Lecture Notes in Networks and Systems

Volume 599

The series "Lecture Notes in Networks and Systems" publishes the latest developments in Networks and Systems—quickly, informally and with high quality. Original research reported in proceedings and post-proceedings represents the core of LNNS.

Volumes published in LNNS embrace all aspects and subfields of, as well as new challenges in, Networks and Systems.

The series contains proceedings and edited volumes in systems and networks, spanning the areas of Cyber-Physical Systems, Autonomous Systems, Sensor Networks, Control Systems, Energy Systems, Automotive Systems, Biological Systems, Vehicular Networking and Connected Vehicles, Aerospace Systems, Automation, Manufacturing, Smart Grids, Nonlinear Systems, Power Systems, Robotics, Social Systems, Economic Systems and other. Of particular value to both the contributors and the readership are the short publication timeframe and the world-wide distribution and exposure which enable both a wide and rapid dissemination of research output.

The series covers the theory, applications, and perspectives on the state of the art and future developments relevant to systems and networks, decision making, control, complex processes and related areas, as embedded in the fields of interdisciplinary and applied sciences, engineering, computer science, physics, economics, social, and life sciences, as well as the paradigms and methodologies behind them.

Indexed by SCOPUS, INSPEC, WTI Frankfurt eG, zbMATH, SCImago.

All books published in the series are submitted for consideration in Web of Science.

For proposals from Asia please contact Aninda Bose (aninda.bose@springer.com).

Nadia Nedjah · Gregorio Martínez Pérez ·
B. B. Gupta
Editors

International Conference on Cyber Security, Privacy and Networking (ICSPN 2022)

Springer

Editors
Nadia Nedjah
Department of Electronics Engineering
and Telecommunications
State University of Rio de Janeiro
Rio de Janeiro, Brazil

Gregorio Martínez Pérez
University of Murcia
Murcia, Spain

B. B. Gupta
Asia University
Taichung, Taiwan

ISSN 2367-3370   ISSN 2367-3389 (electronic)
Lecture Notes in Networks and Systems
ISBN 978-3-031-22017-3   ISBN 978-3-031-22018-0 (eBook)
https://doi.org/10.1007/978-3-031-22018-0

This Springer imprint is published by the registered company Springer Nature Switzerland AG
The registered company address is: Gewerbestrasse 11, 6330 Cham, Switzerland

Preface

The International Conference on Cyber Security, Privacy and Networking (ICSPN 2022), held online, is a forum intended to bring high-quality researchers, practitioners, and students from a variety of fields encompassing interests in massive scale complex data networks and big-data emanating from such networks.

Core topics of interest included security and privacy, authentication, privacy and security models, intelligent data analysis for security, big data intelligence in security and privacy, deep learning in security and privacy, identity and trust management, AI and machine learning for security, data mining for security and privacy, data privacy, etc. The conference welcomes papers of either practical or theoretical nature, presenting research or applications addressing all aspects of security, privacy, and networking, that concerns to organizations and individuals, thus creating new research opportunities. Moreover, the conference program will include various tracks, special sessions, invited talks, presentations delivered by researchers from the international community, and keynote speeches. A total of 115 papers were submitted, from which 38 were accepted as regular papers.

This conference would not have been possible without the support of a large number of individuals. First, we sincerely thank all authors for submitting their high-quality work to the conference. We also thank all Technical Program Committee members and reviewers, sub-reviewers for their willingness to provide timely and detailed reviews of all submissions. Working during the COVID-19 pandemic was especially challenging, and the importance of team work was all the more visible as we worked toward the success of the conference. We also offer our special thanks to the publicity and publication chairs for their dedication in disseminating the call, and encouraging participation in such challenging times, and the preparation of these proceedings. Special thanks are also due to the special tracks chair, finance chair, and the web chair. Lastly, the support and patience of Springer staff members throughout the process are also acknowledged.

Rio de Janeiro, Brazil   Nadia Nedjah
Taichung, Taiwan   B. B. Gupta
Murcia, Spain   Gregorio Martínez Pérez
September 2022

Organization

Organizing Committee

Honorary Chairs

Jie Wu	Temple University, USA
Valentina E. Balas	Aurel Vlaicu University of Arad, Romania
Amiya Nayak	Professor, University of Ottawa, Canada
Xiang Yang	Swinburne Institute of Technology Australia
Zhili Zhou	Nanjing University of Information Science and Technology, NUIST, China
Michael Sheng	Macquarie University, Sydney, Australia

General Chairs

Nadia Nedjah	State University of Rio de Janeiro, Brazil
Gregorio Martínez Pérez	University of Murcia (UMU), Spain
Dragan Peraković	University of Zagreb, Croatia

Program Chairs

Francesco Palmieri	University of Salerno, Italy
B. B. Gupta	Asia University, Taichung, Taiwan

Publicity Chairs

Kwok Tai Chui   The Open University of Hong Kong, Hong Kong
Francesco Colace   University of Salerno, Italy
Wadee Alhalabi   Department of Computer Science, KAU, Saudi Arabia
Milad Taleby Ahvanooey   Nanjing University (NJU), China

Publication Chairs

Deepak Gupta   Founder and CEO, LoginRadius Inc., Canada
Shingo Yamaguchi   Yamaguchi University, Japan
Francisco José García-Peñalvo   University of Salamanca, Spain

Industry Chairs

Srivathsan Srinivasagopalan   AT&T, USA
Suresh Veluru   United Technologies Research Centre Ireland, Ltd.,
                              Ireland
Sugam Sharma   Founder and CEO, eFeed-Hungers.com, USA

Contents

Data Mining Techniques for Intrusion Detection on the Internet of Things Field

Marco Carratù, Francesco Colace, Angelo Lorusso, Antonio Pietrosanto,
Domenico Santaniello, and Carmine Valentino[✉]

DIIn, University of Salerno, Fisciano, SA, Italy
{mcarratu,fcolace,alorusso,pietrosanto,dsantaniello,cvalentino}@unisa.it

Abstract. Over the years, the Internet of Things (IoT) paradigm has acquired great importance due to various application possibilities. The need for Intrusion Detection System (IDS) arises related to the widespread of smart tools connected to each other. This paper aims to present a methodology based on data mining techniques to improve the protection of the connection in an Internet of Things application. In particular, this paper exploits machine learning techniques and Recommender Systems. The K-Nearest Neighbor method and a Context-Aware Recommender System allow the identification of attacks. A multiclassification module based on binary perceptron classifiers with a one-versus-one strategy allows the identification of the attack typology. The obtained numerical results are promising.

Keywords: Internet of Things · Intrusion detection system · Data mining · Machine learning · Classification · Recommender Systems

1 Introduction

Over the years, the Internet of things (IoT) [1,2] paradigm has empowered people to improve their daily life and develop a lot of new possibilities. The acronym IoT refers to a set of connected smart tools able to exchange information with each other. This information allows providing services to users that can interact with them. The services and the interaction allow applying IoT to various fields such as smart cities [3], smart buildings [4], smart homes [5], Industry 4.0 [6], or cultural heritage fields [7,8].

The possibility of managing smart sensors through a central device connected to the Internet and of being able to transfer sensitive information introduces a security problem. The cyber-security [9–12] issue arose with the diffusion of the Internet to the public and implied the rise of cyber crimes, which consist of illicit activity to achieve criminal purposes through a computer or a computing device. Cyber crimes can be classified into various categories, such as cyberstalking, cyber terrorism, phishing, and cybersquatting, and are achieved through a cyber attack, indeed the attempt of non-authorized access. The cyber attack, strictly related to cyber crimes, can be divided into two typologies: insider or

N. Nedjah et al. (Eds.): ICSPN 2021, LNNS 599, pp. 1–10, 2023.
https://doi.org/10.1007/978-3-031-22018-0_1

external attack. Moreover, cyber-attacks can be performed by an organization with specific objectives, such as money or cyber-espionage, or they can be performed by an individual that aims for revenge or recognition. Cyber-attacks exploit malware (malicious software) [13–17] that can assume various forms, such as adware, spyware, virus, worms, and trojan horse.

The evolution of malware increases the network vulnerability and raises the security requirement on the Internet of Things paradigm to guarantee sensitive data protection [18]. The introduction of Intrusion Detection Systems aims to identify data traffic anomalies and illicit activities through the network [19]. Because of the evolution of malware and its variety, there are four main Intrusion Detection techniques [20–22]:

- Signature-based approaches [23] require a dataset that collects the main features of attacks to identify patterns and similarities with the possible new attacks. These approaches return a low level of false alarms. On the other hand, the principal limit of these approaches consists of the inability to recognize new attack typologies that the database does not include.
- Anomaly-based approaches [23] exploit every change of the usual behaviors in the network to identify possible attacks. The principal limit of anomaly-based techniques consists of the high percentage of false alarms caused by identifying anomalies even in new behaviors that are not illegal. These approaches can be further divided into:
  - Statistical-based approaches [20];
  - Knowledge-based approaches [20];
  - Anomaly-based approaches through machine learning techniques [20].
- Specification-based approaches, similar to anomaly-based ones, aim to identify behaviors that are different from usual ones. But the main difference with anomaly-based approaches consists of using roles and thresholds determined by experts. Indeed, the ability to identify attacks does not exploit automatic techniques but the knowledge of humans that allows the system to understand behavior changes.
- Hybrid approaches take advantage of Signature-based, anomaly-based, and specification-based techniques to improve the limits of the singular method and identify attacks.

Developing reliable detection techniques allows for protecting sensitive data in the Internet of Things paradigm. In particular, this paper aims to create a detection technique based on two data mining methods: K-Nearest Neighbor, Context-Aware Recommender Systems (CARSs), and the Perceptron [24] algorithm for multiclassification.

The K-Nearest Neighbor (KNN) [25] represents an instance-based machine learning technique that exploits similarity measures to make predictions. It falls into the classification algorithms and takes advantage of a straightforward implementation phase. The similarity measure, that KNN exploits, is the Minkowski distance [26]:

$$d_p(x, y) = \left(\sum_{i=1}^{n} |x_i - y_i|^p \right)^{\frac{1}{p}} \qquad x, y \in R^n \;()$$

This distance becomes the Manhattan distance when $p = 1$, and the Euclidean one when $p = 2$.

Recommender Systems (RSs) [27–29], instead, are data filtering and analysis tools able to provide suggestions to users according to their preferences. In particular, CARSs [30] exploit context [31,32] to improve the provided forecasts, where the context is defined as any information able to influence system entities (users and items) and their interaction with the system.

These two tools will be exploited to develop an intrusion detection [35] method in an Internet of Things application. Instead, the multiclassification through the Perceptron algorithm allows identifying the attack typology. In particular, this paper is structured as follows: Sect. 2 contains the related works, Sect. 3 describes the proposed approach and presents the related architecture, Sect. 4 introduces the experimental results, and, finally, Sect. 5 contains conclusions and future improvements.

2   Related Works

As mentioned above, the widespread of sensors for the acquisition of data and to provide services arose the Cyber-Security problem. Because of this evolution, various Intrusion Detection Systems (IDSs) were developed on the Internet of Things (IoT) [14,33] paradigm to ensure sensitive data protection.

Reference [36] presents an approach based on the selection of features and ensemble learning. The proposed framework consists of four phases. The data preprocessing phase allows transforming raw data into a suitable format; the dimensionality reduction phase enables to select of proper features through the CFS-BA method based on the Correlation-based selection features (CFS) and the Bat algorithm (BA); the training phase focuses on the training of Random Forest, Random by penalizing attributes, and the decision tree algorithm C4.5; the final phase consists of the training of the ensemble through the average of probabilities.

Another IDS is provided by [37], where a recursive neural network allows identifying Denial of Services. In particular, the authors exploit a modified version of the backpropagation algorithm adapted to the specific application case. Moreover, the data pre-elaboration acquired significant importance to the neural network training.

In the IoT field, the intelligent intrusion detection system IMIDS [38] represents an approach based on a Convolutional Neural Network (CNN). The IMIDS components are the Network capture block, The Feature extraction block, and the Train/load model block. This first aims to capture raw network packets and extract network flows, the second allows to identify of relevant features and makes them compatible with the detection model, and the last consists of the CNN training to make predictions.

Reference [39], instead, contains an IDS based on an approach defined Multi-level Graph Approach. This approach exploits an Ontology, that allows managing knowledge through data, the Context Dimension Tree, an acyclic direct graph that represents contextual information and allows for the information specification, and a Bayesian Network that grants to make predictions. This approach takes advantage of the Ontology and the Context Dimension Tree for the construction of the Bayesian Network structure according to data.

These approaches are evidence of the different methodologies applied to the IoT paradigm for data protection.

In the next section, the proposed approach is going to be described to define an Intrusion Detection System based on data mining techniques.

3   The Proposed Approach

As mentioned above, this section focuses on the proposed approach description based on two data mining approaches, the K-Nearest Neighbor and a Context-Aware Hybrid Recommender System, to identify attacks and the multiclassification through the Perceptron algorithm to understand the attack typology.

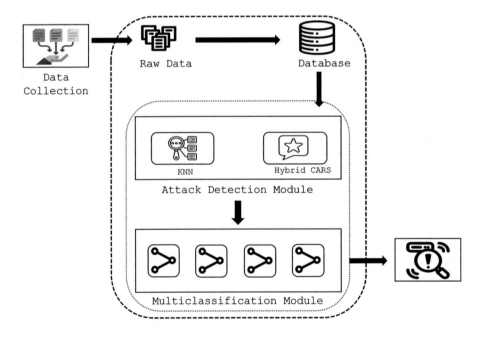

Fig. 1. The system architecture

Figure 1 summarizes the system architecture that needs the data collection to feed the intrusion detection system. The collected data are not suitable for the inference engine, and it requires a pre-elaboration phase to obtain a database that contains suitable data for the training phase. In particular, data features, binary labels related to attacks and usual behaviors, and labels associated with attack typologies are collected. Then, data can be exploited by the Attack Detection Module that consists of the K-Nearest Neighbor method and a Hybrid Context-Aware Recommender System.

The built database contains data composed of features and labels. This database structure allows the application of the KNN machine learning technique. Through the choice of the parameter k, which will be determined through a validation phase on data, the predictions of new data labels will be calculated through the mean of the k-nearest neighbors' labels:

$$y^{new} = \frac{1}{k} \sum_{i=1}^{k} y_{j_i} \, ()$$

The other data mining technique exploited by the proposed approach is a Hybrid Context-Aware Recommender System [40]. This CARS is based on two distinct recommendation techniques: a collaborative filtering approach that takes advantage of matrix factorization [41], and a content-based one that exploits the Gram-Schmidt orthogonalization [42].

In particular, the labels of data are exploited by the Collaborative Filtering methods to predict the attack probabilities. The interaction matrix represents the interaction between the IP addresses and the control devices that manages a subset of sensors. The predictions are calculated through the formula

$$\widehat{y}_{uic}^{new} = \overline{y} + b_{uc} + b_i + p_u q_i^T \, ()$$

where $\overline{y}$ represents the mean of all labels, b_{uc} is the bias related to the IP address u in the location context c that consists of the virtual location referred to as the IP address, b_i represents the bias related to the control device, and $p_u q_i^T$ expresses the affinity between the IP address u and the control device i. This affinity is obtained through the Singular Value Decomposition (SVD) and the acquisition of the formula elements happens through the Stochastic Gradient Descent techniques.

The Content-Based techniques, instead, exploits the features contained in the database. These features are normalized in the pre-elaboration phase, and through the Gram-Schmidt orthogonalization, the forecasts predictions for attack can be achieved. The Gram-Schmidt orthogonalization is described in Algorithm 1 and allows to respect for some properties related to the Singular Value Decomposition [42].

The forecasts obtained with the two recommendation methodologies are ensemble through the mean of the provided forecasts.

Finally, KNN and Hybrid Recommender Systems outputs are ensemble through a learning procedure that allows identifying the weighted mean coefficients.

Then the Multiclassification Module allows identifying the attack typologies through a one-versus-one strategy. Then $n(n-1)/2$ Classifiers have to be trained, where n consists of the number of attack typologies that have to be classified.

Algorithm 1: Gram-Schmidt orthogonalization
1: Input: matrix
$A = (a_1, \ldots, a_l) \in R^{s \times l} : s \leq l$
2: $w_1 \leftarrow a_1$
3: **For** $i = 2, \ldots, l$ **do**
4: $w_i \leftarrow a_i - \sum_{h=1}^{i-1} \frac{<a_i, w_h>}{<w_h, w_h>} w_h$
5: Output: $\overline{A} = (w_1, \ldots, w_l) \in R^{s \times l}$

4   Experimental Phase

This section describes the experimental phase that allows for validating the proposed approach. To collect data, two Raspberries Pi4 were installed in the Icaro Laboratory and the ICT Center of the University of Salerno. These Raspberries were connected to various typologies of sensors summarized in Table 1.

Table 1. Sensors linked to Raspberries Pi4 and related controlled parameters

Parameter	Sensor
Air indoor/outdoor temperature	BME680 sensor module
Air indoor/outdoor humidity	BME680 sensor module
Luminosity	SI1145 digital UV index/IR/visible light sensor
Voltage	Voltage sensor
Light control	Dimmer
Ventilation	Fun motor

Moreover, Raspberries transmit sensor data to a central device that allows the collection of it in raw form in a CSV (comma-separated values) file. Then, through the Python library Pandas, the pre-elaboration phase is applied.

To simulate various attack typologies a bot was developed to try internal or external attacks. This bot was implemented through Ruby, a high-level programming language, and generated the attack typology and the time of the attack through pseudo-random numbers. The attack typologies are four and consist of:

- Distributed Denial of Services (DDoS);
- Denial of Service (DoS);
- Data Manipulation (DM);
- Storage Attack (SA).

The amount of data consists of 64800 where 51840 are usual behaviors and 12960 are attacks (20%). This data was collected for six months acquiring the data every 5 min and trying to attack the system 3 times per hour.

The results related to the Attack Detection Module ability are promising. In particular, the Recall value confirms the Attack Detection Module skill to avoid false alarms.

Table 2. Numerical results related to the ability of the Attack Detection Module to identify attacks over the entire database.

	Precision	Recall	F1-score
Attack Detection Module	0.9013	0.9433	0.9218

Table 3, instead, contains the numerical results related to the ability to identify the attack typology of the multiclassification done through the perceptron algorithm with the one-versus-one strategy. The training of the Attack Detection Module takes advantage of 70% of the dataset that contains both usual behavior and attacks (65%–5% respectively). Instead, the Multiclassification Module takes advantage of 10% of the dataset, which contains only attacks. Then the validation is done on 20% of the dataset. To improve the reliability of the results the Cross Validation k-fold is exploited.

Table 3. Numerical results of the proposed approach

	Precision	Recall	F1-score
The proposed approach	0.8892	0.9231	0.9058

The numerical results are still promising and return a recall value that confirms the ability of the proposed approach to avoid false alarms.

5   Conclusions and Future Works

In this paper, an approach to identifying attacks is developed. In particular, data mining methodologies are exploited to improve data security. The K-Nearest algorithm and a Hybrid Context-Aware Recommender System allow the identification of attacks; instead, the Perceptron algorithm is exploited for multiclassification with a one-versus-one strategy. An ad-hoc database was collected to validate the proposed approach, and the obtained results are promising.

Future works related to this approach include improving the Attack Detection Module and the Multiclassification Module. Both of them will take advantage of two different Neural Networks. Moreover, more data will be collected, and new attack typologies will be introduced to the database.

References

1. Chettri, L., Bera, R.: A comprehensive survey on internet of things (IoT) toward 5G wireless systems. IEEE Internet Things J. **7**(1) (2020). https://doi.org/10.1109/JIOT.2019.2948888
2. Casillo, M., Colace, F., Lorusso, A., Marongiu, F., Santaniello, D.: An IoT-based system for expert user supporting to monitor, manage and protect cultural heritage buildings. In: Studies in Computational Intelligence, vol. 1030 (2022). https://doi.org/10.1007/978-3-030-96737-6_8
3. Zanella, A., Bui, N., Castellani, A., Vangelista, L., Zorzi, M.: Internet of things for smart cities. IEEE Internet Things J. **1**(1), 22–32 (2014)
4. Minoli, D., Sohraby, K., Occhiogrosso, B.: IoT considerations, requirements, and architectures for smart buildings—energy optimization and next-generation building management systems. IEEE Internet Things J. **4**(1) (2017). https://doi.org/10.1109/JIOT.2017.2647881
5. Marikyan, D., Papagiannidis, S., Alamanos, E.: A systematic review of the smart home literature: a user perspective. Technol. Forecast. Soc. Change **138** (2019). https://doi.org/10.1016/j.techfore.2018.08.015
6. Lelli, F.: Interoperability of the time of Industry 4.0 and the internet of things. Future Internet **11**(2) (2019). https://doi.org/10.3390/fi11020036
7. Chianese, A., Piccialli, F., Jung, J.E.: The internet of cultural things: towards a smart cultural heritage (2017). https://doi.org/10.1109/SITIS.2016.83
8. Jara, A.J., Sun, Y., Song, H., Bie, R., Genooud, D., Bocchi, Y.: Internet of things for cultural heritage of smart cities and smart regions (2015). https://doi.org/10.1109/WAINA.2015.169
9. Ten, C.W., Manimaran, G., Liu, C.C.: Cybersecurity for critical infrastructures: attack and defense modeling. IEEE Trans. Syst. Man Cybern. Part A Syst. Hum. **40**(4) (2010). https://doi.org/10.1109/TSMCA.2010.2048028
10. Jang-Jaccard, J., Nepal, S.: A survey of emerging threats in cybersecurity. J. Comput. Syst. Sci. **80**(5) (2014). https://doi.org/10.1016/j.jcss.2014.02.005
11. Buczak, A.L., Guven, E.: A survey of data mining and machine learning methods for cyber security intrusion detection. IEEE Commun. Surv. Tutor. **18**(2) (2016). https://doi.org/10.1109/COMST.2015.2494502
12. Castiglione, A., Palmieri, F., Colace, F., Lombardi, M., Santaniello, D., D'Aniello, G.: Securing the internet of vehicles through lightweight block ciphers. Pattern Recognit. Lett. **135** (2020). https://doi.org/10.1016/j.patrec.2020.04.038
13. Aslan, O., Samet, R.: A comprehensive review on malware detection approaches. IEEE Access **8** (2020). https://doi.org/10.1109/ACCESS.2019.2963724
14. Sharma, R., Sharma, T.P., Sharma, A.K.: Detecting and preventing misbehaving intruders in the internet of vehicles. Int. J. Cloud Appl. Comput. (IJCAC) **12**(1), 1–21 (2022)
15. Ling, Z., Hao, Z.J.: An intrusion detection system based on normalized mutual information antibodies feature selection and adaptive quantum artificial immune system. Int. J. Semant. Web Inf. Syst. (IJSWIS) **18**(1), 1–25 (2022)
16. Gibert, D., Mateu, C., Planes, J.: The rise of machine learning for detection and classification of malware: research developments, trends and challenges. J. Netw. Comput. Appl. **153** (2020). https://doi.org/10.1016/j.jnca.2019.102526
17. Egele, M., Scholte, T., Kirda, E., Kruegel, C.: A survey on automated dynamic malware-analysis techniques and tools. ACM Comput. Surv. **44**(2) (2012). https://doi.org/10.1145/2089125.2089126

18. Chui, K.T., et al.: Handling data heterogeneity in electricity load disaggregation via optimized complete ensemble empirical mode decomposition and wavelet packet transform. Sensors **21**(9), 3133 (2021)
19. Ling, Z., Hao, Z.J.: An intrusion detection system based on normalized mutual information antibodies feature selection and adaptive quantum artificial immune system. Int. J. Semant. Web Inf. Syst. (IJSWIS) **18**(1), 1–25 (2022)
20. Khraisat, A., Gondal, I., Vamplew, P., Kamruzzaman, J.: Survey of intrusion detection systems: techniques, datasets and challenges. Cybersecurity **2**(1), 1–22 (2019). https://doi.org/10.1186/s42400-019-0038-7
21. Zarpelão, B.B., Miani, R.S., Kawakani, C.T., de Alvarenga, S.C.: A survey of intrusion detection in internet of things. J. Netw. Comput. Appl. **84** (2017). https://doi.org/10.1016/j.jnca.2017.02.009
22. Lu, J., et al.: Blockchain-based secure data storage protocol for sensors in the industrial internet of things. IEEE Trans. Ind. Inform. **18**(8), 5422–5431 (2021)
23. Lokman, S.-F., Othman, A.T., Abu-Bakar, M.-H.: Intrusion detection system for automotive controller area network (CAN) bus system: a review. EURASIP J. Wireless Commun. Netw. **2019**(1), 1–17 (2019). https://doi.org/10.1186/s13638-019-1484-3
24. Hernández, G., Zamora, E., Sossa, H., Téllez, G., Furlán, F.: Hybrid neural networks for big data classification. Neurocomputing **390** (2020). https://doi.org/10.1016/j.neucom.2019.08.095
25. Taheri, R., Ghahramani, M., Javidan, R., Shojafar, M., Pooranian, Z., Conti, M.: Similarity-based Android malware detection using Hamming distance of static binary features. Future Gener. Comput. Syst. **105** (2020). https://doi.org/10.1016/j.future.2019.11.034
26. Gao, X., Li, G.: A KNN model based on Manhattan distance to identify the SNARE proteins. IEEE Access **8** (2020). https://doi.org/10.1109/ACCESS.2020.3003086
27. Bobadilla, J., Ortega, F., Hernando, A., Gutiérrez, A.: Recommender systems survey. Knowl.-Based Syst. **46** (2013). https://doi.org/10.1016/j.knosys.2013.03.012
28. Ricci, F., Shapira, B., Rokach, L.: Recommender systems: introduction and challenges. In: Recommender Systems Handbook, 2nd edn. (2015). https://doi.org/10.1007/978-1-4899-7637-6_1
29. Carbone, M., Colace, F., Lombardi, M., Marongiu, F., Santaniello, D., Valentino, C.: An adaptive learning path builder based on a context aware recommender system. In: Proceedings—Frontiers in Education Conference, FIE, vol. 2021, Oct 2021. https://doi.org/10.1109/FIE49875.2021.9637465
30. Adomavicius, G., Mobasher, B., Ricci, F., Tuzhilin, A.: Context-aware recommender systems. AI Mag. **32**(3) (2011). https://doi.org/10.1609/aimag.v32i3.2364
31. Abowd, G.D., Dey, A.K., Brown, P.J., Davies, N., Smith, M., Steggles, P.: Towards a better understanding of context and context-awareness. In: Lecture Notes in Computer Science (including subseries Lecture Notes in Artificial Intelligence and Lecture Notes in Bioinformatics), vol. 1707 (1999). https://doi.org/10.1007/3-540-48157-5_29
32. Annunziata, G., Colace, F., de Santo, M., Lemma, S., Lombardi, M.: AppPoggiomarino: a context aware app for e-citizenship. In: ICEIS 2016—Proceedings of the 18th International Conference on Enterprise Information Systems, vol. 2 (2016). https://doi.org/10.5220/0005825202730281
33. Cvitić, I., Peraković, D., Periša, M., Gupta, B.: Ensemble machine learning approach for classification of IoT devices in smart home. Int. J. Mach. Learn. Cybern. **12**(11), 3179–3202 (2021). https://doi.org/10.1007/s13042-020-01241-0

34. Tewari, A., et al.: A lightweight mutual authentication approach for RFID tags in IoT devices. Int. J. Netw. Virt. Org. **18**(2), 97–111 (2018)
35. Ling, Z., Hao, Z.J.: Intrusion detection using normalized mutual information feature selection and parallel quantum genetic algorithm. Int. J. Semant. Web Inf. Syst. (IJSWIS) **18**(1), 1–24 (2022)
36. Zhou, Y., Cheng, G., Jiang, S., Dai, M.: Building an efficient intrusion detection system based on feature selection and ensemble classifier. Comput. Netw. **174** (2020). https://doi.org/10.1016/j.comnet.2020.107247
37. Almiani, M., AbuGhazleh, A., Al-Rahayfeh, A., Atiewi, S., Razaque, A.: Deep recurrent neural network for IoT intrusion detection system. Simul. Model. Pract. Theory **101** (2020). https://doi.org/10.1016/j.simpat.2019.102031
38. Le, K.H., Nguyen, M.H., Tran, T.D., Tran, N.D.: IMIDS: an intelligent intrusion detection system against cyber threats in IoT. Electronics (Switzerland) **11** (4) (2022). https://doi.org/10.3390/electronics11040524
39. Colace, F., Khan, M., Lombardi, M., Santaniello, D.: A multigraph approach for supporting computer network monitoring systems. In: Advances in Intelligent Systems and Computing, vol. 1184 (2021). https://doi.org/10.1007/978-981-15-5859-7_46
40. Gunti, P., et al.: Data mining approaches for sentiment analysis in online social networks (OSNs). In: Data Mining Approaches for Big Data and Sentiment Analysis in Social Media, pp. 116–141. IGI Global (2022)
41. Casillo, M., et al.: Context aware recommender systems: a novel approach based on matrix factorization and contextual bias. Electronics (Switzerland) **11**(7) (2022). https://doi.org/10.3390/electronics11071003
42. Casillo, M., Conte, D., Lombardi, M., Santaniello, D., Troiano, A., Valentino, C.: A content-based recommender system for hidden cultural heritage sites enhancing. In: Lecture Notes in Networks and Systems, vol. 217 (2022). https://doi.org/10.1007/978-981-16-2102-4_9

Detecting Rumors Transformed from Hong Kong Copypasta

Yin-Chun Fung[1]([✉]), Lap-Kei Lee[1], Kwok Tai Chui[1], Ian Cheuk-Yin Lee[1], Morris Tsz-On Chan[1], Jake Ka-Lok Cheung[1], Marco Kwan-Long Lam[1], Nga-In Wu[2], and Markus Lu[3]

[1] School of Science and Technology, Hong Kong Metropolitan University, Ho Man Tin, Kowloon, Hong Kong SAR, China
ycfung@study.hkmu.edu.hk,{lklee,jktchui}@hkmu.edu.hk
[2] College of Professional and Continuing Education, The Hong Kong Polytechnic University, Kowloon, Hong Kong SAR, China
ngain.wu@cpce-polyu.edu.hk
[3] Hong Kong International School, Hong Kong SAR, China

Abstract. A copypasta is a piece of text that is copied and pasted in online forums and social networking sites (SNSs) repeatedly, usually for a humorous or mocking purpose. In recent years, copypasta is also used to spread rumors and false information, which damages not only the reputation of individuals or organizations but also misleads many netizens. This paper presents a tool for Hong Kong netizens to detect text messages that are copypasta or their variants (by transforming an existing copypasta with new subjects and events). We exploit the Encyclopedia of Virtual Communities in Hong Kong (EVCHK), which contains a database of 315 commonly occurred copypasta in Hong Kong, and a CNN model to determine whether a text message is a copypasta or its variant with an accuracy rate of around 98%. We also showed a prototype of a Google Chrome browser extension that provides a user-friendly interface for netizens to identify copypasta and their variants on a selected text message directly (e.g., in an online forum or SNS). This tool can show the source of the corresponding copypasta and highlight their differences (if it is a variant). From a survey, users agreed that our tool can effectively help them to identify copypasta and hence help stop the spreading of this kind of online rumor.

Keywords: Rumor detection · Copypasta · Natural language processing

© The Author(s), under exclusive license to Springer Nature Switzerland AG 2023
N. Nedjah et al. (Eds.): ICSPN 2021, LNNS 599, pp. 11–23, 2023.
https://doi.org/10.1007/978-3-031-22018-0_2

1   Introduction

Nowadays, many express their opinions by publishing or reposting articles on the Internet. According to Leung [1], online discussion forums are the preferred social medium for gaining recognition; blogs and social networking sites (SNSs) [2] like Facebook are normally used for social and psychological needs and the need for affection. The popularity of these online platforms has also led to the rapid spreading of rumors, which is false information created by, for example, exaggeration, tampering, or mismatching, and can mislead the readers and even have a negative impact on public events [3].

A copypasta is a piece of text that is copied and pasted repeatedly around the Internet and can usually be seen in online discussion forums and social networking sites for humorous or mocking purposes [5,6]. Some articles may be transformed or adapted from existing copypasta by replacing the subjects and/or events with new ones. These copypastas (and their variants) are usually funny or satirical, and this can bring happiness to their readers. Yet these copypastas may evolve into rumors because some true believers would believe that the content of the copypasta is true [7]. For example, in November 2019, a copypasta "A fierce fight broke out at the top of the government", which is transformed from the copypasta about a fight between Netherlands national football team players in 2012, appeared in a popular forum "LIHKG" in Hong Kong, and the then Chief Secretary for Administration of the Hong Kong government dismissed the rumor on his official Facebook page [9]. Such a rumor can be regarded as rumors created by tampering [4,10]. Twitter [14], one of the popular SNSs, has also updated its security policies to combat false information caused by copypasta [11].

Copypasta may be easily identified by netizens who are frequent users of online forums and SNSs. Yet, many other netizens fail to identify copypastas from credible and authentic texts. These rumors are adapted from different articles and events or are made from imagination. This culture has become very common for netizens worldwide. Some netizens may think that the transformation to copypastas will make them more interesting. Some others however maliciously adapted them to achieve purposes like defamation. At present, there is no relevant law in Hong Kong to regulate this kind of behavior,[1] so it is becoming more common in Hong Kong.

The Encyclopedia of Virtual Communities in Hong Kong (EVCHK)[2] is a website with a collection of more than 12,000 entries on the Internet culture in contemporary Hong Kong, and it is operated by a community of volunteer editors in a similar fashion to Wikipedia [6]. EVCHK contains a database of 315 common copypastas in Hong Kong, which may be helpful for netizens to find out manually whether a text is a copypasta, a transformed variant, or neither of them. Such checking of online articles or text messages however would require a lot of time and energy.

---

[1] https://www.info.gov.hk/gia/general/202003/18/P2020031800422.htm.
[2] https://evchk.fandom.com/.

To mitigate the problem, one direction is to educate the netizens to raise their cybersecurity awareness, e.g., [13]; another direction is to develop tools that can automatically detect whether a text is a rumor. There have been many rumor detection algorithms in the Natural Language Processing (NLP) research community (see the surveys [15, 16] and the references therein). While sentiment analysis, e.g., [17, 18], and intent identification, e.g., [19, 20], are well-studied NLP problems and have many applications, e.g., chatbots [13, 22], rumor detection may involve the use of these and more textual features for the machine learning algorithms and most of the existing works can only identify a rumor without offering an explanation why it is a rumor.

This paper aims to develop a tool that can identify copypasta (and its variant) in Hong Kong and provide explanations for why texts are identified as copypasta. Our contributions include the following:

- There is no publicly available dataset for copypasta detection. We collected copypasta (including their variants) and non-copypasta (i.e., text messages that are not copypasta) from different websites, including EVCHK, online forums, and news media in Hong Kong, and created a dataset for Hong Kong copypasta detection.
- We divided our dataset for training and testing, respectively. We trained machine learning models based on CNN and RNN, and found that CNN performs better on copypasta detection. The accuracy is around 98% on the testing dataset.
- We developed a prototype of a Google Chrome browser extension that provides a user-friendly interface for netizens to identify copypasta and their variants on a selected text message directly (e.g., in an online forum or SNS). This tool can show the source of the corresponding copypasta and highlight their differences (if it is a variant).
- We conducted a survey on 45 users and focus group interviews with some participants to show that our tool can effectively help them to identify copypasta and help stop the spreading of this kind of online rumor.

Organization of the paper. Section 2 reviews some existing works on rumor detection. Section 3 gives the detailed design of our copypasta detection tool. Section 4 presents an evaluation of the tool on 45 participants. Section 5 concludes the paper and proposes some future work directions.

2   Existing Works

The Encyclopedia of Virtual Communities in Hong Kong (EVCHK) is one of the most popular websites for netizens to find out the meaning and reference of net slang and copypasta in Hong Kong, as information in EVCHK is well-organized into different categories by a community of volunteer editors. Users can often find the relevant pages in EVCHK using search engines like Google Search. Yet those with lower ICT literacy may not able to find the desired information using appropriate keywords using the search function in EVCHK and search engines.

Table 1. Comparison of existing tools on Hong Kong copypasta detection.

Solution	Snopes websites	Slang detection [16]	HKBU FactCheck Service	TweetCred	Our tool
Detects Hong Kong copypasta	✗	✗	✓	✗	✓
Provides automatic detection	✗	✓	✗	✓	✓
Shows the source of the copypasta	✗	✗	✗	✗	✓

Slang detection and identification. There are many researchers that are working on the detection of rumors using different approaches [12,21]; such as author in [23] proposed data mining approaches for sentiment analysis in online social networks. In another work, authors [8] proposed a model for Fake News Detection Using Multi-Channel Deep Neural Networks. However, to the best of our knowledge, there is no existing work focusing on copypasta detection. A closely related work is the problem of slang detection and identification, proposed by Pei et al. [24]. They used RNN to identify the exact positions of slang keywords to detect the presence of slang in a sentence. However, it locates the slang in a sentence only and cannot be adapted to identify copypasta.

"Snopes" website.[3] It is a famous and popular website for users to fact-check the source of articles, news, and copypasta (see Fig. 1 (left)). It provides a Fact Check Rating System that is credible to users. However, it focuses on English content for US news media and websites, so users using other languages cannot utilize their fact-checking service.

HKBU FactCheck Service.[4] This service (see Fig. 1 (right)) is developed by the School of Communication of the Hong Kong Baptist University (HKBU). Like the "Snopes" website, it provides a Fact Check Rating System to fact-check the article and news and it includes more Hong Kong local articles and news. Users can make a request to fact-check an article on the website, but fact-checking is only done manually and would take a long time. It only has around 10 fact-checking results per month. Identifying copypasta is also not a focus of this service.

TweetCred. It is a tool developed by Gupta et al. [25], which is a Google Chrome browser extension that provides a credibility rating for each tweet. The tool uses a supervised automated ranking algorithm to evaluate the credibility of a tweet such that users can determine whether a tweet is a rumor or not. However, it can only be used for tweets on Twitter and cannot identify copypasta.

---

[3] https://www.snopes.com/.
[4] https://factcheck.hkbu.edu.hk/.

Fig. 1. Screen captures of the Snopes website (left) and HKBU FactCheck Service (right)

3 Design of Our Copypasta Detection Tool

3.1 Overall System Architecture

Our solution can be split into two parts: the server side and the client side. The client side is a JavaScript extension that makes API calls to the server side, which is a web service written in Python.

Figure 2 shows the overall system architecture of our tool. The Google Chrome browser extension (i.e., the Client) gets the text selected by the user and sends an API request to the server. The server contains a web service (the backend application) which has an underlying machine learning detection model trained with a knowledge base of copypasta and non-copypasta to predict the probability that a text is a copypasta (or its variant) and then provide the source of the copypasta if the text is identified as a copypasta. The browser extension receives the result from the web service and then displays the analysis result to the user.

Figure 3 shows the workflow of the browser extension. The text input can be text passages selected by the user on the browser (which is referred to as *user post*) or text inputted by the user directly on the browser extension (which is referred to as *user input*). We employed a very simple text preprocessing strategy by removing all non-Chinese characters and symbols (e.g., punctuation marks and emojis) on the user post and user input.

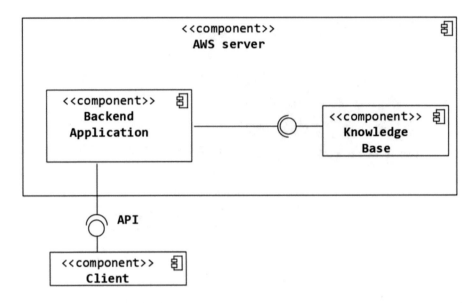

Fig. 2. Overall system architecture of our tool.

3.2   The Machine Learning Models

The detection model is trained using neural networks. Two models have been tested and compared (the comparison result is given in Sect. 4).

The first model is a convolutional neural network (CNN), which consists of the word embedding, convolution, pooling, and output layers. The word embedding layer is used to make the dimension vector of the data dictionary. The convolution layer is for detecting features. The pooling layer is for reducing data dimensions and for generalizing the feature. The output is the fully connected layer with dropout and SoftMax output.

The second model is a recurrent neural network (RNN). Long Short-Term Memory (LSTM, which is a type of RNN) was adopted. The neural network consists of an embedding, a fully connected layer, a sequence pool, and an output layer. There are one LSTM operation and two max sequence pool operation. The output layer is using SoftMax with a size of 2.

3.3   User Interface of the Browser Extension

As the tool is designed for Hong Kong netizens and most of them use Cantonese (a dialect of the Chinese language), the browser extension uses only Cantonese for the user interface. Once the browser extension is installed on the Google Chrome browser, we can select some text passages and then right-click on the selected text to detect whether the selected text is copypasta or not (Fig. 4). Alternatively, we can start the browser extension directly on the browser's menu bar, which displays a text box for user text input (Fig. 5 (left)) and a submit

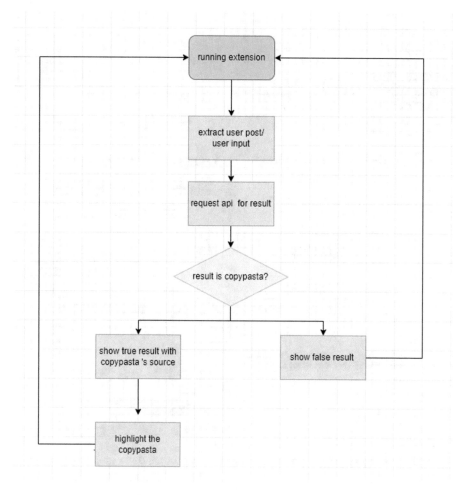

Fig. 3. Workflow of the browser extension.

button. The detection result page is shown in Fig. 5 (right). A detection summary with the copypasta probability is shown. There are three levels of probability: definitely a copypasta ($\geq$ 80%), maybe a copypasta ($\geq$ 50% and < 80%), and not a copypasta (< 50%). For the first two levels, a link to the EVCHK page for the source of the copypasta will be given right below the probability, and the text that matches the copypasta source will be highlighted at the bottom of the result page.

4 Evaluation

Our evaluation focuses on the performance of the machine learning models and evaluation on 45 users.

Fig. 4. Right-click on the selected text to detect copypasta using the browser extension.

Fig. 5. User interface of the browser extension: user input (left) and detection result (right).

4.1 Performance of the Machine Learning Models

Our created dataset. We gathered 233 out of the 315 copypastas from EVCHK as the source. We have collected 15,963 texts from Google News and the popular Hong Kong online forum LIHKG[5] using web scraping techniques and classified them into copypasta and non-copypasta with references to the sources.

Comparisons between CNN and RNN. We trained a CNN model and an RNN model using the same training dataset and performed comparison and evaluation of their performance using the same testing dataset. Table 3 shows the results. Both models have high accuracy and recall, but the CNN model is better than the RNN model on all indicators. As most of the text is non-copypasta in the dataset, precision is more concerned. Since the precision of the

---

[5] https://lihkg.com/.

Table 2. No. of text message samples in our created dataset.

Item	Training	Testing	Total
Copypasta	409	288	697
Non-copypasta	5,446	10,020	15,266
Source of copypasta	–	–	233

CNN model is 0.6911, it is regarded as having a better performance in detecting copypasta.

Table 3. Training results with different models.

	CNN model	RNN model
True positive	273	268
True negative	9898	9772
False positive	122	248
False negative	15	20
Accuracy	0.9867	0.9740
Precision	0.6911	0.5194
Recall	0.9479	0.9306
F1-score	0.7994	0.6667

Comparisons on text preprocessing strategies. We also conducted an experiment on text preprocessing strategy will yield a better result. The strategies are: (1) no preprocessing, (2) remove stop words, and (3) keep only Chinese characters. We used PyCantonese [26] to remove the stop words in Strategy (2). As shown in Table 4, Strategy (3) gives a higher F1 score.

4.2 User Evaluation

We invited 45 undergraduate students in Hong Kong to test our tool and complete a Google Form survey. All participants are experienced netizens and native Cantonese speakers. The survey consists of 6 questions on a 5-point Likert scale (1: disagree, 2: partially disagree, 3: neutral, 4: partially agree, 5: agree). Table 5 shows the survey results.

In the survey, more than half of the participants gave positive feedback on the user interface design (Question 1). Our tool receives an average mark of 4 in Questions 2 to 4, which reflects that the tool is useful for the participants to identify copypasta and rumors. The majority of participants agree that the tool increases their awareness of copypasta and would recommend it to others.

Table 4. Training results with different text preprocessing strategies.

	No preprocessing	Remove stop words	Keep only Chinese characters
True positive	275	272	273
True negative	9871	9850	9898
False positive	149	170	122
False negative	13	16	15
Accuracy	0.9843	0.9820	0.9867
Precision	0.6486	0.6154	0.6911
Recall	0.9549	0.9444	0.9479
F1-score	0.7725	0.7452	0.7994

Table 5. Survey result.

Item	1	2	3	4	5
1. Our user interface is attractive	0 (0%)	6 (13.3%)	3 (6.7%)	11 (24.4%)	25 (55.6%)
2. The extension is useful for determining whether a text passage is a copypasta	1 (2.2%)	2 (4.4%)	12 (26.7%)	14 (31.1%)	16% (35.6%)
3. The extension can help me identify rumors or fake news	0 (0%)	5 (11.1%)	15 (33.3%)	14 (31.1%)	11 (24.4%)
4. The extension detection result returns efficiently related information when the text is a copypasta	0 (0%)	5 (11.1%)	9 (20%)	15 (33.3%)	16 (35.6%)
5. The extension increases my awareness of copypasta	0 (0%)	11 (24.4%)	7 (15.6%)	12 (26.7%)	15 (33.3%)
6. I would recommend this extension to others	0 (0%)	6 (13.3%)	12 (26.7%)	13 (28.9%)	14 (31.1%)

Among the 45 participants, we invited 4 of them to join a focus group interview so as to get more qualitative responses from them. They think the extension is creative because they have never seen an extension that detects the copypasta before. However, they also found that the tool cannot detect non-Chinese copypasta and Chinese copypastas translated from other languages which can be a future work direction. They also expressed concerns that the tool cannot detect the new copypasta, as it requires human effort to update the knowledge base on the server side.

5   Conclusion and Future Work

In this paper, we collected copypasta and non-copypasta from different websites, forums, and news in Hong Kong and created a dataset for Hong Kong copypasta. We trained CNN and RNN prediction models and found that CNN performs better than RNN in copypasta detection. We also showed how the tool can be implemented to facilitate usage: The prediction model is deployed to a server, and a Google Chrome extension can be designed to communicate with the server for users to detect whether selected text is a copypasta or not, and get more information about the identified copypasta (if any). A user survey of our tool showed that our detection tool is straightforward to use and is useful to detect copypasta. Users can learn more about Hong Kong copypasta and reduce their chance of believing some rumors from copypasta. This would help stop the spreading of online rumors from copypasta.

Limitations and future works. Our tool cannot detect new copypasta that is not in our dataset. As a future work direction, the extension may provide a reporting feature for users to report new copypastas when they found them missing in our detection result. When the dataset is sufficient, the model can be trained to classify the source. Another future work direction is to make the extension fully automatic; the extension may work with a script running in the background that grabs the text of the web page and checks whether they are copypasta automatically.

References

1. Leung, L.: Generational differences in content generation in social media: the roles of the gratifications sought and of narcissism. Comput. Hum. Behav. **29**(3), 997–1006 (2013)
2. Sahoo, S.R., et al.: Security issues and challenges in online social networks (OSNs) based on user perspective. In: Computer and Cyber Security, pp. 591–606 (2018)
3. Sahoo, S.R., et al.: Spammer detection approaches in online social network (OSNs): a survey. In: Sustainable Management of Manufacturing Systems in Industry 4.0, pp. 159–180. Springer, Cham (2022)
4. Chen, J., Wu, Z., Yang, Z., Xie, H., Wang, F.L., Liu, W.: Multimodal fusion network with contrary latent topic memory for rumor detection. IEEE Multimed. **29**(1), 104–113 (2022)
5. Riddick, S., Shivener, R.: Affective spamming on Twitch: rhetorics of an emote-only audience in a presidential inauguration livestream. Comput. Compos. **64**, 102711 (2022)
6. Lam, C.: How digital platforms facilitate parody: online humour in the construction of Hong Kong identity. Comedy Stud. **13**(1), 101–114 (2022)
7. Zannettou, S., Sirivianos, M., Blackburn, J., Kourtellis, N.: The web of false information: rumors, fake news, hoaxes, clickbait, and various other shenanigans. J. Data Inf. Qual. **11**(3), 1–37 (2019)
8. Tembhurne, J.V., Almin, M.M., Diwan, T.: Mc-DNN: fake news detection using multi-channel deep neural networks. Int. J. Semant. Web Inf. Syst. (IJSWIS) **18**(1), 1–20 (2022)
9. Facebook Page of the Hong Kong Chief Secretary for Administration's Office. https://www.facebook.com/CSOGOV/posts/420752431929437. Accessed 2022/07/01

10. Srinivasan, S., Dhinesh Babu, L.D.: A parallel neural network approach for faster rumor identification in online social networks. Int. J. Semant. Web Inf. Syst. (IJSWIS) **15**(4), 69–89 (2019)
11. Avery, D.: Twitter updates security policy to combat spam tweets and 'copypasta'. https://www.cnet.com/news/social-media/twitter-updates-security-policy-to-combat-spam-tweets-and-copypasta/. Accessed 2022/07/01
12. Chiang, T.A., Che, Z.H., Huang, Y.L., Tsai, C.Y.: Using an ontology-based neural network and DEA to discover deficiencies of hotel services. Int. J. Semant. Web Inf. Syst. (IJSWIS) **18**(1), 1–19 (2022)
13. Fung, Y.C., Lee, L.K.: A chatbot for promoting cybersecurity awareness. In: Agrawal, D.P., Nedjah, N., Gupta, B.B., Martinez Perez, G. (eds.) Cyber Security, Privacy and Networking. LNNS, vol. 370, pp. 379–387. Springer, Singapore (2022)
14. Sahoo, S.R., et al.: Hybrid approach for detection of malicious profiles in twitter. Comput. Electr. Eng. **76**, 65–81 (2019). ISSN 0045-7906. https://doi.org/10.1016/j.compeleceng.2019.03.003
15. Meel, P., Vishwakarma, D.K.: Fake news, rumor, information pollution in social media and web: a contemporary survey of state-of-the-arts, challenges and opportunities. Expert Syst. Appl. **153**, 112986 (2020)
16. Rani, N., Das, P., Bhardwaj, A.K.: Rumor, misinformation among web: a contemporary review of rumor detection techniques during different web waves. Concurr. Comput. Pract. Exp. **34**(1), e6479 (2022)
17. Fung, Y.C., Lee, L.K., Chui, K.T., Cheung, G.H.K., Tang, C.H., Wong, S.M.: Sentiment analysis and summarization of Facebook posts on news media. In: Data Mining Approaches for Big Data and Sentiment Analysis in Social Media, pp. 142–154. IGI Global (2022)
18. Lee, L.K., Chui, K.T., Wang, J., Fung, Y.C., Tan, Z.: An improved cross-domain sentiment analysis based on a semi-supervised convolutional neural network. In: Data Mining Approaches for Big Data and Sentiment Analysis in Social Media, pp. 155–170. IGI Global (2022)
19. Liu, Y., Liu, H., Wong, L.P., Lee, L.K., Zhang, H., Hao, T.: A hybrid neural network RBERT-C based on pre-trained RoBERTa and CNN for user intent classification. In: International Conference on Neural Computing for Advanced Applications, pp. 306–319. Springer, Singapore (2020)
20. Liu, H., Liu, Y., Wong, L.P., Lee, L.K., Hao, T.: A hybrid neural network BERT-cap based on pre-trained language model and capsule network for user intent classification. Complexity **2020**, 8858852 (2020)
21. Appati, J.K., Nartey, P.K., Yaokumah, W., Abdulai, J.D.: A systematic review of fingerprint recognition system development. Int. J. Softw. Sci. Comput. Intell. (IJSSCI) **14**(1), 1–17 (2022)
22. Lee, L.K., Fung, Y.C., Pun, Y.W., Wong, K.K., Yu, M.T.Y., Wu, N.I.: Using a multiplatform chatbot as an online tutor in a university course. In: 2020 International Symposium on Educational Technology (ISET), pp. 53–56. IEEE (2020)
23. Gunti, P., et al.: Data mining approaches for sentiment analysis in online social networks (OSNs). In: Data Mining Approaches for Big Data and Sentiment Analysis in Social Media, pp. 116–141. IGI Global (2022)
24. Pei, Z., Sun, Z., Xu, Y.: Slang detection and identification. In: Proceedings of the 23rd Conference on Computational Natural Language Learning (CoNLL), pp. 881–889 (2019)
25. Gupta, A., Kumaraguru, P., Castillo, C., Meier, P.: TweetCred: real-time credibility assessment of content on Twitter. In: International Conference on Social Informatics 2014, pp. 228–243. Springer, Cham (2014)

26. Lee, J.L., Chen, L., Lam, C., Lau, C.M., Tsui, T.H.: PyCantonese: Cantonese linguistics and NLP in python. In: Proceedings of the 13th Language Resources and Evaluation Conference, pp. 6607–6611. European Language Resources Association (2022)

Predictive Model Building for Pain Intensity Using Machine Learning Approach

Ahmad Al-Qerem[1(⊠)], Batool Alarmouty[1], Ahmad Nabot[1],
and Mohammad Al-Qerem[2]

[1] Zarqa University, Zarqa, Jordan
ahmad_qerm@zu.edu.jo anabot@zu.edu.jo
[2] Al-Ahliyya Amman University, Amman, Jordan
M.alqerem@ammanu.edu.jo

Abstract. When the patient's body is compromised in any way, they will likely be in a lot of pain. If the caregiver is aware of the level of pain that the patient is experiencing, they will be better able to formulate the most appropriate treatment plan and provide the most appropriate medication. The visual analogue scale, often known as a VAS, is the approach that is used the most frequently to evaluate pain, and it is entirely dependent on patient reporting. Due to the fact that this kind of scale is ineffective when dealing with traumatic experiences or infants, it became necessary to devise a mechanism that could automatically recognize the severity of pain. On a dataset of multi-biopotential signals corresponding to varying degrees of discomfort, we evaluated the performance of the random forest and support vector machine classifiers.

Keywords: Pain intensity · Predictive model · Machine learning

1 Introduction

Pain is an important warning sign that something is amiss somewhere else in the body and should be checked out by a medical professional. Reference [1], it is an annoying and complex feeling that can be affected by a variety of circumstances including age and gender [10], and it is anything that disturbs the patient, causing him to feel uncomfortable. Pain can manifest itself in a variety of several ways and intensities, such as a burn, ashes, or a pinprick in a specific location on your body. There are two different kinds of pain: acute pain, which comes on abruptly and can be caused by a number of different things like an accident or sickness; this sort of pain needs to be detected as soon as possible or the situation will get worse. It is possible for a patient to experience chronic pain for months or years after suffering a severe injury that was left untreated, an infection, or an continuing disease such as cancer. Although this type of pain cannot always be

Supported by Zarqa University.

eliminated, doctors can attempt to handle it so that the patient gets better feeling. Untreated pain can have a substantial influence on the patient as well as the patient's family; it can make the patient irritable and impede his ability to concentrate. Character, bring on trouble sleeping, and have the potential to bring on sadness. It is necessary for the doctor to have an accurate assessment of the patient's level of pain in order to properly diagnose the source of the discomfort, devise an effective treatment strategy, and select the appropriate medication to provide to the patient. In most cases, it is up to the patient to rate the level of discomfort that he is experiencing on one of several scales that have been developed specifically for this function. There are many different scales used to determine the level of pain that a patient is facing; however, the most common scale is called the Visual Analogue Scale (VAS). This scale asks the patient to select their pain level on a scale from 0 to 10, with 0 meaning there is no pain and 10 meaning the most excruciating pain imaginable [4]. The VAS scale is widely utilized due to its ease of use; however, it does have some drawbacks, including the fact that patients who suffer from mental illness, infants, and traumas are unable to use it; additionally, the effectiveness of the scale is dependent on the patient's prior experience with pain; and finally, some patients will feign pain in order to obtain painkillers.

In an effort to find a solution to this issue, researchers have begun to automate the process of recognizing the degree of pain based on a number of parameters, including facial expression and biopotential signals. E-signals created by the electrochemical actions of a particular cell type during physiological processes in the body are called "biopotential signals." These signals are nothing more than that. Electrodes, a specific form of sensor, can be used to measure these signals by sticking them to the skin's surface. T1 represents the lowest level of pain and T4 represents the highest level of pain. Our primary goal in this study is to determine how biopotential signals acquired from electromyography and skin conductance level and electrocardiogram can be used to categorize four levels of pain. Following the application of feature selection to the dataset, the best result that we obtained using random forest was 86.9% accuracy, and the best result that we obtained using the support vector machine was 89.1% accuracy.

The remainder of the paper is organized as follows: we begin by introducing the related work that has been done in the field; next, a brief description of the dataset that we used; next, an explanation of the feature selection methods and the classifiers that we used; and finally, the conclusion of this paper.

2   Related Work

Researchers recently proposed different machine learning based techniques for the solution of many real-world problems [2,3,15]. Gruss et al. [6] used the same data used in this paper, after applying automatic feature selection methods on the data, they used the Support Vector Machine (SVM) classifier obtaining 79.29% accuracy for T1 and 90.94% for T4. Chu et al. [5] used Blood Volume Pulse (BVP), Electrocardiogram (ECG) and Skin Conductance Level (SCL).

The genetic algorithm was applied for feature selection, then a single-signal model that make the prediction based on a single physiological signal. Each signal was fed in the Linear Discriminant Algorithm (LDA), the SCL showed the best result with 68.93 accuracy while the worst results were from the ECG with 53.01 accuracy. Then multi-signal model was tested using the LDA, K-Nearest Neighbor (KNN) and SVM, the best result was obtained from the LDA algorithm with average accuracy of 75%. Jiang et al. [9] used the Heart Rate (HR), Breath Rate (BR), Galvanic Skin Response (GSR), and facial surface Electromyogram physiological signals. After applying the Artificial Neural Network (ANN) to classify the pain to three levels (no pain, mild pain, or severe pain), this classification obtained 83.3% average accuracy. Mamata et al. [7] proposed secure data communication for cloud based healthcare systems.

3   Dataset

The dataset that was used for this work consisted of 85 healthy people of varying ages and genders who were subjected to unpleasant heat stimuli. These stimuli were administered by connecting the ATS thermode to the right forearm of each participant [6]. The participants were asked to report when they reached both the pain sill and the pain tolerance sill. The pain sill is the point at which a subject begins to start feeling painful rather than just the sensation of heat. Both of these records were saved as T1 and T4, and the other levels were made in between T2 and T3. Please refer to the figure below. Following the recording of the temperatures that result in each of the four categories of discomfort, each subject was then subjected to these temperatures in a random order 20 times, with breaks in between each stimulus; as a result, each subject experienced a total of 80 stimuli during the experiment. The following list is comprised of the patients' biopotential signals that were preserved during this experiment.

4   Feature Selection Process

Feature selection refers to the process of choosing, from among the total set of features that are included in the dataset, the subset of characteristics that are most pertinent to the prediction model and provide the most helpful information for it. This process is known as feature selection. During this process, features that are less relevant, redundant, noisy, or completely useless are discarded without the loss of any important information [13]. The process of feature selection is seen as being of critical significance since it cuts down on memory storage, training time, and computing cost while simultaneously improving the predictive model's overall performance [8,12]. The feature selection process can be broken down into three primary groups: First, there are the wrapper approaches that think of the feature selection as a search problem. This means that it will experiment with a variety of alternative combinations of features and then use a predictive model to assess the quality of each combination. Second, there are the filter methods that deal with the feature selection as a pre-processing step. These methods use statistical methods to associate each feature with a score, rank the features based on this score, and then use the features that have the highest score in the chosen predictive model. Last but not least, we have the embedded approaches, which include carrying out the process of feature selection concurrently with the development of the predictive model. During this portion of the experiment, we are going to make use of one of the wrapper methods. The wrapper methods generally consist of two phases [12]: first, it generates a subset of features, and then it evaluates the combination of characteristics that are in the subset. [Create] and [Evaluate] are both in bold. It will continue to cycle through these two processes until either the optimal combination is found or the predetermined number of iterations has been reached. Exhaustive search, also known as brute force, is an example of a wrapper method that guarantees finding the best subset of features by trying every possible combination of features. However, it is rarely used because it requires a significant amount of computational power, which makes it impractical for datasets that contain a large number of features. For this reason, more sophisticated approaches are applied in order to achieve better results. Using a correlation measure that computes the strength of the linear relationship between features, we began the process of feature selection in our paper by removing the features that depend on others. If the correlation is too high between two features, then it indicates that the features are dependent on one another; accordingly, we eliminated one of the features that had a correlation value of more than 0.9 or less than −0.9. The Boruta algorithm was then used in our process. The Boruta technique was initially presented to the public in 2010 by Jankowski et al. [11]. This algorithm is an advancement on the practice of utilizing random forest classification for the purpose of feature selection. Constructing a random forest classifier involves mixing numerous decision trees, where each decision tree stands in for a separate classifier. Different subsets of the whole objects and features that were included in the dataset are utilized by each tree. After computing a score for each feature based on the classification results of each tree [14], after which the features that provide the least amount of information are removed, and the process is then restarted. In the Boruta algorithm, every feature is copied and then its values are mixed up in a random manner; the resulting new features are referred to as shadow features. The combined characteristics of the original features and the shadow features are then fed into the random forest; for further information, see Table 1.

Table 1. Shadow features

Original features		Shadow features		Label
F1	F2	F1	F2	
57.47	79.02	73.61	83.12	2
73.61	141.96	63.09	79.02	0
60.01	83.12	57.47	79.97	1
63.09	79.97	60.01	141.96	3

During each repetition, the values that are included within the shadow features are mixed up, and a score is calculated for each individual feature. If the score of a feature is higher than the maximum score of the shadow features, then the feature is regarded as important; however, if the score of a feature is lower than or equal to the score of its shadow feature, then the feature is regarded as unimportant, and as a result, the feature is removed and replaced with its shuffled version. Figure 1 presents the boruta algorithm. The process is finished either when all of the characteristics have been categorised or when a predefined number of iterations have been completed.

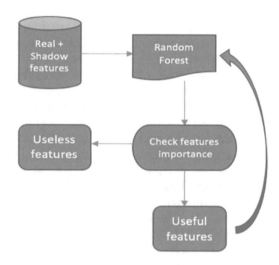

Fig. 1. Boruta algorithm

In the boruta algorithm, we specified a random forest classifier with a depth equal to 5, ran the process for a total of one hundred iterations, and set the perc value so that it would be equal to 85 if the maximum function could be reached. The process of selecting features is illustrated in Fig. 2.

5   Classification

Identifying the pain intensity level require a predictive model, after selecting the most important features we used two types of classification models. But before applying the classifiers on the data we need to do some normalization on it. We scaled the data using the standard scaler in which the distribution of each feature became centered around 0 with a standard deviation of 1. Then we grouped the data based on the pain level to have the ability to apply binary classification for each pain level vs. the no pain class. Then and for evaluation purposes we split the data randomly before training the model into training and test sets with 75% for the training set and the remaining for the testing set. After normalizing

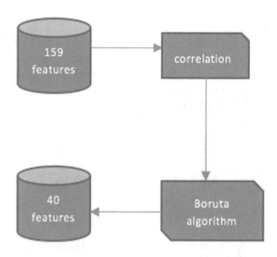

Fig. 2. Steps in the features selection process

and splitting the dataset, its ready to be used for the predictions. The process of dataset preparation for classification and evaluation described in Fig. 3.

Two predictive model were tested on the data. The first model is the Random Forest classifier which used several decision tree classifiers, in which each tree takes a subset from the data and predict the class which considered this tree vote, then the random forest aggregates these votes to decide the final class. We initialized the random forest classifier with 30 decision trees and measured the quality of split in each tree using the entropy to measure the information gain. The second model is the SVM classifier, that tries to split the classes by finding the hyperplane with the largest margin between the classes. Since our attributes and the class label have a non-linear relation we used the Radial Basis Function (RBF), the error rate that allows the hyperplane to ignore some training points (misclassify them) in trade of getting a higher margin was specified to 2.4, as for the gamma that determine the influence that training example can reach we use the auto parameter which divide 1 by the number of features in the dataset. For comparison purposes we tested the SVM classifier using the features used in the original paper, after applying the same pre-processing we used on our model, we found that using the boruta algorithm for feature selection shows an enhancement on the results. Figure 4 shows a comparison of the result for the three models based on accuracy for each pain level.

Both classifiers had the same accuracy for identifying the first level of pain, as for the second level of pain the random forest gave better result, while the support vector machine had the advantage for the third and fourth levels of pain, SVM accuracy drops when predicting the second level, but have better overall results than random forest.

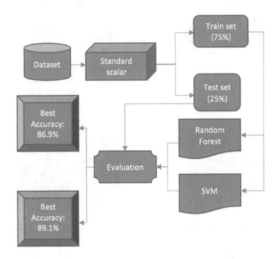

Fig. 3. Process of dataset preparation for classification and evaluation

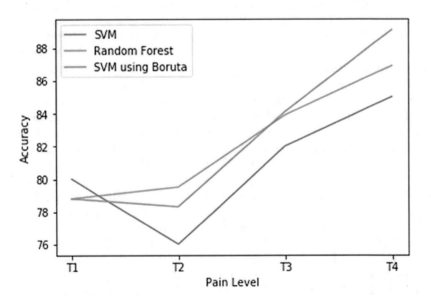

Fig. 4. Comparison between classifiers based on the accuracy

6   Conclusion

It is essential to determine the level of pain a patient is experiencing in order to devise the most effective treatment strategy for that patient; nevertheless, not everyone is able to articulate the degree to which they are experiencing pain. Therefore, it became essential to automate this procedure, as estimating the intensity of pain might provide a helpful assessment for the caregiver to use in therapy. In this paper we applied the boruta algorithm to select the most important features from the features extracted from biopotential signals of different parts of the body, obtained from heat stimuli experiments on healthy people, then compared the results between the random forest and support vector machine classifiers for pain intensity recognition, based on our experiments SVM showed better results than the random forest.

Acknowledgments. This research is funded by the Deanship of Research and Graduate Studies at Zarqa University/Jordan.

References

1. Al-Qerem, A.: An efficient machine-learning model based on data augmentation for pain intensity recognition. Egypt. Inform. J. **21**(4), 241–257 (2020)
2. Almomani, A., et al.: Phishing website detection with semantic features based on machine learning classifiers: a comparative study. Int. J. Semant. Web Inf. Syst. (IJSWIS) **18**(1), 1–24 (2022)
3. Bhardwaj, A., Kaushik, K.: Predictive analytics-based cybersecurity framework for cloud infrastructure. Int. J. Cloud Appl. Comput. (IJCAC) **12**(1), 1–20 (2022)
4. Bodian, C.A., Freedman, G., Hossain, S., Eisenkraft, J.B., Beilin, Y.: The visual analog scale for pain: clinical significance in postoperative patients. J. Am. Soc. Anesthesiol. **95**(6), 1356–1361 (2001)
5. Chu, Y., Zhao, X., Han, J., Su, Y.: Physiological signal-based method for measurement of pain intensity. Front. Neurosci. **11**, 279 (2017)
6. Gruss, S., et al.: Pain intensity recognition rates via biopotential feature patterns with support vector machines. PLoS ONE **10**(10), e0140330 (2015)
7. Gupta, B.B., Li, K.C., Leung, V.C., Psannis, K.E., Yamaguchi, S., et al.: Blockchain-assisted secure fine-grained searchable encryption for a cloud-based healthcare cyber-physical system. IEEE/CAA J. Autom. Sin. **8**(12), 1877–1890 (2021)
8. Guyon, I., Elisseeff, A.: An introduction to variable and feature selection. J. Mach. Learn. Res. **3**, 1157–1182 (2003)
9. Jiang, M., et al.: Acute pain intensity monitoring with the classification of multiple physiological parameters. J. Clin. Monit. Comput. **33**(3), 493–507 (2019)
10. Kumar, K.H., Elavarasi, P.: Definition of pain and classification of pain disorders. J. Adv. Clin. Res. Insights **3**(3), 87–90 (2016)
11. Kursa, M.B., Jankowski, A., Rudnicki, W.R.: Boruta—a system for feature selection. Fundam. Inform. **101**(4), 271–285 (2010)
12. Li, J., et al.: Feature selection: a data perspective. ACM Comput. Surv. (CSUR) **50**(6), 1–45 (2017)

13. Mostert, W., Malan, K., Engelbrecht, A.: Filter versus wrapper feature selection based on problem landscape features. In: Proceedings of the Genetic and Evolutionary Computation Conference Companion, pp. 1489–1496 (2018)
14. Rudnicki, W.R., Kierczak, M., Koronacki, J., Komorowski, J.: A statistical method for determining importance of variables in an information system. In: International Conference on Rough Sets and Current Trends in Computing, pp. 557–566. Springer (2006)
15. Vijayakumar, P., Rajkumar, S., et al.: Deep reinforcement learning-based pedestrian and independent vehicle safety fortification using intelligent perception. Int. J. Softw. Sci. Comput. Intell. (IJSSCI) 14(1), 1–33 (2022)

Analysis of N-Way K-Shot Malware Detection Using Few-Shot Learning

Kwok Tai Chui[1], Brij B. Gupta[2,3(✉)], Lap-Kei Lee[4], and Miguel Torres-Ruiz[5]

[1] Hong Kong Metropolitan University (HKMU), Kowloon, Hong Kong, China
`jktchui@hkmu.edu.hk`
[2] Department of Computer Science and Information Engineering, International Center for AI and Cyber Security Research and Innovations, Asia University, Taichung 41354, Taiwan
`bbgupta@asia.edu.tw`
[3] Lebanese American University, Beirut 1102, Lebanon
[4] Department of Electronic Engineering and Computer Science, School of Science and Technology, Hong Kong Metropolitan University, Kowloon, Hong Kong, China
`jktchui@hkmu.edu.hk`
[5] Instituto Politécnico Nacional, Centro de Investigacion en Computacion, UPALM-Zacatenco, Mexico City 07320, Mexico
`mtorresru@ipn.mx`

Abstract. Solving machine learning problems with small-scale training datasets becomes an emergent research area to fill the opposite end of big data applications. Attention is drawn to malware detection using few-shot learning, which is typically formulated as N-way K-shot problems. The aims are to reduce the effort in data collection, learn the rare cases, reduce the model complexity, and increase the accuracy of the detection model. The performance of the malware detection model is analyzed with the variation in the number of ways and the number of shots. This facilitates the understanding on the design and formulation of three algorithms namely relation network, prototypical network, and relation network for N-way K-shot problems. Two benchmark datasets are selected for the performance evaluation. Results reveal the general characteristics of the performance of malware detection model with fixed ways and varying shots, and varying ways and fixed shots based on the trends of results of 30 scenarios.

Keywords: Few-shot learning · Imbalanced dataset · Malware detection · N-way K-shot · Small-scale dataset

1 Introduction

It is well demonstrated in numerous research studies that the machine learning models can enhance the performance with more training data, typically the models take advantages from deep learning algorithms and big data techniques [1–3]. However, practical scenarios may experience the issues of small-scale datasets [4],

N. Nedjah et al. (Eds.): ICSPN 2021, LNNS 599, pp. 33–44, 2023.
https://doi.org/10.1007/978-3-031-22018-0_4

insufficient computational resources [6], and restriction in model complexity [7]. To solve these scenarios via machine learning algorithm, few-shot learning (FSL) has become a cutting-edge technique [8]. Generating additional training samples via generative adversarial network may only help relieving the issue of imbalanced dataset in a small extent [9]. Downsampling of the majority classes is an alternative approach, however, it sacrifices some truth samples [10].

Different Cyber attacks such as botnet attack [20], DDoS attack [12,28,33] and Malware attacks [16] such as spyware, Trojan horses, worms, and viruses can wreak havoc on human beings (privacy, financial loss, etc.), computer networks, servers, and computers. For robust and secured computer systems, the incidents of malware attacks are minorities compared with normal samples. The number of training data of malware attacks is limited that drives the research initiative to formulate the malware detection problem using FSL.

The rest of the paper is organized as follows. Section 1 presents the research topic, related works and their research limitations, and research contributions of our work. Section 2 shares the design and formulation of the proposed FSL algorithm. Section 3 details the performance evaluation of the proposed algorithm and the comparison with existing works. Section 4 concludes the research work with the potential future research directions.

1.1 Literature Review

Researchers proposed many machine learning techniques for cyber-security domains [5,14]. Such as, Siamese convolutional neural network with FSL was proposed for the anomaly detection in cyber-physical system [11]. The model achieved precision of 90.6% which outperformed existing methods such as Naïve Bayes, Siamese convolutional autoencoder, random forest, one-shot support vector machine, and time series analysis by 0.778–21.4%. Deep neural network and convolutional neural network with FSL were presented [13]. It was remarkably reducing the number of training data to only 1% of the whole dataset, compared with typically 20% setting. Simultaneously, the accuracy of the model was enhanced by 3.28–32.8% compared with existing methods such as recurrent neural network, support vector machine, multi-layer perceptron, random tree, random forest, Naïve Bayes, and J48. In [15], malware detection was achieved by a sample adaptation-based dynamic prototype network. Six settings of N-way K-shot including 5-way 5-shot, 5-way 10-shot, 10-way 5-shot, 10-way 10-shot, 5-way 1-shot, and 10-way 1-shot were evaluated with ranges of accuracy from 62.8% to 90.3%. Supervised infinite mixture prototypes with FSL was proposed for the malware detection [17]. The model embedded with 1D convolution neural network and bi-directional long short-term memory. The novelty of the proposed algorithm was able to recognize new classes that were not trained. Four settings of N-way K-shot namely 5-way 5-shot, 5-way 10-shot, 10-way 5-shot, and 10-way 10-shot, were selected to evaluate the performance of the algorithm. The ranges of accuracy were from 89.2% to 94.2%.

1.2   Research Limitations

The existing works have encountered some key limitations which are explained as follows.

- The performance evaluation omitted the scenarios of N-way K-shot [11,13] or shallowly analyzed on a few scenarios [15,17];
- Lack of studies on the N-way varying-shot problem where the available samples in different ways are differed; and
- There is room for improvement in the accuracy of few-shot learning as limited samples are available.

1.3   Research Contributions

To address the abovementioned limitations, three algorithms namely relation network, prototypical network, and matching network are employed. The key contributions are summarized as follows.

- Extensive analysis on 30 scenarios of N-way K-shot problems; and
- Discussion on the design and formulation of three malware detection algorithms namely relation network, prototypical network, and matching network for N-way K-shot problem.

2   Methodology

The methodology is comprised of the formulations of relation network, prototypical network, and matching network for FSL.

2.1   Relation Network

The original idea of relation network was proposed by a highly cited research work in 2018 [18]. The formulation follows an episode-based learning by choosing N classes (N-way) from training dataset with K labelled samples (K-shot) from each class. Define the sample set $X_{sample} = \{x_i, y_i\} \, \forall i \in [1, m]$ with $m = K \times N$. The remainder of the samples becomes query set $X_{query} = \{x_j, y_j\} \, \forall j \in [1, n]$.

The K-shot learning is extended from one-shot learning. The formulation begins with one-shot learning. A relation network is constructed by two modules namely embedding module f_e and relation module f_r. Sampling x_i and x_j where the outputs are passing into f_e to form feature spaces $f_e(x_i)$ and $f_e(x_j)$. These feature spaces are merged via concatenation $C(f_e(x_i), f_e(x_j))$. The concatenated feature space is passed into f_r. A relation score r_{ij} is thus computed as the similarity between x_i and x_j. A basic example of N-way 1-shot problem will obtain N r_{ij} which are defined as follows.

$$r_{ij} = f_r\left(C\left(f_e(x_i), f_e(x_j)\right)\right) \, \forall i \in [1, N] \tag{1}$$

Extending to N-way K-shot problem, the outputs of f_e perform element-wise sum. This class-level feature map is merge red with the abovementioned feature map. Overall, the number of r_{ij} is fixed for any-shot learning. The mean square error loss is defined as:

$$e, r = \underset{e,r}{argmin} \sum_{i=1}^{m} \sum_{j=1}^{n} (r_{ij} - 1 (y_i == y_j))^2 \tag{2}$$

2.2 Prototypical Network

Prototypical network was proposed in 2017 [19]. It is tailored for the FSL problems in which the model generalizes to any new class being not trained. In FSL, denote P labeled examples $X_{labeled} = \{(x_1, y_1), \ldots, (x_P, y_P)\}$ of a support set. $x_i \in R^N$ is the N-dimensional feature vector of a sample with corresponding label $y_i \in \{1, \ldots, K\}$, and $X_{labeled,k}$ is the set of labeled examples of class k.

The algorithm aims at calculating a prototype for every class with an embedding module $f_e : R^N \rightarrow R^M$ with hyperparameter e to control the learning. The prototype serves as the vector of the averaged support points in the corresponding class. The M-dimensional prototype p_k is defined as follows:

$$p_k = \frac{1}{|X_{labeled,k}|} \sum_{(x_i, x_y) \in X_{labeled,k}} f_e(x_i) \tag{3}$$

A distribution for data sample x using softmax activation function over distance to the prototypes in the embedding space:

$$f_e(y = k|x) = \frac{e^{(-d(f_e(x), p_k))}}{\sum_{k'} e^{(-d(f_e(x), p_{k'}))}} \tag{4}$$

with distance $d : R^M \times R^M \rightarrow [0, +\infty)$. Finally, the goal is to solve the minimization problem:

$$\min J(e) = -log f_e(y = k|x) \tag{5}$$

where $J(e)$ is the negative log-probability.

2.3 Matching Network

Matching network was originally formulated for one-shot learning problem [21]. It aims at mapping some samples n of input-output pairs $X_{map} = \{(x_i, y_i), \ldots, (x_n, y_n)\} \forall i \in [1, \ldots, n]$ to a classifier $C_{X_{map}}(\hat{x})$. Denote $\hat{x}$ as a testing sample with probability distribution $P(\hat{y}|\hat{x}, X_{map})$ over some outputs $\hat{y}$:

$$P(\hat{y} | \hat{x}, X_{map}) = \sum_{i=1}^{n} \alpha(\hat{x}, x_i) y_i \tag{6}$$

where $\alpha(.,.)$ is an attention scheme to fully specify $C_{X_{map}}(\widehat{x})$. Typically, it can be formulated as:

$$\alpha\left(\widehat{x}, x_i\right) = \frac{e^{\cos_{dis}(f_{e1}(\widehat{x}), f_{e2}(x_i))}}{\sum_{j=1}^{n} e^{\cos_{dis}(f_{e1}(\widehat{x}), f_{e2}(x_i))}} \tag{7}$$

where $\cos_{dis}(.,.)$ is the cosine distance function and f_{e1} and f_{e2} are some embedding functions. For simplicity, choose $f_{e1} = f_{e2}$.

The objective function of the matching network is formulated as follows:

$$\beta = \underset{\beta}{argmax} F_{A \sim T}\left[F_{X_{map} \sim T, B \sim A}\left[\sum_{(x,y) \in B} \log P_\beta\left(y | x, X_{map}\right)\right]\right] \tag{8}$$

where β is the hyperparameter, A is a sample from task T, A is then utilized to sample X_{map} and a batch B.

3   Performance Evaluation and Analysis

The benchmark datasets for the performance evaluation of the three FSL approaches are firstly discussed. It is followed by the performance evaluation of these approaches. A comparison is also made with existing works.

3.1   Benchmark Datasets for Malware Detection

Two benchmark datasets namely VirusTotal [22] and API-based malware detection system (APIMDS) [23] are chosen. Former dataset comprises of more than 500 classes of malware whereas the latter dataset has more than 20 classes. It is worth noting that k-fold cross-validation with small values of k, k = 2 is chosen as the datasets may have classes with few samples [24]. This is not identical to more common settings of k = 5 [25] and k = 10 [26].

The major settings of the evaluation are illustrated as follows.

- Meta-training stage contains some epochs where each epoch has 50 episodes, being randomly sampled from the training dataset;
- Each episode comprises of N classes and K support samples per class to form the N-way K-shot problem; and
- 30 scenarios are defined by varying the number of ways (5, 6, 7, 8, 9, 10) and the number of shots (1, 2, 3, 4, 5).

3.2   Performance Evaluation of the Three FSL Approaches

Using relation network, the accuracies of the malware detection model in 30 scenarios using two benchmark datasets are shown in Fig. 1. Key observations are drawn as follows.

- With the increase in the number of ways (N-Way), the malware detection problem becomes more complex that leads to the decrease in accuracies.
- With the increase in the number of shots (K-Shot), the accuracy of the malware detection model increases as more information is available during model training.
- Consider VirusTotal dataset, the ranges of accuracies of the N-Way 1-Shot are 62–71.3%, N-Way 2-Shot are 68.5–77%, N-Way 3-Shot are 73.8–81.9%, N-Way 4-Shot are 77.9–86.1%, and N-Way 5-Shot setting 81.4–88.6%. On the other hand, the ranges of accuracies of the 5-Way K-Shot are 71.3–88.6%, 6-Way K-Shot are 70.1–87.7%, 7-Way K-Shot are 68.6–86.4%, 8-Way K-Shot are 66.4–84.8%, 9-Way K-Shot are 64.3–83.1%, and 10-Way K-Shot are 62–81.4%.
- Consider APIMDS dataset, the ranges of accuracies of the N-Way 1-Shot are 60.3–68.5%, N-Way 2-Shot are 66.7–74.8%, N-Way 3-Shot are 71.9–79.8%, N-Way 4-Shot are 75.7–83%, and N-Way 5-Shot are 80–86.1%. On the other hand, the ranges of accuracies of the 5-Way K-Shot are 68.5–86.1%, 6-Way K-Shot are 67.4–85.3%, 7-Way K-Shot are 66.2–84.3%, 8-Way K-Shot are 64.3–83%, 9-Way K-Shot are 62.5–81.6%, 10-Way K-Shot are 60.3–80%.

Using prototypical network, the accuracies of the malware detection model in 30 scenarios using two benchmark datasets are shown in Fig. 2. The first two observations align with those in the analysis with relation network. Other key observations are summarized as follows.

- Consider VirusTotal dataset, the ranges of accuracies of the N-Way 1-Shot are 62.6–69.5%, N-Way 2-Shot are 68.8–75.8%, N-Way 3-Shot are 73.8–81.1%, N-Way 4-Shot are 75.4–84.9%, and N-Way 5-Shot setting 81.8–87.2%. On the other hand, the ranges of accuracies of the 5-Way K-Shot are 69.5–87.2%, 6-Way K-Shot are 68.3–86.5%, 7-Way K-Shot are 66.9–85.6%, 8-Way K-Shot are 65.6–84.5%, 9-Way K-Shot are 64.3–83.3%, and 10-Way K-Shot are 62.6–81.8%.
- Consider APIMDS dataset, the ranges of accuracies of the N-Way 1-Shot are 59.4–67.2%, N-Way 2-Shot are 66.2–73.4%, N-Way 3-Shot are 71–78.2%, N-Way 4-Shot are 75.6–81.9%, and N-Way 5-Shot are 79.1–85%. On the other hand, the ranges of accuracies of the 5-Way K-Shot are 67.2–85%, 6-Way K-Shot are 66.1–84%, 7-Way K-Shot are 64.6–82.8%, 8-Way K-Shot are 63–81.4%, 9-Way K-Shot are 61.4–80.2%, 10-Way K-Shot are 59.4–79.1%.

Using matching network, the accuracies of the malware detection model in 30 scenarios using two benchmark datasets are shown in Fig. 3. The first two observations align with those in the analysis with relation network and prototypical network. Other key observations are summarized as follows.

- Consider VirusTotal dataset, the ranges of accuracies of the N-Way 1-Shot are 60.4–67.7%, N-Way 2-Shot are 66.8–73.9%, N-Way 3-Shot are 72.2–78.6%, N-Way 4-Shot are 76.9–82.9%, and N-Way 5-Shot setting 80.6–86.3%. On the other hand, the ranges of accuracies of the 5-Way K-Shot are 67.7–86.3%, 6-Way K-Shot are 66.7–85.5%, 7-Way K-Shot are 65.5–84.5%, 8-Way K-Shot

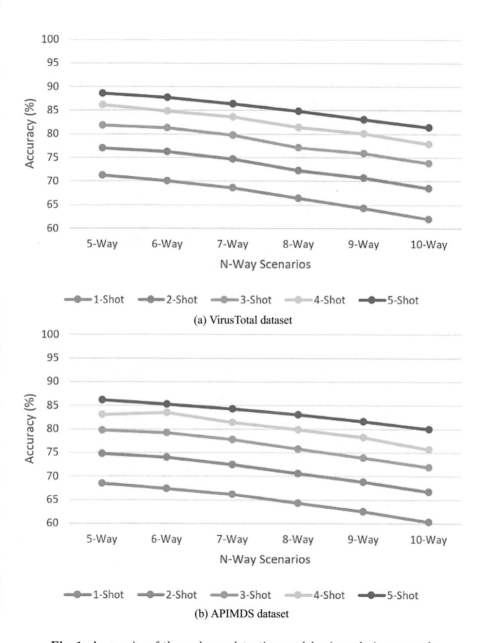

(a) VirusTotal dataset

(b) APIMDS dataset

Fig. 1. Accuracies of the malware detection model using relation network

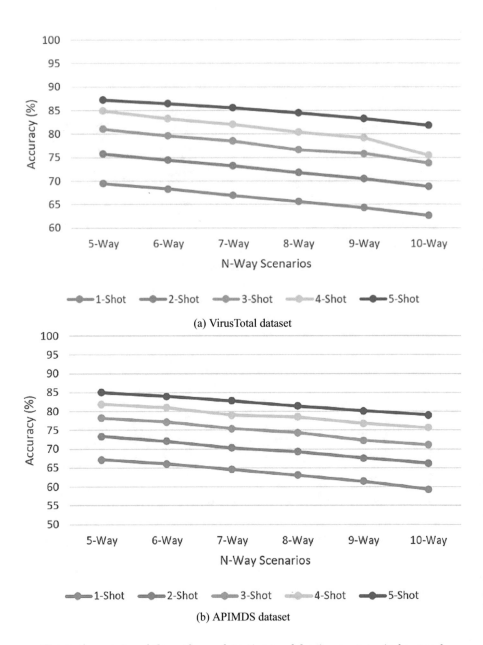

(a) VirusTotal dataset

(b) APIMDS dataset

Fig. 2. Accuracies of the malware detection model using prototypical network

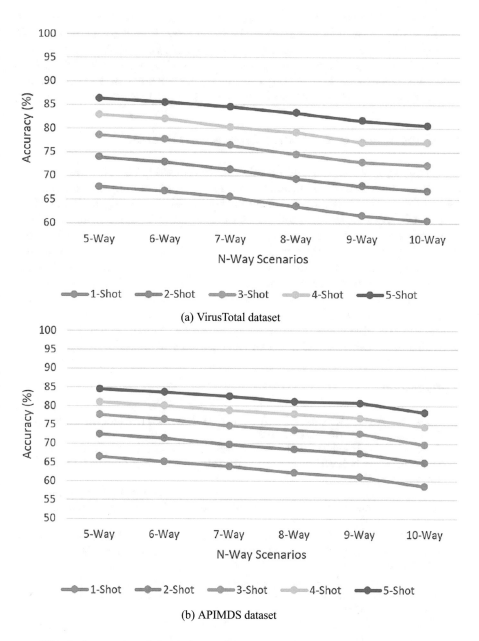

Fig. 3. Accuracies of the malware detection model using matching network.

are 63.5–83.2%, 9-Way K-Shot are 61.6–81.6%, and 10-Way K-Shot are 60.4–80.6%.
– Consider APIMDS dataset, the ranges of accuracies of the N-Way 1-Shot are 58.7–66.5%, N-Way 2-Shot are 64.9–72.4%, N-Way 3-Shot are 69.7–77.7%, N-Way 4-Shot are 74.4–81%, and N-Way 5-Shot are 78.3–84.5%. On the other hand, the ranges of accuracies of the 5-Way K-Shot are 66.5–84.5%, 6-Way K-Shot are 65.2–83.6%, 7-Way K-Shot are 63.9–82.5%, 8-Way K-Shot are 62.3–81.2%, 9-Way K-Shot are 61.1–80.8%, 10-Way K-Shot are 58.7–78.3%.

4   Conclusion and Future Research Directions

In this paper, three algorithms namely relation network, prototypical network, and matching network are proposed for malware detection. 30 scenarios are defined for varying settings of N-Way K-shot few-shot learning problems. Two benchmark datasets are selected for the performance evaluation and analysis. Common characteristics are drawn for all algorithms on the trends of varying ways and fixed shot, and fixed ways and varying shots. The ranges of accuracies of the malware detection model in 30 scenarios are 62–88.6% for relation network, 62.6–87.2% for prototypical network, and 60.4–86.3% for matching network.

Future research directions are recommended (i) generating additional training data in all classes [27,29]; (ii) merging the algorithms via optimization algorithms to take advantages from different networks [30,31]; and (iii) borrowing knowledge from other domains via transfer learning [32,34,35].

References

1. Dong, S., Wang, P., Abbas, K.: A survey on deep learning and its applications. Comput. Sci. Rev. **40**, 100379 (2021)
2. Sreedevi, A.G., Harshitha, T.N., Sugumaran, V., Shankar, P.: Application of cognitive computing in healthcare, cybersecurity, big data and IoT: a literature review. Inf. Process. Manage. **59**(2), 102888 (2022)
3. Almomani, A., et al.: Phishing website detection with semantic features based on machine learning classifiers: a comparative study. Int. J. Semant. Web Inf. Syst. (IJSWIS) **18**(1), 1–24 (2022)
4. Liang, W., Hu, Y., Zhou, X., Pan, Y., Kevin, I., Wang, K.: Variational few-shot learning for microservice-oriented intrusion detection in distributed industrial IoT. IEEE Trans. Ind. Inform. **18**(8), 5087–5095 (2022)
5. Ammi, M., Adedugbe, O., Benkhelifa, E.: Taxonomical challenges for cyber incident response threat intelligence: a review. Int. J. Cloud Appl. Comput. (IJCAC) **12**(1), 1–14 (2022)
6. Kim, J., Chi, S.: A few-shot learning approach for database-free vision-based monitoring on construction sites. Autom. Constr. **124**, 103566 (2021)
7. Hu, X., Chu, L., Pei, J., Liu, W., Bian, J.: Model complexity of deep learning: a survey. Knowl. Inf. Syst. **63**(10), 2585–2619 (2021)
8. Duan, R., Li, D., Tong, Q., Yang, T., Liu, X., Liu, X.: A survey of few-shot learning: an effective method for intrusion detection. Secur. Commun. Netw. **2021** (2021)

9. Li, Y., Xu, F., Lee, C.G.: Self-supervised meta learning generative adversarial network for few-shot fault diagnosis of hoisting system with limited data. IEEE Trans. Ind. Inform. (Early Access)
10. Duan, M., Liu, D., Chen, X., Liu, R., Tan, Y., Liang, L.: Self-balancing federated learning with global imbalanced data in mobile systems. IEEE Trans. Parallel Distrib. Syst. **32**(1), 59–71 (2021)
11. Zhou, X., Liang, W., Shimizu, S., Ma, J., Jin, Q.: Siamese neural network based few-shot learning for anomaly detection in industrial cyber-physical systems. IEEE Trans. Ind. Inform. **17**(8), 5790–5798 (2021)
12. Singh, A., et al.: Distributed denial-of-service (DDoS) attacks and defense mechanisms in various web-enabled computing platforms: issues, challenges, and future research directions. Int. J. Semant. Web Inf. Syst. (IJSWIS) **18**(1), 1–43 (2022)
13. Yu, Y., Bian, N.: An intrusion detection method using few-shot learning. IEEE Access **8**, 49730–49740 (2020)
14. Bhardwaj, A., Kaushik, K.: Predictive analytics-based cybersecurity framework for cloud infrastructure. Int. J. Cloud Appl. Comput. (IJCAC) **12**(1), 1–20 (2022)
15. Chai, Y., Du, L., Qiu, J., Yin, L., Tian, Z.: Dynamic prototype network based on sample adaptation for few-shot malware detection. IEEE Trans. Knowl. Data Eng. (Early Access)
16. Gaurav, A., et al.: A comprehensive survey on machine learning approaches for malware detection in IoT-based enterprise information system. Enterpr. Inf. Syst. 1–25
17. Wang, P., Tang, Z., Wang, J.: A novel few-shot malware classification approach for unknown family recognition with multi-prototype modeling. Comput. Secur. **106**, 102273 (2021)
18. Sung, F., Yang, Y., Zhang, L., Xiang, T., Torr, P.H., Hospedales, T.M.: Learning to compare: relation network for few-shot learning. In: Proceedings of the IEEE Conference on Computer Vision and Pattern Recognition, pp. 1199–1208. IEEE, USA (2018)
19. Snell, J., Swersky, K., Zemel, R.: Prototypical networks for few-shot learning. In: Advances in Neural Information Processing Systems, pp. 1–11. Curran Associates Inc., USA (2017)
20. Pan, X., Yamaguchi, S., Kageyama, T., Kamilin, M.H.B.: Machine-learning-based white-hat worm launcher in botnet defense system. Int. J. Softw. Sci. Comput. Intell. (IJSSCI) **14**(1), 1–14 (2022)
21. Vinyals, O., Blundell, C., Lillicrap, T., Wierstra, D.: Matching networks for one shot learning. In: Advances in Neural Information Processing Systems, pp. 1–9. Curran Associates Inc., Spain (2016)
22. Virus Total: Virustotal-Free Online Virus, Malware and Url Scanner. Available online: https://www.virustotal.com/en
23. Ki, Y., Kim, E., Kim, H.K.: A novel approach to detect malware based on API call sequence analysis. Int. J. Distrib. Sens. Netw. **11**(6), 1–9 (2015)
24. Zou, C., Wang, G., Li, R.: Consistent selection of the number of change-points via sample-splitting. Ann. Stat. **48**(1), 413–439 (2020)
25. Chui, K.T., et al.: An MRI scans-based Alzheimer's disease detection via convolutional neural network and transfer learning. Diagnostics **12**(7), 1–14 (2022)
26. Chui, K.T.: Driver stress recognition for smart transportation: applying multiobjective genetic algorithm for improving fuzzy c-means clustering with reduced time and model complexity. Sustain. Comput. Inf. Syst. **35**, 1–11 (2022)
27. Gao, N., et al.: Generative adversarial networks for spatio-temporal data: a survey. ACM Trans. Intell. Syst. Technol. **13**(2), 1–25 (2022)

28. Mishra, A., et al.: A comparative study of distributed denial of service attacks, intrusion tolerance and mitigation techniques. In: 2011 European Intelligence and Security Informatics Conference, Sept 2011, pp. 286–289. IEEE (2011)

29. Sajun, A.R., Zualkernan, I.: Survey on implementations of generative adversarial networks for semi-supervised learning. Appl. Sci. **12**(3), 1718 (2022)

30. Katoch, S., Chauhan, S.S., Kumar, V.: A review on genetic algorithm: past, present, and future. Multimed. Tools Appl. **80**(5), 8091–8126 (2021)

31. Chui, K.T., et al.: Extended-range prediction model using NSGA-III optimized RNN-GRU-LSTM for driver stress and drowsiness. Sensors **21**(19), 6412 (2021)

32. Kumar, S.: MCFT-CNN: malware classification with fine-tune convolution neural networks using traditional and transfer learning in internet of things. Future Gener. Comput. Syst. **125**, 334–351 (2021)

33. Gupta, B.B., Misra, M., Joshi, R.: An ISP level solution to combat DDoS attacks using combined statistical based approach. Int. J. Inf. Assur. Secur. (JIAS) **3** (2012)

34. Chui, K.T., et al.: Transfer learning-based multi-scale denoising convolutional neural network for prostate cancer detection. Cancers **14**(15), 3687 (2022)

35. Gupta, S., et al.: Detection, avoidance, and attack pattern mechanisms in modern web application vulnerabilities: present and future challenges. Int. J. Cloud Appl. Comput. (IJCAC) **7**(3), 1–43 (2017)

Efficient Feature Selection Approach for Detection of Phishing URL of COVID-19 Era

Md Saif Ali and Ankit Kumar Jain[✉]

National Institute of Technology, Kurukshetra, Kurukshetra, Haryana 136119, India
ankitjain@nitkkr.ac.in

Abstract. Cybercrime is a growing concern, particularly in this COVID-19 era. The COVID-19 outbreak has shown the significant impact potential of such crises on our daily lives worldwide. Phishing is a social engineering crime that can cause financial and reputational damages such as data loss, personal identity theft, money loss, financial account credential theft, etc., to people and organizations. In the recent outbreak of the COVID-19 pandemic, many companies and organizations have changed their working conditions, moved to an online environment workspace, and implemented the Work From Home (WFH) business model that increases the phishing attacks vectors and risk of breaching internal data. In this paper, we have extracted nine efficient features from the URLs and applied seven different Machine Learning algorithms to recognize phishing URLs. Machine learning algorithms are often used to detect phishing attacks more accurately before affecting users. The obtained result concludes that the Random Forest model provides the best and highest accuracy of 95.2%.

Keywords: COVID-19 · Phishing attacks · Machine learning

1 Introduction

On Dec 2019, a virus named SARS-COV2 was discovered in Wuhan, china. It is very contiguous and has quickly spread at an unprecedented rate around the more than 210 countries in the world [1]. COVID-19 has affected millions of people worldwide, and its long-term impact remains to be seen. Due to the COVID-19 pandemic, governments have enforced strict lockdown, border closures, quarantining of infected people, and social distancing [2,4]. The COVID-19 outbreak has shown the significant impact potential of such crises on our daily lives worldwide. With physical lockdowns becoming more common, cybercrime has become more popular. The arrival of COVID-19 was a significant factor in any discussion about 2020 development. However, COVID-19 relates to cybercrime in its proper context. The COVID-19 cyberattack showed that cybercrime is still essentially the same, but criminals change the narrative to fool victims.

© The Author(s), under exclusive license to Springer Nature Switzerland AG 2023
N. Nedjah et al. (Eds.): ICSPN 2021, LNNS 599, pp. 45–56, 2023.
https://doi.org/10.1007/978-3-031-22018-0_5

The difference with COVID-19 is that since the virus was stopped from spreading due to physical restrictions, many people are now working from home and using remote access to business resources, which makes them more vulnerable to attack [3,6]. More recently, scammers have been tricking their victims using the COVID-19 pandemic. Since the WHO reported that COVID-19 had caused an infodemic that is beneficial for phishers, attackers took advantage of people's anxiety about getting COVID-19 and the urgency to hunt for information connected to the virus [5,8]. Cybercriminals frequently use disasters and notable events for their personal gain. Because the COVID-19 pandemic began, fraudsters have targeted employees, healthcare facilities, and even the general public with many themed phishing and malware assaults [7,11]. Phishing is a social engineering crime that can cause financial and reputational damages such as data loss, personal identity theft, money loss, financial account credential theft, etc., to people and organizations [9,13]. In the recent outbreak of the COVID-19 pandemic, many companies and organizations experienced changes in their working condition due to COVID-19 Pandemic and shifting to an online workspace. Companies, organizations, and institutions worldwide have executed the Work From Home (WFH) working model that increases the phishing attacks and risk of a breach in sensitive internal data [10,17].

2   Related Work

The Covid-19 epidemic has tragically led to a sharp rise in different types of cybercrime all over the globe. Because of the COVID-19 pandemic, cybercriminals are attacking government organizations, Retail and E-commerce, healthcare organizations, educational organizations, and other industries. The urgent requirement is comprehensively analyzing cyberattacks, including their signs and effects. We see a need to design an efficient Machine Learning classifier that can efficiently detect phishing URLs.

A machine learning model is trained using predetermined features, and the model then declares whether or not a URL is phishing [12,19,22]. Recently, several researchers have proposed machine learning methods for phishing URL identification that use a massive database of legitimate and phishing websites. The features relating to the URL, page content, DNS, etc., are extracted to create the new data set with the chosen feature set. After preprocessing, Machine learning is applied to these massive datasets.

In Zahra et al. [14], authors analyzed various COVID-19 themed cyberattacks and the impact of COVID-19 on cyber security and proposed a novel fuzzy logic and data mining based phishing detection system with 3 layers, 6 segments, and 30 components synchronized with each other with accuracy rate 98.19%. Basit et al. [15], proposed a novel ensemble machine learning model to detect phishing attacks. ANN, KNN, and Decision Tree (C4.5) are used in ensemble models with Random Forest Classifier (RFC) to detect phishing attacks with 97.33% accuracy. Islam et al. [16], proposed a Machine Learning based framework to identify phishing domains having COVID-related keywords. Authors collect

7849 domains from DomainTools and WhoisDS and apply 5 lexical features to detect phishing domains. The proposed framework has a 99.2% accuracy rate. Wang [18], proposed a conventional machine and deep learning model to detect COVID-19-related domain names. CNN-LSTM and CNN-Bi-LSTM showed the best accuracy rate of 98.70% among all the models. In Viktor et al. [20] proposed an efficient approach for detecting covid-19 related domain names. Using 10 lexical-based features and applying batch learning and online learning classification to detect covid-19 related domain names. Boyle et al. [21], developed an anti-phishing browser extension called "MailTrot" to detect phishing e-mail in the era of COVID-19. Developed approach having high accuracy and high level of usability for end users. Alsaidi et al. [23] proposed 4 Machine Learning algorithms (Logistics Regression, Decision Tree, Support Vector Machine, and Naive Bayes) on small datasets to detect fake news during the COVID-19 epidemic. Naive Baye's has a 94.6% accuracy high among all four machine learning algorithms. Xenakis et al. [24] proposed a novel approach to detect phishing e-mails. Proposed approaches use a combination of Natural Language Processing (TF-IDF, word2vec, and BERT) and Machine Learning (Logistic Regression, Decision Tree, Random Forest, Gradient Boosting Tress, and Naive Bayes) to detect phishing e-mails. A combination of word2vec with Random Forest algorithms gives better accuracy on balanced datasets, and word2vec with Logistic Regression gives better results on imbalanced datasets.

Jafar et al. [25], proposed a phishing URL detection system by using Gated Recurrent Unit (GRU) with an accuracy of 98.30%. The proposed system is able to detect phishing URLs quickly with a higher accuracy rate. Afandi et al. [26], developed a COVID-19 themed phishing detection system that can detect phishing URLs based on the hyperlink. The proposed system uses the KNN algorithm and URL-based features to classify phishing URLs with 97.80% and 99.60% accuracy for dataset 1 and dataset 2, respectively. Almomani et al. [27], proposed 16 Machine Learning classifiers using 48 semantic features on two different data sets. Proposed approaches are more effective in detecting web and spear phishing attacks effectively. Gradient Boosting and Random Forest have the best accuracy of 97% among 16 Machine Learning Classifiers. Alrefaai [28], used Machine Learning approaches (Decision Tree, Random Forest, XG Boost, Multilayer Perceptron, KNN, Naive Bayes, Adaboost, and Gradient Boosting) to detect phishing attacks. The proposed approach uses Kaggle data sets, which have 86 features and 11,430 URLs, and divides datasets in Training and Testing with a ratio of 50:50, and achieves 96.6% accuracy using XG Boost Model.

3   Materials and Methods

The principal aim of our proposed approach is to distinguish phishing URLs from Legitimate ones. In our proposed method, we firstly collect a dataset from The Majestic Million Top 1 million websites and Spycloud COVID-19 Themed dataset and finalize the 9 different Address-Bar based Features. After finalizing the features, we extracted the features and used them for testing and training purposes. The proposed approach takes 8 different machine learning algorithms, i.e. Decision Tree, Random Forest, Logistic Regression, K-Nearest Neighbors, Support Vector Machine, Naive Bayes, Gradient Boosting Classifier, and Multi-Layer Perceptron, to classify phishing URLs. Figure 1 demonstrates the system architecture of our proposed approach for detecting phishing URLs. In the end, we get Machine Learning algorithm results, and performance is evaluated for each and every Machine Learning algorithm in order to get more pre-eminent algorithms for our proposed novel approach for detecting phishing URLs.

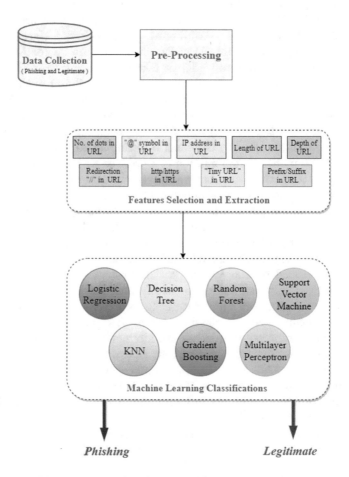

Fig. 1. Proposed architecture for phishing detection

3.1   Dataset Collection

We used the publicly available dataset downloaded from The Majestic Million Top 1 million websites [29] and Spycloud COVID-19 Themed dataset [30]. The dataset contains 5000 legitimate URLs collected from The Majestic Millions and 5000 Phishing URLs collected from Spycloud COVID-19 Themed Dataset. We used 9 different features from the dataset, which distinctly identify Malicious (-1) and benign (1) websites.

Table 1. Distribution of legitimate and phishing URLs

Data set	Description
The Majestic Millions Top 1 millions websites	5000 legitimate URLs
Spycloud COVID-19 Themed dataset	5000 phishing URLs

3.2 Pre-processing

In Pre-processing phase, raw data is turned into a format that machine learning algorithms can understand and evaluate. Data cleaning, Data integration, Data transformation, smoothing, normalization, and aggregation are examples of data pre-processing.

3.3 Features Selection and Extraction

F1 - No of Dots (.) in URL In general, legitimate URLs do not contain more than three dots, but phishing ones may include more than three. The phishing URLs have many sub-domains separate by the dot (.) symbol.

Example: http://coronavirus-com.mail.protection.outlook.com.

It is a phishing website, but some users may believe that they are visiting a legitimate outlook website. Therefore, If a URL contains more than three dots then it is considered a phishing URL.

$$\text{F1} = \text{if} \begin{cases} No.\ of\ dots\ in\ URL\ \geq\ 4\ \rightarrow\ Phishing \\ Otherwise\ \rightarrow\ Legitimate \end{cases} \tag{1}$$

F2 - IP Address in URL An IP address may be found in the domain name palace of a phishing URL.

Example: http://221.143.49.61/anticorona19.com.

Phishers often use IP address in place of the domain name to hide the identity of websites and try to steal personal information from this URL.

$$\text{F2} = \text{if} \begin{cases} Domain\ Part\ has\ an\ IP\ Address\ \rightarrow\ Phishing \\ Otherwise\ \rightarrow\ Legitimate \end{cases} \tag{2}$$

F3 - "@" Symbol in URL The "@" symbol in the URL causes the browser to ignore everything before it, and the actual site address frequently comes after the "@" symbol.

$$F3 = \text{if} \begin{cases} Presence\ of\ ``@"\ symbol\ in\ URL\ \rightarrow\ Phishing \\ Otherwise\ \rightarrow\ Legitimate \end{cases} \quad (3)$$

F4 - Length of URL Phishers frequently choose a lengthy URL to hide the questionable domain name. An average length of a phishing URL is determined to be larger than or equal to 52 characters during feature extraction.

$$F4 = \text{if} \begin{cases} Length\ of\ URL\ >\ 54 \rightarrow\ Phishing \\ Otherwise\ \rightarrow\ Legitimate \end{cases} \quad (4)$$

F5 - Depth of URL Count the number of subpages in the input URL and determine the depth of the URL based on the '/' symbol. A website containing more than three '/' symbols is considered phishing or otherwise legitimate.

$$F5 = \text{if} \begin{cases} Occurance\ of\ ``/"\ in\ the\ URL\ >3\ \rightarrow\ Phishing \\ Otherwise\ \rightarrow\ Legitimate \end{cases} \quad (5)$$

F6 - Redirection "//" in URL The user will move to another website if the URL path contains the symbol "//". This feature checks whether the URL contains the "//" character. We calculated that if the URL starts with HTTP, it means that "//" will appear in the sixth position. If the URL uses HTTPS, the "//" should appear in the seventh position. If the last occurrence of "//" in the URL is greater than the seventh position, then it is considered a phishing URL; otherwise Legitimate.

$$F6 = \text{if} \begin{cases} Last\ Occurance\ of\ ``//"\ in\ the\ URL\ >7\ \rightarrow\ Phishing \\ Otherwise\ \rightarrow\ Legitimate \end{cases} \quad (6)$$

F7 - Presence of "HTTP/HTTPS" in Domain Name In the phishing URL, 'HTTP' or 'HTTPS' protocols may appear more than one time. However, in the legitimate URL, 'HTTP' or 'HTTPS' protocols appear only once. This features checks if the domain part of the URL contains the 'HTTP' or 'HTTPS' protocol. Phishers can add the 'HTTP' or 'HTTPS' token to the URL's domain section to trick the user.

$$F7 = \text{if} \begin{cases} Presence\ of\ `http'\ or\ `https'\ in\ Domain\ name\ >0 \rightarrow\ Phishing \\ Otherwise\ \rightarrow\ Legitimate \end{cases}$$

$$(7)$$

F8 - Using URL Shortening Services "TinyURL" URL Shortening is a "worldwide web" method where the URL can be much shorter in length and still leads to the required web page. This redirects a browser to a webpage with a longer URL by using the HTTP protocol. Phishers generally used this to hide malicious domain names. For example, the URL "http://covid-19help. info/" may be shortened to "https://bit.ly/3aT2o9N".

$$F8 = \text{if} \left\{ \begin{array}{ccc} TinyURL & \to & Phishing \\ Otherwise & \to & Legitimate \end{array} \right. \tag{8}$$

F9 - Prefix or Suffix "–" in Domain In legitimate URLs, the dash (–) symbol is rarely used. Phishers frequently add prefixes or suffixes to domain names to give an illusion to users that they are visiting a trustworthy website. If the domain part of the URL includes a hyphen (–), then it is considered a phishing URL, otherwise considered a Legitimate URL.

$$F9 = \text{if} \left\{ \begin{array}{ccc} Presence\ of\ dash(-)\ symbol\ in\ Domain & \to & Phishing \\ Otherwise & \to & Legitimate \end{array} \right. \tag{9}$$

3.4 Machine Learning Algorithms

Linear Regression
A statistic called simple linear regression is used to find any correlations between independent and dependent variables (sometimes called the response or result variables) (also known as explanatory variables, predictors, or characteristics). It's a technique. It is only possible to declare that the outcome variable's (Y) value varies in relation to the characteristic's (X) value. Regression techniques cannot be used to establish a causal relationship between two variables. Regression is one of the most used supervised learning methods in predictive analytics. Regression models must be used with an understanding of both the training dataset's outcome and feature variables [31].

Decision Tree
A group of "divide and conquer" problem-solving techniques known as "classification trees" or decision trees use an inverted tree structure with roots at the top to anticipate the values of outcome variables. The tree begins with a root node that contains all of the data and then employs a clever technique to divide the node (parent node) into numerous branches (and thus create child nodes). Sub-sets of the original data are created. To increase the number of comparable groups on the child nodes, this is done. It is one of the most effective methods for creating business rules using predictive analytics [32].

Random Forest
Because of its effectiveness and scalability, random forest is one of the most widely used ensemble methods in the business. Each decision tree in a random forest is made up of a subset of randomly chosen features without substitution and a bootstrap sample (sampling with substitution), which is a collection of

decision trees (classification trees and regression trees). Normally, decision trees expand deeply (without pruning). To increase the model's accuracy, the estimator or quantity of models employed in the random forest can be changed [33].

Support Vector Machine
A supervised machine learning algorithm is a support vector machine. Used primarily for classification problems but also for regression tasks as well. Plotting each retrieved data value as a point in an n-dimensional space or graph is how the support vector algorithm works. Where "n" stands for the overall number of data pieces, each data point's value is represented in the graph by a unique coordinate. The classification can then be carried out by locating a line or hyperplane that specifically divides and distinguishes the two classes of data after the distribution of the coordinate data [34].

K-Nearest Neighbors
The Nearest Neighbor (KNN) algorithm is a nonparametric lazy learning algorithm used for regression and classification problems. The KNN algorithm detects observations similar to the new observations in the training set. There, the observed values are called neighbors. For better accuracy, you can consider a range of neighborhoods (K) to classify new observations. The new observation class can be expected to be the same as the class to which the majority of neighbors belong [35].

Gradient Boosting Classifier
A gradient boosting classifier is a type of classifier that creates a sequence of weak prediction models, typically in the form of a decision tree. Throughout the learning process, contiguous trees are produced. This method builds the initial model, makes value predictions, and determines the loss or discrepancy between the initial model's output and the actual value. The loss following the first step is then predicted using a second model. You can keep doing this till you are happy with the outcomes. Finding new trees that reduce the loss function repeatedly is the fundamental goal of gradient boosting. The loss function expresses the amount of the model's inaccuracy [36].

Multi-Layer Perceptron
Multilayer Perceptron (MLP) is a feedforward artificial neural network technique that uses backpropagation learning to classify target variables used in supervised learning. MLPs can be applied to complex nonlinear problems and work well with large input data with relatively fast performance. Algorithms tend to achieve the same accuracy ratio even when the data is small [37].

3.5   Experimental Results and Discussion

We have examined our proposed phishing detection mechanism on different Machine Learning classifiers. Initially, we selected 9 Address Bar Features and then extracted these selected features from the dataset, different machine learning including Random Forest, Decision Tree, Logistic Regression, Support Vector Machine, K-Nearest neighbor, Gradient Boosting Classifier, and Multi-Layer Perceptron is, being applied to obtain the performance metrics. We used the Jupyter Notebook platform for classification and used 20% data for testing purposes and the remaining 80% data for training purposes. Table 2 demonstrates the accuracy, precision, recall, and F1 score of our proposed approach on various machine learning algorithms, i.e., Logistic Regression, Random Forest, Decision Tree, SVM (Support Vector Machine), K-Nearest Neighbors, Gradient Boosting Classifier, and Multi-Layer Perceptron.

Table 2. Summary of machine learning classifier results

Algorithms	Accuracy	Precision	Recall	F1 score
Random Forest	95.2%	98.1%	98.6%	95.7%
Gradient Boosting	94.8%	96.9%	97.8%	95.4%
Decision Tree	94.7%	98.2%	98.6%	95.3%
Multi-Layer Perceptron	94.4%	96.4%	98.3%	95.0%
Support Vector Machine	93.9%	94.3%	97.0%	94.6%
K-Nearest Neighbors	93.7%	98.1%	98.1%	94.4%
Logistic Regression	88.6%	88.9%	90.6%	89.8%

After comparing the various performance metrics for different machine learning classifiers, we have evaluated the Random Forest machine learning classifier gives the best result. The Random Forest achieves the better classification result with higher accuracy, i.e., 95.2%. Moreover, we achieved 98.1% precision, 98.6% recall, and 95.7% F1-score with Random Forest Machine Learning Algorithms.

4   Conclusion and Future Work

The covid-19 epidemic has tragically led to a sharp rise in different types of cybercrime all over the globe. Because of the current situation, cybercriminals are attacking government organizations, Retail and E-commerce, healthcare organization, educational organizations, and other industries. The urgent requirement is to conduct a thorough analysis of cyberattacks, including their signs and effects. To address these issues, we have proposed novel machine learning algorithms using nine different features on seven different machine learning classifiers to detect phishing URLs with better and higher accuracy. With the obtained result, it is concluded that the Random Forest model provides the best and highest accuracy. In the future, we aim to further study the new more no. of features in more depth and add these features to our existing model to get better performance. In addition, to get better performance, we also aim to work on Deep Learning and Artificial Intelligence techniques to efficiently detect the phishing websites.

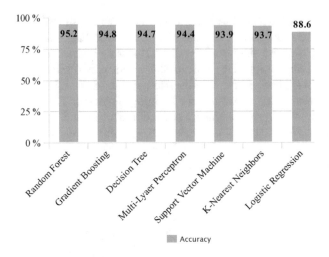

Fig. 2. Comparison of results

References

1. Debata, B., Patnaik, P., Mishra, A.: COVID-19 pandemic! It's impact on people, economy, and environment. J. Public Aff. **20**(4), e2372 (2020)
2. Harper, L., et al.: The impact of COVID-19 on research. J. Pediatr. Urol. **16**(5), 715–716 (2020)
3. Ozili, P.K., Arun, T.: Spillover of COVID-19: impact on the global economy (2020). Available at SSRN 3562570
4. Gupta, S., et al.: PHP-sensor: a prototype method to discover workflow violation and XSS vulnerabilities in PHP web applications. In: Proceedings of the 12th ACM International Conference on Computing Frontiers (CF '15), pp. 1–8. Association for Computing Machinery, New York, NY, USA. Article 59 (2015). https://doi.org/10.1145/2742854.2745719
5. Eian, I.C., Yong, L.K., Li, M.Y.X., Qi, Y.H., Fatima, Z.: Cyber attacks in the era of covid-19 and possible solution domains (2020)
6. Almomani, A., et al.: Phishing website detection with semantic features based on machine learning classifiers: a comparative study. Int. J. Semant. Web Inf. Syst. (IJSWIS) **18**(1), 1–24 (2022)
7. Lallie, H.S., et al.: Cyber security in the age of COVID-19: a timeline and analysis of cyber-crime and cyber-attacks during the pandemic. Comput. Secur. **105**, 102248 (2021)
8. Chui, K.T., et al.: Handling data heterogeneity in electricity load disaggregation via optimized complete ensemble empirical mode decomposition and wavelet packet transform. Sensors **21**(9), 3133 (2021). https://doi.org/10.3390/s21093133
9. What Is Phishing? https://www.phishing.org/what-is-phishing
10. Eian, I.C., Yong, L.K., Li, M.Y.X., Qi, Y.H., Fatima, Z.: Cyber attacks in the era of covid-19 and possible solution domains. Preprints 2020, 2020090630 (2020). https://doi.org/10.20944/preprints202009.0630.v1
11. Lu, J., Shen, J., et al.: Blockchain-based secure data storage protocol for sensors in the industrial internet of things. IEEE Trans. Ind. Inform. **18**(8), 5422–5431 (2022). https://doi.org/10.1109/TII.2021.3112601

12. Jain, A.K., Gupta, B.B.: A machine learning based approach for phishing detection using hyperlinks information. J. Ambient Intell. Humaniz. Comput. **10**(5), 2015–2028 (2018). https://doi.org/10.1007/s12652-018-0798-z
13. Cvitić, I., Peraković, D., Periša, M., Gupta, B.: Ensemble machine learning approach for classification of IoT devices in smart home. Int. J. Mach. Learn. Cybern. **12**(11), 3179–3202 (2021). https://doi.org/10.1007/s13042-020-01241-0
14. Zahra, S.R., Chishti, M.A., Baba, A.I., Wu, F.: Detecting covid-19 chaos driven phishing/malicious URL attacks by a fuzzy logic and data mining based intelligence system. Egypt. Inform. J. **23**(2), 197–214 (2022)
15. Basit, A., Zafar, M., Javed, A.R., Jalil, Z.: A novel ensemble machine learning method to detect phishing attack. In: 2020 IEEE 23rd International Multitopic Conference (INMIC), Nov 2020, pp. 1–5. IEEE (2020)
16. Ispahany, J., Islam, R.: Detecting malicious COVID-19 URLs using machine learning techniques. In: 2021 IEEE International Conference on Pervasive Computing and Communications Workshops and other Affiliated Events (PerCom Workshops), pp. 718–723 (2021). https://doi.org/10.1109/PerComWorkshops51409.2021.9431064
17. Gupta, B.B.: A lightweight mutual authentication approach for RFID tags in IoT devices. Int. J. Netw. Virt. Org. (2016)
18. Wang, Z.: Use of supervised machine learning to detect abuse of COVID-19 related domain names. Comput. Electr. Eng. **100**, 107864 (2022)
19. Sharma, R., Sharma, T.P., Sharma, A.K.: Detecting and preventing misbehaving intruders in the internet of vehicles. Int. J. Cloud Appl. Comput. (IJCAC) **12**(1), 1–21 (2022)
20. Mvula, P.K., Branco, P., Jourdan, G.V., Viktor, H.L.: COVID-19 malicious domain names classification. Expert Syst. Appl. 117553 (2022)
21. Boyle, P., Shepherd, L.A.: MailTrout: a machine learning browser extension for detecting phishing e-mails. In: 34th British HCI Conference 34, July 2021, pp. 104–115
22. Ling, Z., Hao, Z.J.: An intrusion detection system based on normalized mutual information antibodies feature selection and adaptive quantum artificial immune system. Int. J. Semant. Web Inf. Syst. (IJSWIS) **18**(1), 1–25 (2022)
23. Etaiwi, H.A.: Empirical evaluation of machine learning classification algorithms for detecting COVID-19 fake news. Int. J. Adv. Soft Comput. Appl. **14**(1) (2022)
24. Bountakas, P., Koutroumpouchos, K., Xenakis, C.: A comparison of natural language processing and machine learning methods for phishing email detection. In: The 16th International Conference on Availability, Reliability and Security, Aug 2021, pp. 1–12
25. Jafar, M.T., AlFawareh, M., Barhoush, M., AlshiraH, M.H.: Enhanced analysis approach to detect phishing attacks during COVID-19 crisis. Cybern. Inf. Technol. **22**(1), 60–76 (2022)
26. Afandi, N.A., Hamid, I.R.A.: Covid-19 phishing detection based on hyperlink using K-nearest neighbor (KNN) algorithm. Appl. Inf. Technol. Comput. Sci. **2**(2), 287–301 (2021)
27. Almomani, A., et al.: Phishing website detection with semantic features based on machine learning classifiers: a comparative study. Int. J. Semant. Web Inf. Syst. (IJSWIS) **18**(1), 1–24 (2022)
28. Alswailem, A., Alabdullah, B., Alrumayh, N., Alsedrani, A.: Detecting phishing websites using machine learning. In: 2019 2nd International Conference on Computer Applications & Information Security (ICCAIS), May 2019, pp. 1–6. IEEE

29. The Majestic Million. https://majestic.com/reports/majestic-million
30. COVID-19 Themed Domain Dataset. Available at: https://spycloud.com/resource/covid19-domain-dataset/
31. Liner Regression. https://www.techopedia.com/definition/32060/linear-regression
32. Tu, P.L., Chung, J.Y.: A new decision-tree classification algorithm for machine learning. In: TAI'92-Proceedings Fourth International Conference on Tools with Artificial Intelligence, Jan 1992, pp. 370–371. IEEE Computer Society
33. Understanding Random Forest. https://www.analyticsvidhya.com/blog/2021/06/understanding-random-forest/
34. Support Vector Machine. https://machinelearningmedium.com/2018/04/10/support-vector-machines/
35. Understand the KNN Algorithm. https://openclassrooms.com/en/courses/6389626-train-a-supervised-machine-learning-model/6405936-understand-the-knn-algorithm
36. Gradient Boosting Trees for Classification: A Beginner's Guide. https://affine.ai/gradient-boosting-trees-for-classification-a-beginners-guide/
37. Delashmit, W.H., Manry, M.T.: Recent developments in multilayer perceptron neural networks. In: Proceedings of the Seventh Annual Memphis Area Engineering and Science Conference, MAESC, May 2005

Optimal Feature Selection to Improve Vehicular Network Lifetime

Sakshi Garg[1]([✉]), Deepti Mehrotra[1], Sujata Pandey[1], and Hari Mohan Pandey[2]

[1] Amity University, Noida, Uttar Pradesh, India
sakshijyotigarg@gmail.com, spandey@amity.edu
[2] Bournemouth University, Bournemouth, UK

Abstract. The evolution of the Internet of Things (IoT) leads to the ascent of the need to develop a protocol for Low power and Lossy Networks (LLNs). The IETF ROLL working group then proposed an IPv6 routing protocol called RPL in 2012. RPL is in demand because of its adaptability to topology changes and its capacity to identify and evade loops. Although RPL in the recent past was only used for IoT networks. But, contemporary studies show that its applicability can be extended to vehicular networks also. Thus, the domain of the Internet of Vehicles (IoV) for RPL is of significant interest among researchers. Since the network becomes dynamic when RPL is deployed for vehicular networks, the heterogeneous network suffers from extreme packet loss, high latency and repeated transmissions. This reduces the lifetime of the network. The idea behind this article is to simulate such a dynamic environment using RPL and identify the principal features affecting the network lifetime. The network setup is simulated using the Cooja simulator, a dataset is created with multiple network parameters and consequently, the features are selected using the Machine Learning (ML) technique. It is inferred from the experiment that increasing PDR and reducing EC will improve the overall network lifetime of the network.

Keywords: RPL · IoV · Network lifetime · PDR · EC · ML

1   Introduction

Internet of Things (IoT) [1] network permits devices and people to collectively exchange information. IoT is further considered a dynamic network which lead to the evolution of a new domain altogether called the Internet of Vehicles (IoV) [2]. IoV paradigm includes Vehicle-to-Roadside (V2R), Vehicle-to-Vehicle (V2V), Vehicle-to-Infrastructure (V2I) and Vehicle-to-Communication network (V2C). Low power and Lossy Networks (LLNs) based on IoV network face multiple constraints like energy, memory, reliability, heterogeneity, etc. Therefore, it is necessary to optimize LLNs issues concerning IoV. Since several studies [3–5] are available in the literature that has targeted LLN-based IoT network and only a few works [6,7] can be located that have explored IoV for LLNs, it still remains contemporary distress.

© The Author(s), under exclusive license to Springer Nature Switzerland AG 2023
N. Nedjah et al. (Eds.): ICSPN 2021, LNNS 599, pp. 57–68, 2023.
https://doi.org/10.1007/978-3-031-22018-0_6

Routing Protocol for LLN (RPL) was introduced in 2012 by the IETF ROLL working group to tackle these LLN issues [8]. RPL to date is used for IoT networks [9] but recent state-of-art [11] has shown its implementation for IoV networks too. RPL within IoV requires development to address heterogeneity, energy Quality of Service (QoS), scalability, and network QoS requirements. Multiple researchers have suggested RPL optimization to improve network implementation [12–14] concerning IoV networks but to enhance the network performance it is vital to analyze the network at its core.

In this article, an optimal feature selection is proposed to identify the chief features affecting the network lifetime. It is evident from the literature [15] that the higher the network lifetime, the higher the network performance. It is also well established [10,16] that the network lifetime can be boosted by increasing the Packet Delivery Ratio (PDR) and reducing Energy Consumption (EC). Arbitrarily proposing enhancements and optimization to RPL within the IoV network are of little significance on contrary to proposing solutions with optimally chosen network parameters. This motivated us to simulate a mobile environment using the RPL protocol, collect various parameters affecting the network and finally chose the optimal features to improve the network lifetime. The selection is performed on the collected dataset by using Principal Component Analysis (PCA) for dimensionality reduction and Extra Tree Classifier (ETC) to identify those reduced features. The accuracy of the proposed selection is also tested using Machine Learning (ML) technique to justify the efficiency of the selected parameters. Network lifetime is improved by dropping EC and rising PDR. Thus, the parameters touching these two factors are calculated in this study. The results justify that the major contributing factors affecting PDR are Throughput (THPT), Control Traffic Overhead (CTO) and Total Latency (TL). The prime parameters responsible for EC are Expected Transmission Count (ETX), Sent Packets (SP) and TL. It can also be seen that TL is a common factor for both PDR and EC. The accuracy of this selection is 91.756% and 90.116% respectively.

The paper is further structured as follows: the literature study is summarized in Sect. 2, and the particulars of the collected dataset are conferred in Sect. 3. In Sect. 4 the methodology is explained and Sect. 5 examines the conclusions of the study. Finally, the paper is finished with potential future directions in Sect. 6.

2   Related Works

The literature suggests the necessity to improve network lifetime to improve network performance using RPL [17]. Such enhancements will prove beneficial for both IoT and IoV networks. The authors of the paper [18] proposed the use of ETX and EC to enhance network lifetime for the IoT networks by utilizing the Network Interface Average Power metric (NIAP). The use of ML techniques in RPL assessment. The parameters chosen to improve network lifetime are not based on scientific calculations but merely on observations. Also, it is limited to the IoT network implementation, while in this paper, the parameters are optimally chosen based on the ML strategies. In paper [19], author proposed a trust infrastructure based authentication method for clustered vehicular ad hoc networks. In paper [20], the authors proposed a new objective function based on energy, the number of siblings and ETX. The results showed an improvement in network lifetime by lowering delay at 95% of PDR. Their proposal is for IoT networks and metric consideration to form objective function is freehold. Authors in [21] proposed a secure rotting protocols for UAV networks. Authors [22] advised the use of Mobility Energy and Queue Aware-RPL (MEQA-RPL) to increase network lifetime. Their results show improvement in network lifetime but the results could be better if the metric parameter were optimally selected, which is suggested in our study. Authors in this review paper [23] clearly highlighted that increase in PDR and lower EC can elevate network lifetime significantly. Paper [24] also utilized the trade-off between PDR and latency, load balancing and transmission performance to maintain a higher network lifetime. Authors of the paper [25] recommended the use of the ML technique to optimally select features for predictive modelling. Our paper is so conceptualized to select the most optimal network parameters for the vehicular networks and later the accuracy of the selected features is predicted by modelling it using the LR technique to justify the preciseness of the feature selection process. Also, in [26], authors proposed Deep Reinforcement Learning-Based Pedestrian and Independent Vehicle Safety Fortification Using Intelligent Perception.

These studies intended at improvising the RPL performance and mostly optimizing RPL by improving objective function metrics but very few studies have discussed network performance of RPL. This paper covers this gap and tackles the network performance by proposing optimal feature selection to enhance network lifetime for vehicular networks.

3   Dataset

We have accumulated the data by simulating the environment over the Cooja simulator on Contiki OS. This dataset contains multiple factors that may contribute to the network statistics. The experimental setup includes various parameters as shown in Table 1. The data is collected by increasing the number of nodes at continuous intervals to consider scalability as one of the network parameters. Each experiment has been recapped at least 5 times and then an average of those values is taken to reduce any biases. So, the original dataset measures 30×16 (6 node intervals × 5 times repetition × 16 parameters). Further, the dataset is preprocessed and a short illustration of this dataset is shown in Fig. 1.

The chief parameters of the dataset are explained briefly ahead:

- **Nodes**: They represent the density of the simulated network.
- **Routing Metric (RTM)**: A unit calculated by the RPL algorithm to select or reject a path for data transmission.
- **Expected Transmission Count (ETX)**: Total number of transmissions performed o send a packet from source to destination.
- **Energy Consumption (EC)**: Total energy consumed by a mote during the network lifetime.

Nodes	HC	RTM	ETX	EC	PDR	CTO	THPT	SP
10	1.580	1037.417	32.136	1.667	52.432	28904	0.570	49260
20	1.478	1051.614	32.119	2.260	51.546	95096	1.616	139650
30	1.697	1069.849	33.666	2.209	50.725	122970	2.187	188987
40	1.882	1206.850	38.141	2.425	47.619	182845	3.802	328502
50	1.520	1074.142	40.223	2.276	38.043	200934	4.008	346262
60	1.688	1345.975	41.944	2.895	29.458	289742	4.734	409023

Fig. 1. An instance of the collected dataset

- **Packet Delivery Ratio (PDR)**: A ratio of the total number of packets delivered to the total number of packets sent from source to destination.
- **Control Traffic Overhead (CTO)**: The sum of control messages generated by the nodes in the network.
- **Throughput**: The rate of successful packet delivery over the network channel.
- **Sent Packets**: Total number of packets sent from source to destination.
- **Received Packets**: Total packets received during the network lifetime.
- **Lost Packets**: Total packets lost during network lifetime.
- **Total Latency**: Total round trip time the data packet takes to travel.
- **Simulation Time**: Total time the experiment was performed.

4 Methodology

To predict the important features of the network, the steps are shown in Fig. 2 as follows. The model is developed using a python script on a jupyter notebook.

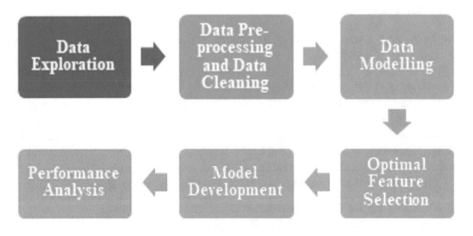

Fig. 2. Classification of metrics and assessment of objective function

4.1   Problem Statement

It is sure that the expansion in network execution and network lifetime with expanding network size, will enhance the network performance and network QoS if it is anticipated and improved. Limited exploration of vehicular networks within IoV prompts the need to evaluate the chief parameters affecting the network. It is apparent from the literature that improving network lifetime can significantly improve network QoS and performance. Hence, it is crucial to identify prime parameters that impact network performance. So, it breeds the necessity to select optimal features that can enhance network performance and provide QoS with great precision.

4.2   Proposed Model

Data Exploration. The data is explored by simulating the dynamic environment using the RPL protocol. To incorporate mobility in RPL, the BonnMotion tool is exploited to generate mobility and the Random Waypoint mobility model is applied to ensure a veritable representation of vehicular network traffic. Further, the data is collected by performing diverse experiments and some parameters are obtained using the Wireshark network analysis tool like Total Latency, Throughput and Control Traffic Overhead. The parameters set for the data collection in the Cooja simulator are shown in Tables 1 and 2.

Table 1. Simulation detail using Cooja

Index	Parameters
Random seed	123,456
Start-up mode delay	65 s
Radio messages	6LoWPAN with pcap analyzer
Propagation model	UDGM with constant distance loss
Mote type	Sky mote
Number of nodes	10, 20, 30, 40, 50, 60
TX ratio	100%
RX ratio	50%
TX range	45 m
INT range	90 m
Total simulation time	24 h

Data Pre-processing and Cleaning. The primary data is cleaned of any Null values. Multiple readings of the same experiments are averaged to obtain holistic values of the parameters. Any parameter that is not contributing to the prediction of the target parameter (Network Lifetime) is dropped. Like, the

Table 2. BonnMotion setup details

Index	Parameter
Number of nodes	10, 20, 30, 40, 50, 60
Mobility model	Random waypoint
X; Y area	100 m ∀ value of nodes
Clustering range	10 m
Pause time	20 s
Number of waypoint	6
Environment	Mobile

simulation time for all the experiments is the same, hence its standard deviation is zero, and therefore it is dropped from the dataset. Redundant parameters like CV RP, CV SP, and CV LP are also ignored. An instance of the refined dataset is shown in Fig. 3.

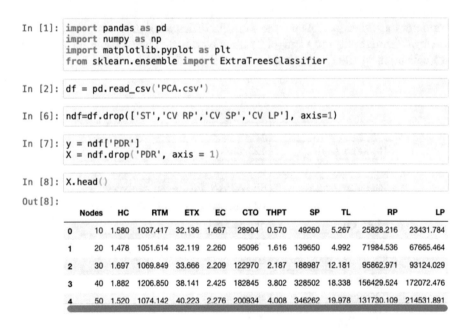

Fig. 3. Example of the pre-processed data

Data Modelling. The data is then modelled using Extra Trees Classifier to identify the significant features with respect to PDR and EC.

Optimal Feature Selection. Our objective is to escalate network lifetime, which is evident from the literature that it can be accomplished by increasing

PDR and reducing EC. For this, we need to identify critical features affecting PDR and EC. Evaluation of varied features in the network increases the network complexity. Calculating all those parameters to estimate the target variable Network Lifetime is needless. Thus, dimensionality reduction is done using Principal Component Analysis (PCA) to select the most optimal features concerning PDR and EC.

Model Development. The obtained features are then modelled using Linear Regression to predict the accuracy of the model for PDR and EC separately.

Performance Analysis. The accuracy of the model is tested for PDR and EC both.

This article commends the need to increase PDR and reduce EC to increase Network Lifetime as an answer to pre-decide the network execution and enhance it for unrivalled performance and better QoS. The essential thought behind fostering this model is the utilization of ML methodology, on account of its capacity to handle enormous information. Since the vehicular network is lively and fanciful, the size of information can't be pre-accessed. Consequently, this model purposes the multi-variate Linear Regression method by choosing optimal features to fabricate the model with high exactness.

5   Results and Discussions

The primary dataset is modelled and important features impacting PDR and EC are plotted using the Extra Trees Classifier. The graph obtained is reflected in Figs. 4 and 5.

Figure 4 shows that TL is the most imperative feature and EC is the least affecting the PDR of the network. However, Fig. 5 shows that TL is the most vital feature and the scalability of nodes is least disturbing the EC of the network. Further, the dataset has numerous parameters and some features might not be essential for PDR or EC. Evaluating compound features repeatedly will increase the complexity of the network. Hence, it is obligatory to reduce features and find the most contributing ones. The dimensionality reduction of the dataset is performed using PCA for PDR and EC both. Most contributing factors are identified out of multiple factors using PCA taking 95% of the components of the dataset parameters. The results reveal that there are three most important features that impact PDR and EC significantly. It can be seen in Figs. 6 and 7.

The most important extracted features for PDR and EC are listed in Table 3.

The accuracy of the selection is predicted using a machine learning technique called Linear Regression (LR) and it is seen that optimally selected features for PDR give 91.756% accuracy while the accuracy score for EC is 90.116%.

This table fairly clarifies the use of the recommended optimally selected features for evaluating the network performance and the key parameters influencing the network lifetime for vehicular networks.

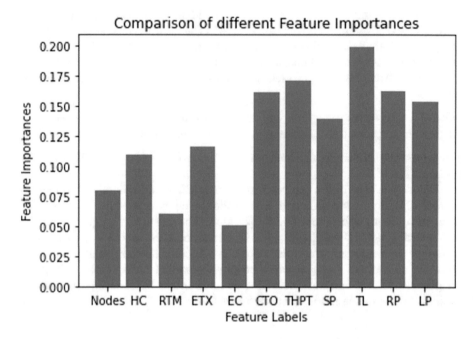

Fig. 4. Feature importance with respect to PDR

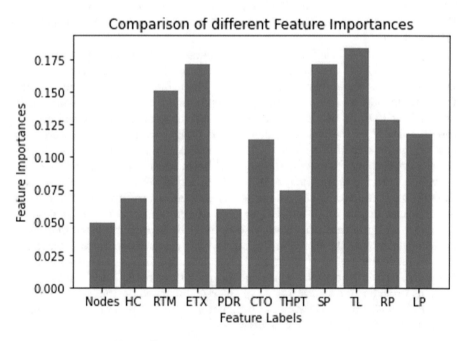

Fig. 5. Feature importance with respect to EC

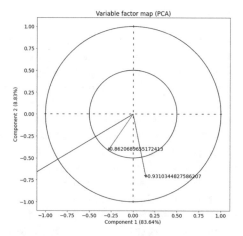

Fig. 6. Extracted features for PDR using PCA variable map

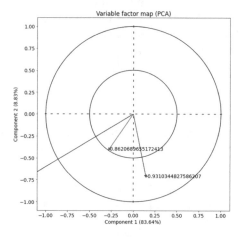

Fig. 7. Extracted features for EC using PCA variable map

Table 3. Optimally selected feature

Network lifetime	Factors affecting PDR and EC
Increase PDR	1. Total Latency (TL)
	2. Throughput (THPT)
	3. Control Traffic Overhead (CTO)
Reduce EC	1. Total Latency (TL)
	2. Sent Packets (SP)
	3. Expected Transmission Count (ETX)

6   Conclusion and Future Directions

This article distinctly underlines the necessity to find optimal features for the IoV network. Predictive analysis is performed using ETC, PCA and LR strategies of ML to select the most optimal parameters for the vehicular network to upsurge network lifetime. The dataset is first simulated by recreating the dynamic environment using the RPL protocol, mobile nodes and RWP mobility model. Then, the dataset is modelled using ETC to find the importance of features for PDR and EC. Further, the dataset is condensed using PCA to reduce dimensionality and thereby network complexity. The dataset is collected using the collect view in the Cooja simulator and processing the simulation log files over the Wireshark network analyzer. The obtained feature through PCA fit into the LR model to obtain the accuracy of the prediction. The accuracy score for the selected features is predicted to justify the efficiency of the proposed parameters to enhance network lifetime for vehicular networks. The accuracy for selected parameters for PDR and EC is equitably great to justify its use for future vehicular networks. The average precision score for PDR is 91.756%, while the average accuracy score for EC is 90.116%. This notion has direct applications to smart city networks, Electronic Toll Collection systems networks, etc. In future, this work can be extended for real data and other mobility models like Random Walk (RWK), Nomadic, Manhattan Grid, etc. to analyze the behavior of optimally selected features on network lifetime concerning those models.

References

1. Safaei, B., et al.: Impacts of mobility models on RPL-based mobile IoT infrastructures: an evaluative comparison and survey. IEEE Access **8**, 167779–167829 (2020)
2. Garg, S., Mehrotra, D., Pandey, H.M., Pandey, S.: Accessible review of internet of vehicle models for intelligent transportation and research gaps for potential future directions. Peer-to-Peer Netw. Appl. **14**(2), 978–1005 (2021). https://doi.org/10.1007/s12083-020-01054-6
3. Bendouda, D., Haffaf, H.: QFM-MRPL: towards a QoS and fault management based of mobile-RPL in IoT for mobile applications. In: 2019 15th International Wireless Communications & Mobile Computing Conference (IWCMC), June 2019, pp. 354–359. IEEE
4. Vaezian, A., Darmani, Y.: MSE-RPL: mobility support enhancement in RPL for IoT mobile applications. IEEE Access (2022)
5. Zarzoor, A.R.: Enhancing IoT performance via using mobility aware for dynamic RPL routing protocol technique (MA-RPL). Int. J. Electron. Telecommun. **68**(2), 187–191 (2022)
6. Garg, S., Mehrotra, D., Pandey, S., Pandey, H.M.: Network efficient topology for low power and lossy networks in smart corridor design using RPL. Int. J. Pervas. Comput. Commun. (2021)
7. Kumar, A., Hariharan, N.: Enhanced mobility based content centric routing in RPL for low power lossy networks in internet of vehicles. In: 2020 3rd International Conference on Intelligent Autonomous Systems (ICoIAS), Feb 2020, pp. 88–92. IEEE

8. Winter, T., et al.: RPL: IPv6 routing protocol for low-power and lossy networks. RFC **6550**, 1–157 (2012)

9. Darabkh, K.A., Al-Akhras, M., Ala'F, K., Jafar, I.F., Jubair, F.: An innovative RPL objective function for broad range of IoT domains utilizing fuzzy logic and multiple metrics. Expert Syst. Appl. 117593 (2022)

10. Yang, C.: A path-clustering driving travel-route excavation. Int. J. Semant. Web Inf. Syst. (IJSWIS) **18**(1), 1–16 (2022)

11. Garg, S., Mehrotra, D., Pandey, H.M., Pandey, S.: Static to dynamic transition of RPL protocol from IoT to IoV in static and mobile environments. Clust. Comput. 1–16 (2022)

12. Seyfollahi, A., Moodi, M., Ghaffari, A.: MFO-RPL: a secure RPL-based routing protocol utilizing moth-flame optimizer for the IoT applications. Comput. Stand. Interfaces **82**, 103622 (2022)

13. Aboubakar, M., Kellil, M., Bouabdallah, A., Roux, P.: Using machine learning to estimate the optimal transmission range for RPL networks. In: NOMS 2020–2020 IEEE/IFIP Network Operations and Management Symposium, Apr 2020, pp. 1–5. IEEE

14. Yang, C.: A path-clustering driving travel-route excavation. Int. J. Semant. Web Inf. Syst. (IJSWIS) **18**(1), 1–16 (2022)

15. Sennan, S., Ramasubbareddy, S., Nayyar, A., Nam, Y., Abouhawwash, M.: LOA-RPL: novel energy-efficient routing protocol for the internet of things using lion optimization algorithm to maximize network lifetime (2021)

16. Cyriac, R., Karuppiah, M.: A comparative study of energy retaining objective functions in RPL for improving network lifetime in IoT environment. Recent Adv. Comput. Sci. Commun. (Formerly Recent Pat. Comput. Sci.) **13**(2), 159–167 (2020)

17. Garg, S., Mehrotra, D., Pandey, S.: A study on RPL protocol with respect to DODAG formation using objective function. In: Soft Computing: Theories and Applications, pp. 633–644. Springer, Singapore (2022)

18. Pereira, H., Moritz, G.L., Souza, R.D., Munaretto, A., Fonseca, M.: Increased network lifetime and load balancing based on network interface average power metric for RPL. IEEE Access **8**, 48686–48696 (2020)

19. Mirsadeghi, F., Rafsanjani, M.K., Gupta, B.B.: A trust infrastructure based authentication method for clustered vehicular ad hoc networks. Peer-to-Peer Netw. Appl. **14**(4), 2537–2553 (2020). https://doi.org/10.1007/s12083-020-01010-4

20. Moradi, S., Javidan, R.: A new objective function for RPL routing protocol in IoT to increase network lifetime. Int. J. Wireless Mob. Comput. **19**(1), 73–79 (2020)

21. Fatemidokht, H., et al.: Efficient and secure routing protocol based on artificial intelligence algorithms with UAV-assisted for vehicular ad hoc networks in intelligent transportation systems. IEEE Trans. Intell. Transp. Syst. **22**(7), 4757–4769 (2021). https://doi.org/10.1109/TITS.2020.3041746

22. Durai, S.: MEQA-RPL: A Mobility Energy and Queue Aware Routing Protocol for Mobile Nodes in Internet of Things (2021)

23. Ghanbari, Z., Navimipour, N.J., Hosseinzadeh, M., Shakeri, H., Darwesh, A.: The applications of the routing protocol for low-power and lossy networks (RPL) on the internet of mobile things. Int. J. Commun. Syst. e5253 (2022)

24. Jamalzadeh, M., Maadani, M., Mahdavi, M.: EC-MOPSO: an edge computing-assisted hybrid cluster and MOPSO-based routing protocol for the internet of vehicles. Ann. Telecommun. 1–13 (2022)

25. Garg, S., Mehrotra, D., Pandey, S., Pandey, H.M.: A convergence time predictive model using machine learning for LLN. In: 2021 IEEE 8th Uttar Pradesh

Section International Conference on Electrical, Electronics and Computer Engineering (UPCON), Nov 2021, pp. 1–5

26. Vijayakumar, P., Rajkumar, S.C.: Deep reinforcement learning-based pedestrian and independent vehicle safety fortification using intelligent perception. Int. J. Softw. Sci. Comput. Intell. (IJSSCI) **14**(1), 1–33 (2022)

Machine Learning Based Two-Tier Security Mechanism for IoT Devices Against DDoS Attacks

Domenico Santaniello[1], Akshat Gaurav[2,3(✉)], Wadee Alhalabi[4],
and Francesco Colace[1]

[1] University of Salerno, Fisciano, Italy
dsantaniello@unisa.it, fcolace@unisa.it
[2] Ronin Institute, Montclair, USA
akshat.gaurav@roninstitute.org
[3] Lebanese American University, Beirut 1102, Lebanon
[4] Department of Computer Science, King Abdulaziz University, Jeddah, Saudi
Arabia
wsalhalabi@kau.edu.sa

Abstract. IoT devices are becoming an increasingly important part of our everyday lives, and their worth is rising with each passing year. Because IoT devices capture and handle all of our personal and private data, they are a primary target for cyber attackers. Due to the limited processing power and memory capacity of IoT devices, it is challenging to apply complicated security algorithms. The development of a lightweight security mechanism for IoT devices is necessary. In this context, we create a two-tier security solution for Internet of Things devices that protects against DDoS attacks, the most well-known kind of cyber assault. The suggested solution makes extensive use of statistical technologies and machine learning techniques at several tiers to effectively recognise DDoS attacks. The suggested technique made advantage of active learning to determine the appropriate attributes for detecting DDoS attack traffic.

Keywords: IoT · Machine learning · DDoS · Entropy

1 Introduction

The proliferation of IoT devices has expanded as a result of increased digitalization. According to some estimates, the Internet of Things will reach 20 billion devices by 2020. However, the security of these devices continues to be a significant issue. Several IoT devices have recently been evaluated and it was discovered that they contain numerous vulnerabilities, making them an easy target for various forms of cyber attack [9,23].

DDoS attacks is more than a decade old cyber attacks that affect the functioning of online websites or servers [19,26]. In DDoS attack, the attacker consumes the resources of the victim and made them unavailable to its potential

N. Nedjah et al. (Eds.): ICSPN 2021, LNNS 599, pp. 69–82, 2023.
https://doi.org/10.1007/978-3-031-22018-0_7

uses [11,13]. After the first recorder DDoS attack in 1993, its impact continues to increase day by day. Currently, DDoS attacks are the major threat to manufacturers and industries that are developing and producing online services [26]. As IoT devices are shearing the information and services on-line, they are also affected by the DDoS attack. Mirai botnet is the most notorious attack that uses more than a million IoT devices to generate a DDoS attack on many popular websites. In addition, it is very difficult to identify bots in the IoT environment because it is easy to hack vulnerable IoT devices. There are many types of DDoS attacks possible in an IoT environment, such as volumetric DDoS attacks and resource-depletion DDoS attacks. According to a study, volumetric DDoS attacks account for a total of 65% of DDoS attacks [15].

Detection of DDoS attacks may be broken down into three categories: attack prevention strategies, defensive tactics, and traceback. Load balancing and honeypots are two methods used by users to mitigate the effects of DDoS attacks. Techniques for finding the source of a DDoS attack are employed in many ways. Most challenging, however, is devising a DDoS protection mechanism capable of both identifying and counteracting a specific DDoS assault. Detecting and distinguishing between a DDoS assault and flash crowd traffic is the goal of a successful DDoS defensive strategy. DDoS attacks may cause significant economic and reputational harm to victims if detection algorithms fail to distinguish legitimate traffic from DDoS traffic. An effective DDoS attack detection system must thus be able to distinguish between the flash crowd and the DDoS attack traffic without filtering the latter out. Researchers are keep on working in the development of a optimant technique for the identification and mitigation of DDoS attacks in IoT network. AS, the nature of the architecture and traffic distribution of the IoT network is different from traditional wired networks; classical DDoS attack techniques are not effective against the IoT network [9]. In this context, recently, many scholars have proposed the use of machine learning and artificial intelligence for the detection of malicious users and normal traffic on the IoT network. There are many ways in which ML and AI can be used for the identification of DDos attacks, such as some research supporting the use of statistical features for the algorithm, identification of malicious traffic. However, this method has some limitations that show that there is still some scope of development [26]. Some researchers develop the integration of ML and AI techniques with other techniques such as SDN [2,4], cloud computing [15,24], blockchain [16], and fog computing to develop an efficient algorithm for detection of DDoS.

However, developing an appropriate system to detect DDoS attacks in an IoT context is challenging, since the characteristics of malicious data are similar to those of normal traffic. As a result, identifying and filtering out attack traffic is very challenging. Additionally, since IoT devices have limited power and memory capacity, it is inefficient to implement a sophisticated threat detection technique on them. In addition to that, most of the time attackers use IP spoofing to hide their real identity; which makes it difficult to backtrack the attacker location. The spoofed traffic is called backscatter traffic [3,10]. Therefore, different types of IDS and IPS are not effective against these types of DDoS attacks. The main limitation of using ML and AI techniques for the detection of DDoS techniques is that the training model requires labeled data sets, which is not always possible [7]. In addition, analysis of large datasets consumes the limited resources of IoT devices. Therefore, a DDoS detection technique that reliably and effectively detects malicious traffic is needed.

In this context, we develop a multilayer DDoS detection model for IoT network, The proposed model has the following contributions:

- The proposed model is implemented in the fog layer and gateway routers for the identification and mitigation of DDoS attacks.
- The proposed system used statistical and ML algorithms for the detection of DDoS attacks.
- For reducing the cost of the ML algorithm, we use active learning technology [7] to identify the most optimal features to minimize the impact of the DDoS attack.

The remainder of the paper is organized as follows: the literature review is presented in Sect. 2. Details about the proposed research methodology are presented in Sect. 3. Section 4 presents the simulation and the analysis of the results. The conclusion and future work is presented in Sect. 5.

2 Related Work

Most of the work in recent time is based on online resources [14]. Hence, DDoS becomes the omnipresent problem for uses and service providers [22]. In order to get more information about the development in the field of DDoS attacks detection, we analyze various research works present in Scopus database from 2002 to 2022. In our study, we include only conferences and journal papers. This section gives a brief overview of the research development that has taken place in the last two decades. As we can see in Fig. 1 there is an exponential growth in the publication of articles after 2016. This shows that as the online presence of users increased, more research was conducted in the field of developing techniques for the protection of user information.

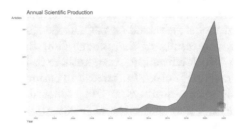

Fig. 1. Paper publication variation

The authors propose various techniques for the detection of ML DDoS attack. The authors in [1] proposed ML techniques for the detection of DDoS attacks; in this model, the authors used feature engineering techniques to obtain the most optimal features of the features sets. Due to the reduced feature size, the DDoS detection time was reduced. In addition to that, the author proved that the KNN algorithm works best with the reduced feature sets. However, the authors in this model are able to reduce the features only to 68%. The selection of optimal features for the detection of DDoS attack attracts many researchers. Recently, the author in [8] proposed an intrusion detection system based on selection of characteristics. The proposed IDS is tested on combination of many datasets such as CAIDA07, CIC-IDS-2017, wordcup 1998, and SlowDos2016. The author used information gain to rank the features and selected the most optimal features from the set of features. The proposed IDS analyzes the traffic in two phase; due to which the detection rate of the proposed model is 99%. Author in [27] proposed a statistical approach to secure health care services from DDoS attacks during COVID-19 pandemic. In another work, author [6] proposed a model for DDoS Detection Framework in Resource Constrained IoT domain.

The most challenging part in developing the DDoS attack traffic is the identification of the source of attack [12,21]. But most of the time attackers use spoofed packets for the DDoS attack, which is called backscatter traffic. Therefore, researchers are working on techniques for the identification and detection of DDoS attacks due to backscatter traffic. In this context, the authors develop a model based on ML in 2014, that analyses the performance of open source IDS and supervised ML technique against backscatter traffic. The experiment proves that the ML techniques can effectively detect backscatter traffic without the requirements of IP address and port number.

The collaborative technique is one of the possible and effective techniques proposed by researchers for the detection of DDoS attacks. In this technique, the individual nodes analyze the local traffic and then that information is used to analyze the traffic performance globally. In this context, the authors of [5] develop an additive technique for the detection of DDoS attack. In this proposed model, all nodes have the ability to analyze the network traffic locally; due to this, the proposed framework works effectively against distributed DDoS attacks. In another approach, the authors [25] develop a hybrid deep learning approach for the detection of DDoS attacks. In this proposed approach, autoencoders are used to select the most effective features for the identification of DDoS attacks. The proposed model is trained on CICDDoS2019 dataset and gives 98% F-1 score. Therefore, the results of this model show that a hybrid technique can detect the DDoS attack efficiently. Also, authors in [17] proposed Intrusion Detection Using Normalized Mutual Information Feature Selection and Parallel Quantum Genetic Algorithm. Authors in [20] proposed a model for classification based machine learning for detection of DDoS attack in cloud computing. In another work, the authors [18] proposed a CNN and entropy-based DDoS attack detection technique in the SDN network. The authors in this projected model are able to obtain an accuracy of 98.98%. This showed that combination of statistical method and ML/DL methods can give good results.

3   Proposed Methodology

This section presents detailed information on our proposed methodology. The basic architecture of our proposed framework is presented in Fig. 2. Our proposed framework is divided into two layers that combine to detect DDoS attack traffic. The first part of the proposed approach is implemented on the gateway rooters that perform the preliminary check of the DDoS traffic. The second stage of the framework is implemented on the fog layer. The two layers collaboratively work together to identify the DDoS attack.

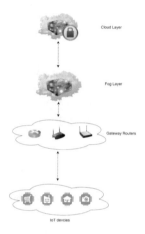

Fig. 2. Proposed methodology

3.1   First Stage

This stage provides the first layer of protection against DDoS attacks. At this stage, gateway routers use the statistical technique to identify DDoS attacks. The topological representation is presented in Fig. 3. The IoT devices are generating the packets at a regular interval of time and these packets are passed to the gateway routers. The gateway routers forwarded the packets to the fog layer for further processing.

Consider that there is a Gateway router ($\mathbb{G}$) that is connected to n IoT devices $\{t_i, t_2, t_3, \ldots, t_n\}$. These IoT devices transmit traffic at different rates ($\{r_1, r_2, r_3, \ldots, r_n\}$). We analyze their traffic for the period of time t and calculate the probability and entropy using the following equations.

$$\mathbb{P}_j = \frac{x_j}{\sum_{j=1}^{n} x_j} \tag{1}$$

where $\mathbb{P}$ is the probability of distribution and x is the packets captured in a time window t. Probability of distribution is an important factor for measuring the distribution of traffic characteristics. In order to obtain more information about the traffic, we calculate the entropy with the help of Eq. 1.

$$\mathbb{H}_j = -\sum_{i=1}^{n} \mathbb{P}_i \times \ln \mathbb{P}_j \tag{2}$$

where $\mathbb{H}$ represents the entropy, for the time interval t. We normalized the entropy according to Eq. 3.

$$\mathbb{H}_i = \alpha \mathbb{H}_i; \ 0 < \alpha \le 1 \tag{3}$$

The block diagram for stage 1 is represented in Fig. 4. The detailed work of this stage is represented in Algorithm 1. The important steps of the algorithm are as follows.

- For every time window ΔT, the gateway router analyzes the incoming traffic.
- If the packets belong to the blacklist, then the packets are discarded.
- If not, then calculate the probability of the distribution and entropy.
- Compare the entropy with normal traffic.
- If entropy is higher, then a warning is sent to the fog layer.

Algorithm 1: Algorithm to analyze the coming packets

Input: Incoming Traffic
Output: Check the traffic behavior
Begin
for Every time window (ΔT) **do**
    if *If packet belongs to the blacklist* **then**
      | Discard packet
    end
    else
      $\mathbb{IP}_i \to A_k$, IP address is extracted
      Calculate Probability $\mathbb{P}_j$;
      Calculate Entropy $\mathbb{H}$;
      Normalize the entropy $\alpha \mathbb{H}$
      Analysis the packets
      if *Entropy is high* **then**
        | send the warning to the fog layer
      end
      else
        | forward the packets
        | Update Database
      end
    end
end
End

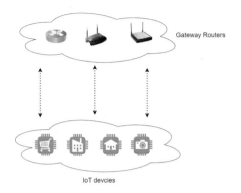

Fig. 3. Stage 1

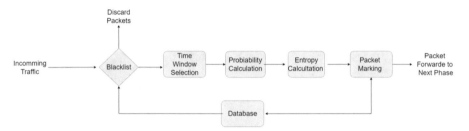

Fig. 4. Block diagram of Stage 1

3.2 Second Stage

The second stage is implemented in the fog layer. The fog layer is located between the highway route and the cloud layer. The purpose of fog layer is to move the processing power near the nodes. As the fog layer is close to the physical nodes it can process the traffic quickly and filter out the malicious traffic efficiently. The basic topological representation of this layer is presented in Fig. 5. This stage used the concepts of machine learning to analyze the incoming traffic.

In traditional machine learning technique, the database is processed and labeled; then that labeled data set is used to train the machine learning model. However, this processing and labeling of the data set makes the overall performance of the machine learning model slow and complex. Therefore, in order to improve the speed and effectiveness of the machine learning model, we use the active learning technique [7] to process the data set. The functioning of the active learning technique is represented in Fig. 6. In this active learning technique, the data set is partitioned into small subsets to remove redundancy and reduce the training time. In active learning, the unlabeled data set is analyzed, and the most optimal patterns are used to train the machine learning model.

Feature Section The first step in the development of the machine learning model is the selection of optimal features. Most of the time, the data set consists

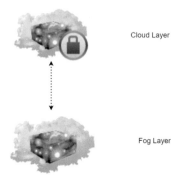

Fig. 5. Stage 2

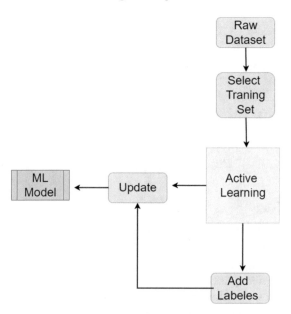

Fig. 6. Active learning model

of a large number of features; hence, a reduction of the feature set is required. The statistical parameters of all marked packets are extracted in the fog layer, and then the feature section algorithm is used to select the best set of features. The detailed procedure for the selection of features is represented in Algorithm 2. The explanation of each feature is as follows:

- The final set of features is selected ($\mathbb{F}$).
- All the features are grouped into small groups ($\mathbb{G} \leftarrow \{g_1, g_2, \ldots, g_n\}$).
- Using the ranking algorithm, the best features of each group are selected ($\{f_a, f_b, f_c, \ldots, f_n\}$).

– The most common features of the feature groups are stored in the final feature set ($\mathbb{F} \leftarrow \{f_a \cup f_b \cup f_c \cup \cdots \cup f_n\}$).

Algorithm 2: Algorithm to select best features

Input: All features
Output: Select best features
Begin
$\mathbb{F} \leftarrow \varnothing$
for Every time window (ΔT) **do**
$\quad$ Group the features into different groups
$\quad$ $\mathbb{G} \in \{g_1, g_2, \ldots, g_n\} \leftarrow f_\alpha \in \{f_1, f_2, \ldots, f_n\}$
$\quad$ Select the features that are common in all the groups
$\quad$ $\mathbb{F} \leftarrow \{f_a \cup f_b \cup \cdots \cup f_n\}$
$\quad$ Add the most optimal features in the final list
end
End

Feature Ranking Algorithm There are many techniques through which the optimal features can be selected, such as genetic algorithm, greedy algorithm. However, we used the correlation ran algorithm to rank the features. The formula for calculating the Pearson correlation is given in Eq. 4.

$$C = \frac{(N \sum(a \times b) - (\sum a) \times (\sum b))}{\sqrt{(N \sum(x^2) - (\sum x)^2) \times (N \sum(b)^2 - (\sum b)^2)}} \tag{4}$$

The correlation coefficient (C) can take any values between -1 and $+1$ and represents the correlation between two variables a and b. Positive values increase in correlation, and negative values represent lack of correlation between the variables. In our model, for every k features, we calculate the correlation coefficient $k - 1$. Finally, compare all the correlation coefficients in a group to find the optimal features.

Classifier In this section, we give information about the machine learning model used in our framework for the detection of DDoS attacks.

– SVM: The support vector machine (SVM), invented by Vapnik et al., is a machine learning method used in regression and pattern recognition approaches. Input data is transformed into a high-dimensional linear separable feature space by using the kernel mapping technique in an SVM. In addition to linking predetermined kernels such as Gaussian and polynomial, the decision function of an SVM is proportional to the number of SVs and their respective weights.

– This technique creates a treelike structure by periodically dividing the data set according to a criteria that maximize data separation. The Decision Node and the Leaf Node are the only two parts of a decision tree.
– Random Forest: Classification and regression techniques based on ensemble learning, such as random forest (RF), are ideal for problems that need data sorting. The algorithm was devised by Breiman and Cutler. Decision trees are used to anticipate outcomes in RF. Multi-decision trees are trained and then used to forecast classes based on the votes cast for each tree during the training phase. This is the class that has the biggest margin function (MF).
– Gradient Boosting: Gradient boosting is a machine learning approach that has been shown to be very successful. The terms "bias" and "variance" are often used to describe the faults that might occur in machine learning systems. Gradient boosting, one of the boosting approaches, minimizes the bias error in the model. For example, the gradient boost approach may be used to generate an additive model with the lowest possible loss function. As a result, the gradient boosting approach continually increases the number of decision trees that reduce the loss function. Using the shrinkage parameter *beta*, also known as the learning rate, the gradient boosting approach works better when the contribution of the new decision tree is reduced at each iteration step. Smaller steps are better than larger ones when it comes to gradient boosting, according to the theory of the shrinking approach.

4   Results and Discussions

4.1   Data Set Prepossessing

For the simulation, we used the open source software OMNET++. In this situation, the attack packets overwhelm the target with a large volume of erroneous traffic. As a result, the attacker node generates packets every second, whereas the legitimate nodes create packets every five seconds in this case. We collected all log data throughout the simulation's 200 s runtime. We use a generic routing protocol to replicate our approach, as it is not protocol-specific. Because there are so many of them, we eliminated them from the dataset during the preparation phase. Figure 7a and b show the change in entropy over time during and after a DDoS attack. We can clearly see that the entropy values at the time of normal traffic are constant. However, at the time of DDoS attack, there is a change in the extract values. Finally, we divided the data into two sets: one for training and the other for testing. Due to the partitioned data, the suggested approach can be thoroughly evaluated on both sets.

4.2   Analysis

In this subsection, we analyze the performance of our proposed model using the following parameters:

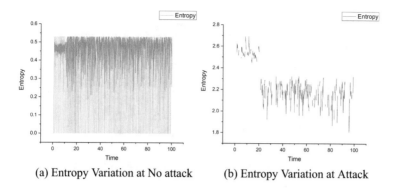

(a) Entropy Variation at No attack   (b) Entropy Variation at Attack

Fig. 7. Dataset representation

- **Precision** - Forwarded packets' percentage of legitimate packets is determined by this metric.

$$Precision\,(P) = \frac{\Delta T}{\Delta T + \Delta F} \tag{5}$$

- **Recall** - If the technique is implemented, it measures the percentage of legitimate packets that are accepted.

$$Recall\,(R) = \frac{\Delta T}{\Delta T + \Delta \hat{F}} \tag{6}$$

- **Accuracy** - It evaluates the suitability of our proposed course of action.

$$Accuracy = \frac{\Delta T + \Delta \hat{T}}{Total\,Packets} \tag{7}$$

- **F-1 Score** - It evaluates the effectiveness of the method.

$$F-1\,score = 2 \times \frac{P \times R}{P + R} \tag{8}$$

where δT is True positive, δF is false positive, $\delta \hat{T}$ is true negative, and $\delta \hat{F}$ is false negative.

For our proposed technique, we now assess multiple machine learning algorithms to see which one is the most suitable. Each of the machine learning algorithms is evaluated for its performance in accuracy, precision, recall, and f-1 score. Figure 8 clearly illustrate that the SVM-based approach has the greatest accuracy and recall rate. As a result, we can state that the suggested LR-based strategy is able to recognise DDoS attacks.

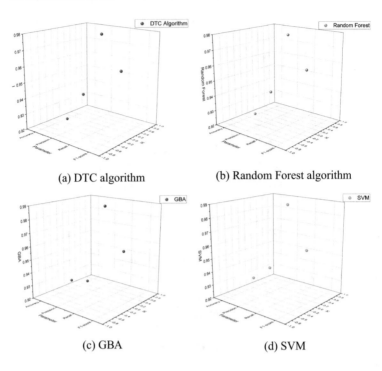

(a) DTC algorithm   (b) Random Forest algorithm

(c) GBA   (d) SVM

Fig. 8. Statistical parameters calculation

5   Conclusion

Automated and networked gadgets are transforming the world into an interconnected network of devices that communicate with each other. Sensors and
actuators will be attached to both living and nonliving "things" in the Internet of Things (IoT) environment, and all of them will be part of the Internet,
not just computers and smartphones. IoT devices and related technologies face
significant security issues that play an important role in their deployment. In
this paper, we proposed a DDoS attack detection model for IoT devices. The
proposed model is implemented on gateway routers and fog nodes, and both
devices collaboratively detect the DDoS attack. As the resources of the gateway
routers are limited, they used the entropy-based technique for the identification
of DDoS attack traffic. If the gateway router detects malicious traffic, it marks
the packets, and in the second stage the fog nodes analyze the marked packets using the machine learning technique. In the proposed approach, we used an
active learning technique that will reduce the complexity of the machine learning
technique. In the future, we will plan to test the framework on different datasets.

References

1. Aamir, M., Zaidi, S.M.A.: DDoS attack detection with feature engineering and machine learning: the framework and performance evaluation. Int. J. Inf. Secur. **18**(6), 761–785 (2019). https://doi.org/10.1007/s10207-019-00434-1

2. Alshamrani, A., Chowdhary, A., Pisharody, S., Lu, D., Huang, D.: A defense system for defeating DDoS attacks in SDN based networks. In: Proceedings of the 15th ACM International Symposium on Mobility Management and Wireless Access, pp. 83–92. ACM, Miami, FL, USA (2017). https://doi.org/10.1145/3132062.3132074

3. Balkanli, E., Alves, J., Zincir-Heywood, A.: Supervised learning to detect DDOS attacks (2014). https://doi.org/10.1109/CICYBS.2014.7013367

4. Barki, L., Shidling, A., Meti, N., Narayan, D.G., Mulla, M.M.: Detection of distributed denial of service attacks in software defined networks. In: 2016 International Conference on Advances in Computing, Communications and Informatics (ICACCI), pp. 2576–2581. IEEE, Jaipur, India (2016). https://doi.org/10.1109/ICACCI.2016.7732445, http://ieeexplore.ieee.org/document/7732445/

5. Berral, J., Poggi, N., Alonso, J., Gavaldà, R., Torres, J., Parashar, M.: Adaptive distributed mechanism against flooding network attacks based on machine learning, pp. 43–49 (2008). https://doi.org/10.1145/1456377.1456389

6. Chaudhary, P., et al.: DDOS detection framework in resource constrained internet of things domain. In: 2019 IEEE 8th Global Conference on Consumer Electronics (GCCE), pp. 675–678. IEEE (2019)

7. Deka, R.K., Bhattacharyya, D.K., Kalita, J.K.: Active learning to detect DDoS attack using ranked features. Comput. Commun. **145**, 203–222 (2019). https://doi.org/10.1016/j.comcom.2019.06.010, https://linkinghub.elsevier.com/retrieve/pii/S0140366419303858

8. Dhingra, A., Sachdeva, M.: Detection of denial of service using a cascaded multi-classifier. Int. J. Comput. Sci. Eng. **24**(4), 405–416 (2021). https://doi.org/10.1504/IJCSE.2021.117028

9. Doshi, R., Apthorpe, N., Feamster, N.: Machine learning DDoS detection for consumer internet of things devices. In: 2018 IEEE Security and Privacy Workshops (SPW), May 2018, pp. 29–35. https://doi.org/10.1109/SPW.2018.00013, arXiv:1804.04159

10. Furutani, N., Ban, T., Nakazato, J., Shimamura, J., Kitazono, J., Ozawa, S.: Detection of DDoS backscatter based on traffic features of darknet TCP packets. In: 2014 Ninth Asia Joint Conference on Information Security, pp. 39–43. IEEE, Wuhan, China (2014). https://doi.org/10.1109/AsiaJCIS.2014.23, http://ieeexplore.ieee.org/document/7023237/

11. Gaurav, A., et al.: Filtering of distributed denial of services (DDOS) attacks in cloud computing environment. In: 2021 IEEE International Conference on Communications Workshops (ICC Workshops), pp. 1–6. IEEE (2021)

12. Gaurav, A., et al.: A novel approach for DDOS attacks detection in covid-19 scenario for small entrepreneurs. Technol. Forecast. Soc. Change **177**, 121554 (2022)

13. Gulihar, P., et al.: Cooperative mechanisms for defending distributed denial of service (DDOS) attacks. In: Handbook of Computer Networks and Cyber Security, pp. 421–443. Springer (2020)

14. Gupta, B., Gupta, S., Gangwar, S., Kumar, M., Meena, P.: Cross-site scripting (XSS) abuse and defense: exploitation on several testing bed environments and its defense. J. Inf. Priv. Secur. **11**(2), 118–136 (2015)

15. He, Z., Zhang, T., Lee, R.B.: Machine learning based DDoS attack detection from source side in cloud, pp. 114–120. IEEE (2017)
16. Kumar, P., Kumar, R., Gupta, G.P., Tripathi, R.: A distributed framework for detecting DDoS attacks in smart contract-based blockchain-IoT systems by leveraging fog computing. Trans. Emerg. Telecommun. Technol. **32**(6) (2021). https://doi.org/10.1002/ett.4112
17. Ling, Z., Hao, Z.J.: Intrusion detection using normalized mutual information feature selection and parallel quantum genetic algorithm. Int. J. Semant. Web Inf. Syst. (IJSWIS) **18**(1), 1–24 (2022)
18. Liu, Y., Zhi, T., Shen, M., Wang, L., Li, Y., Wan, M.: Software-defined DDOS detection with information entropy analysis and optimized deep learning. Future Gener. Comput. Syst. **129**, 99–114 (2022). https://doi.org/10.1016/j.future.2021.11.009
19. Mishra, A., et al.: A comparative study of distributed denial of service attacks, intrusion tolerance and mitigation techniques. In: 2011 European Intelligence and Security Informatics Conference, pp. 286–289. IEEE (2011)
20. Mishra, A., et al.: Classification based machine learning for detection of DDOS attack in cloud computing. In: 2021 IEEE International Conference on Consumer Electronics (ICCE), pp. 1–4. IEEE (2021)
21. Pan, X., Yamaguchi, S., Kageyama, T., Kamilin, M.H.B.: Machine-learning-based white-hat worm launcher in botnet defense system. Int. J. Softw. Sci. Comput. Intell. (IJSSCI) **14**(1), 1–14 (2022)
22. Singh, A., et al.: Distributed denial-of-service (DDOS) attacks and defense mechanisms in various web-enabled computing platforms: issues, challenges, and future research directions. Int. J. Semant. Web Inf. Syst. (IJSWIS) **18**(1), 1–43 (2022)
23. Tewari, A., et al.: A lightweight mutual authentication approach for RFID tags in IoT devices. Int. J. Netw. Virt. Org. **18**(2), 97–111 (2018)
24. Virupakshar, K.B., Asundi, M., Channal, K., Shettar, P., Patil, S., Narayan, D.: Distributed denial of service (DDoS) attacks detection system for OpenStack-based private cloud. Procedia Comput. Sci. **167**, 2297–2307 (2020). https://doi.org/10.1016/j.procs.2020.03.282, https://linkinghub.elsevier.com/retrieve/pii/S1877050920307481
25. Wei, Y., Jang-Jaccard, J., Sabrina, F., Singh, A., Xu, W., Camtepe, S.: AE-MLP: a hybrid deep learning approach for DDOS detection and classification. IEEE Access **9**, 146810–146821 (2021). https://doi.org/10.1109/ACCESS.2021.3123791
26. Yuan, X., Li, C., Li, X.: DeepDefense: identifying DDoS attack via deep learning. In: 2017 IEEE International Conference on Smart Computing (SMARTCOMP), May 2017, pp. 1–8. IEEE, Hong Kong, China (2017). https://doi.org/10.1109/SMARTCOMP.2017.7946998, http://ieeexplore.ieee.org/document/7946998/
27. Zhou, Z., et al.: A statistical approach to secure health care services from DDOS attacks during covid-19 pandemic. Neural Comput. Appl. 1–14 (2021)

An Analysis of Machine Learning Algorithms for Smart Healthcare Systems

Mai Alduailij[1], Anupama Mishra[2(✉)], Ikhlas Fuad Zamzami[3],
and Konstantinos Psannis[4]

[1] Department of Information Systems, College of Computer and Information
Sciences, Princess Nourah Bint Abdulrahman University, Riyadh, Saudi Arabia
MAAlduailij@pnu.edu.sa
[2] Department of Computer Science and Engineering, Himalayan School of Science
and Technology, Swami Rama Himalayan University, Dehradun, India
anupamamishra@srhu.edu.in
[3] Faculty of Business, King Abdulaziz University, Rabigh, Saudi Arabia
ifzamzami@kau.edu.sa
[4] University of Macedonia, Thessaloniki, Greece
kpsannis@uom.edu.gr

Abstract. The epidemic taught us to keep digital health records. It
also showed us how wearable observing gear, video conferences, and AI-
powered chatbots may give good treatment remotely. Real-time data
from health care devices across the globe helped attack and track
the infection. Biomedical imaging, sensors, and machine learning have
improved health in recent years. Medical care and biomedical sciences
are now information science sectors that demand enhanced data mining
approaches. Biomedical data have high dimensionality, class irregularity,
and few tests. AI uses information and computations to imitate how people learn, continually improving its accuracy. ML is an important part
of information science. Calculations utilising measurable methods reveal
essential information about information mining operations. This paper
explains and compares ML algorithms that can detect diseases sooner.
We summarise ML's algorithms and processes to extract information for
a data-driven society.

Keywords: Machine learning · Smart health systems · Quantum
algorithms

1 Introduction

In contrast to the 1980s and 1990s, clinical technology has progressed significantly. The medical-care industry is at its peak, thanks to the new age and technology. An ageing population has a number of repercussions, including shifting
patient expectations, a shift in lifestyle choices, and, as a result, the never-ending
cycle of innovation. So, what exactly is AI? It is indeed a system's ability to
carry out behaviors or cognitive processes which we associate with intelligence in

N. Nedjah et al. (Eds.): ICSPN 2021, LNNS 599, pp. 83–91, 2023.
https://doi.org/10.1007/978-3-031-22018-0_8

humans. AI has the ability to boost clinical outcomes along with primary health-care efficiency and productivity. It could also end up making health providers' job much easier by collaborating with them and improve their efficiency [1, 2]. What is ML, basically? Artificial intelligence relies heavily on machine learning [3]. It's all about getting computers to do things without being explicitly programmed. It's also a tool and technology that we can use with your data to know information. It use computer algorithms to determine how to do a task based on the generalization of data or examples, and it can learn to improve itself over time by gaining experience from previous data. Today, machine learning is everywhere, whether it's in business, industry, research, or government, and it's used to solve some of the world's most pressing problems. It's a crucial piece of technology in the financial, medical, commercial, and scientific fields. ML has had an important part in many aspects of health care, including the recent medical practices, treatment of chronic illnesses, as well as the processing of patient data and records. It is understood that hospitals, clinics, and other healthcare organizations around the globe are gradually starting to recognize the necessity for digitization and integration within administrative processes. In recent years, scientists and students have joined the sector of cancer diagnosis and treatment. Combining cognitive computation with genomic tumor sequencing is one of the future approaches. This uses machine learning to create diagnostics and clinical therapies. These robotic arms are more precise and reliable than humans. Moving to Deep Learning, here is what it really means and the way it's helping the healthcare industry. Deep learning is one of a component in a family of machine learning inspired by the human brain and also methods supported by artificial neural networks with representation learning. Deep learning networks along with the extraction of features in an unsupervised manner might improve the performance of classification solutions. Since for some diseases and symptoms the data used for classification and study is scarce, generative adversial networks can be used to generate more datasets [4]. Transfer learning techniques can also be considered when studying different diseases with similar symptomatology [5–9]. Our traditional understanding is based on our life routines, but that is not the fundamental underlying process of nature. Our surroundings is just the development of quantum mechanics, which is the underlying and more fundamental mechanics. Our everyday intuition does not fit quantum phenomena. Indeed, these basic mechanisms remained concealed from us. We just began to notice this feature of nature in the previous century. We generated ideas and mathematical tools from our famous experts as the investigation progressed. Quantum theory, being a probabilistic theory, has sparked several philosophical discussions [10]. Many quantum phenomena, such as wave function collapse, quantum tunnelling, quantum superposition, and so on, continue to interest us. Our comprehension of reality's underlying quantum nature remains a mystery. Quantum technologies aspire to take use of these fundamental rules for our own technological benefit. In the past decades or so, for the application which are based on quantum mechanical laws made advancement by huge strides, with the purpose of replacing or coexisting with conventional systems. Machine learning is the process of enabling a machine to educate or learn like human from the methods used to manage data. The conventional machine learning technique assists in the categorization of images, identification of patterns and speech, and management of enormous volumes of data, among other things, by using its sub module of deep learning. However, a vast amount of data is currently being generated [11]. To handle, organize, and categories vast data, new techniques are necessary. In the great majority of cases, traditional machine learning is capable of finding patterns in data [12,13]. Traditional machine learning algorithms can discover patterns in data in most cases, but they can't tackle exceptional situations that need a large volume of data in a short amount of time. Large database businesses are conscious of their constraints, and they are seeking solutions, including quantum machine learning. Quantum machine learning [14–19] solves problems by employing quantum phenomena such as superposition and entanglement. Quantum computers employ numerous superposition states 0 and 1 to allow them to apply any processing strategy a similar time, giving them a competitive edge on standard machine learning methods. We build quantum algorithms to execute conventional algorithms on a quantum computer using quantum machine learning approaches. This method may be used to categories, analyses, and evaluate data. Quantum algorithms are employed, as well as supervised and unsupervised learning approaches.

2   Machine Learning in BioMedical

Many researchers proposed different machine learning based healthcare techniques [20–23]. Such as, author in [17] proposed Accelerating compute intensive medical imaging segmentation algorithms using hybrid CPU-GPU implementations. Also, author in [18], proposed Blockchain-Assisted Secure Fine-Grained Searchable Encryption for a Cloud-Based Healthcare Cyber-Physical System. In another work, authors [15] proposed an MRI Scans-Based Alzheimer's Disease Detection via Convolutional Neural Network and Transfer Learning Classification is a method for labelling issue domain instances. Author in [24] create a model of Pre-Trained Convolutional Neural Networks for Breast Cancer Detection Using Ultrasound Images. In classification, a class label is predicted for a given sample of input data. For example, a person's illness. Classification modelling demands large training datasets with multiple inputs and outputs. Training data will be used to map input data samples to class labels. The training dataset must be indicative of the issue and contain many class label samples. Textual values must be transformed to numeric numbers before being submitted into the system. Textual variables like "spam" and "not spam" must be converted to numbers before modelling. Label encoding assigns unique integers to class labels, such as "spam" = 0 and "no spam" = 1. We used publicly source datasets of breast cancer, heart failure, and diabetes to test many Machine algorithms [25,26]. We tested the accuracy of machine learning systems for predicting breast cancer, with M = 1 indicating positive cases and B = 0 indicating negative. A digital image of a needle aspirate is utilised to compute breast mass attributes.

They show cell nuclei in three dimensions. The table gives feature info. Mean, standard error, and mean of three highest values (worst) were determined for each image. 30 characteristics. Worst Radius is field 23, Mean Radius is field 3. All feature values get four significant digits. Correlation spans from −1 to 1, with 1 signifying strong linkage and −1 indicating unrelatedness. The pair plot illustrates attribute fluctuations relative to the desired value. All feature values are recorded using four digits. The dataset comprises 569 occurrences and 32 features used to train the prediction model. The correlation values are represented in percentages, with factors that have a substantial influence represented in a brighter shade and features that are highly connected in a deeper shade. Before implementing the machine learning method, the dataset is preprocessed and standardized (Figs. 1 and 2). Popular algorithms that can be used for binary classifications are Logistics Regression, KNN and Gaussian Naive Bayes.

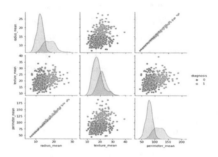

Fig. 1. Pair plot between features (top 3 features in this case only)

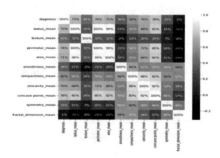

Fig. 2. Correlation matrix of features

3   Selected and Applied Classifiers

3.1   Logistic Regression

Although the name implies regression, Logistic regression is a supervised machine learning algorithm used to predict class or occurrence likelihood. When data is linearly separable and dichotomous, this approach is used. Logistic regression is used for binary classification. Binary classification predicts discrete two-class output variables. Logistic regression is a binary Classification Algorithm used to detect spam. Diabetes forecasting, predicting if a consumer will buy a product or switch to a rival, and determining if a user will click on a marketing link are examples. Logistic regression is a popular Machine Learning technique for two-class categorization. Logistic regression is used to predict. Early identification controls coronavirus, heart disease, and diabetes (Fig. 3). The recommended approach might help clinicians prescribe appropriate therapeutic medicines, according to a comparison [14]. Table 1 shows the outcomes of training and testing the above datasets in Logistic regression. Logistic regression accuracy is 95.10.

Table 1. Results from Logistic regression

Precision	Recall	f1-score	Support
0.97	0.96	0.96	90
0.93	0.94	0.93	53

3.2   KNN (K Nearest Neighbors)

KNN stands for K-nearest neighbors and this machine learning algorithm prediction technique that uses distance calculations between pairs of data. It is also known as lazy learning algorithm as keep all the available instances i.e. training data and make prediction or classify based on a how similar new data to the stores one. Initially "K" need to define K is the nearest neighbour to measure distance from. When training data with known target values is available, the method is employed in classification issues. The algorithm provides a class to each record based on the majority of its k closest neighbours. The method assigns the item the average value among its k closest neighbours for solving a prediction. KNN is a non-parametric classification approach. The core principle is that known facts are arranged in a space defined by the chosen attributes. The algorithm compares the classes of the previous k data to determine the new data's class when it gets new data. Results are displayed in Table 2. KNN accuracy is (96%).

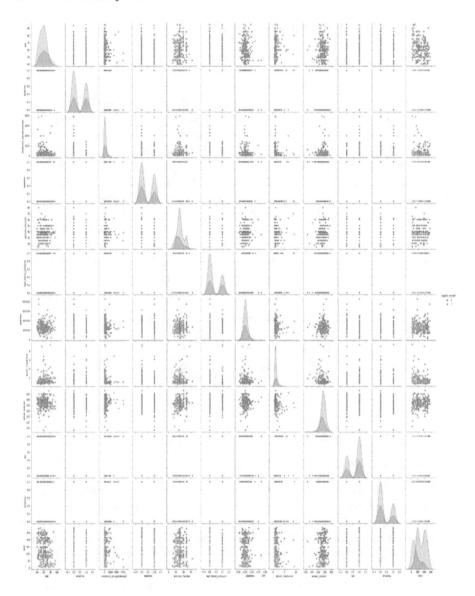

Fig. 3. Pair plot of different feature of diabetes datasets

3.3 Gaussian Naive Bayes

Based on Bayes' Theorem, Naive Bayes is a classification algorithm that assumes predictors are independent. Naive Bayes classifiers presume that the importance features in a class has relation with presence of any other feature to look at it from different perspective. Consider and example of an Orange. If it's Orange in Color, round, and has diameter around 4 inch. Though these qualities are inter-

Table 2. Results from KNN

Precision	Recall	f1-score	Support
0.94	0.99	0.96	67
0.98	0.91	0.95	47

dependent or dependent on the existence of other features, each one contributes to the possibility that this could an Orange, That is why this is called as 'Naive.' The Naive Bayes model is useful with large data sets and its also easy to build. It can outperform other algorithms quite easily. Bayesian Networks (BNs), also described as causal probabilistic models or belief networks, are directed a cyclic graphical (DAG) models. Table 3 displays the results.

Gaussian Naive Bayes model accuracy (in %): 66.66.

Table 3. Results from Gaussian Naive Bayes

Precision	Recall	f1-score	Support
0.69	0.88	0.77	48
0.57	0.30	0.39	27

4   Performance Metrics and Result Discussion

Every trained model's performance is determined by how close the observation is to actual events [27, 28]. With labelled data, we can evaluate any model. Required some evaluation metrics to calculate the performance of any algorithms, the general metric of evaluation used in any algorithm is precision. Certain useful rumour identification metrics, apart from this, are accuracy, recall and F1-score. During trying to predict values toward labels, four bins were received that are FP, FN, TP, and TN. False Positive is a non-rumoured event that is estimated as a rumour, and False Negative is estimated as a non-rumoured event. True Positive is rumoured event is estimated as rumour, True Negative is non-rumoured event is estimated as non-rumour.

5   Conclusion

In this paper we have seen, the importance of machine learning in the field of health care. We have applied different classifiers over datasets and received different accuracy and performance of the classifiers. Before applying the classifiers, we also found the important features based on their co-relation which has been shown through figures.

References

1. Chicco, D., Jurman, G.: Machine learning can predict survival of patients with heart failure from serum creatinine and ejection fraction alone. BMC Med. Inform. Decis. Mak. **20**, 16 (2020)
2. Mishra, V., Singh, Y., Kumar Rath, S.: Breast cancer detection from thermograms using feature extraction and machine learning techniques. In: Proceedings of the IEEE 5th International Conference for Convergence in Technology, Bombay, India, Mar 2019
3. Almomani, A., et al.: Phishing website detection with semantic features based on machine learning classifiers: a comparative study. Int. J. Semant. Web Inf. Syst. (IJSWIS) **18**(1), 1–24 (2022)
4. Kumar, N., et al.: Efficient automated disease diagnosis using machine learning models. J. Healthc. Eng. Article ID 9983652, 13 pages (2021)
5. Smith, J.W., Everhart, J.E., Dickson, W.C., Knowler, W.C., Johannes, R.S.: Using the ADAP learning algorithm to forecast the onset of diabetes mellitus. In: Proceedings of the Symposium on Computer Applications and Medical Care, pp. 261–265. IEEE Computer Society Press (1988)
6. Zou, L., Sun, J., Gao, M., Wan, W., Gupta, B.B.: A novel coverless information hiding method based on the average pixel value of the sub-images. Multimed. Tools Appl. **78**(7), 7965–7980 (2018). https://doi.org/10.1007/s11042-018-6444-0
7. Gupta, B.B., Misra, M., Joshi, R.C.: An ISP level solution to combat DDoS attacks using combined statistical based approach (2012). arXiv preprint arXiv:1203.2400
8. Tăutan, A.-M., Ionescu, B., Santarnecchi, E.: Artificial intelligence in neurodegenerative diseases: a review of available tools with a focus on machine learning techniques. Artif. Intell. Med. **117**, 102081 (2021)
9. Nielsen, M.A.: Neural Networks and Deep Learning, vol. 25. Determination Press, San Francisco, CA, USA (2015)
10. Lee, M.T., Suh, I.: Understanding the effects of environment, social, and governance conduct on financial performance: arguments for a process and integrated modelling approach. Sustain. Technol. Entrepr. **1**(1), 100004 (2022). https://doi.org/10.1016/j.stae.2022.100004
11. Onyebuchi, A., et al.: Business demand for a cloud enterprise data warehouse in electronic healthcare computing: issues and developments in e-healthcare cloud computing. Int. J. Cloud Appl. Comput. (IJCAC) **12**, 1–22 (2022). https://doi.org/10.4018/IJCAC.297098
12. Neural Designer. Available online: https://www.neuraldesigner.com/blog/what_is_advanced_analytics. Accessed 7 June 2019
13. Pan, X., Yamaguchi, S., Kageyama, T., Kamilin, M.H.B.: Machine-learning-based white-hat worm launcher in botnet defense system. Int. J. Softw. Sci. Comput. Intell. (IJSSCI) **14**, 1–14 (2022). https://doi.org/10.4018/IJSSCI.291713
14. Schuld, M., Sinayskiy, I., Petruccione, F.: An introduction to quantum machine learning. arXiv:1409.3097v1
15. Chui, K.T., et al.: An MRI scans-based Alzheimer's disease detection via convolutional neural network and transfer learning. Diagnostics **12**(7), 1531 (2022). https://doi.org/10.3390/diagnostics12071531
16. Sarrab, M., Alshohoumi, F.: Assisted-fog-based framework for IoT-based healthcare data preservation. Int. J. Cloud Appl. Comput. (IJCAC) **11**, 1–16 (2021). https://doi.org/10.4018/IJCAC.2021040101

17. Alsmirat, M.A., Jararweh, Y., Al-Ayyoub, M., Shehab, M.A., Gupta, B.B.: Accelerating compute intensive medical imaging segmentation algorithms using hybrid CPU-GPU implementations. Multimed. Tools Appl. **76**(3), 3537–3555 (2016). https://doi.org/10.1007/s11042-016-3884-2

18. Mamta, et al.: Blockchain-assisted secure fine-grained searchable encryption for a cloud-based healthcare cyber-physical system. IEEE/CAA J. Autom. Sin. **8**(12), 1877–1890 (2021). https://doi.org/10.1109/JAS.2021.1004003

19. Martínez, J.M.G., Carracedo, P., Gorgues Comas, D., Siemens, C.H.: An analysis of the blockchain and COVID-19 research landscape using a bibliometric study. Sustain. Technol. Entrepreneur. **1**(1), 100006 (2022). https://doi.org/10.1016/j.stae.2022.100006

20. Al-Ayyoub, M., et al.: Accelerating 3D medical volume segmentation using GPUs. Multimed. Tools Appl. **77**(4), 4939–4958 (2018)

21. Liu, Y., et al.: Survey on atrial fibrillation detection from a single-lead ECG wave for internet of medical things. Comput. Commun. **178**, 245–258 (2021). ISSN 0140-3664. https://doi.org/10.1016/j.comcom.2021.08.002

22. Zhou, Z., Gaurav, A., Gupta, B.B., Hamdi, H., Nedjah, N.: A statistical approach to secure health care services from DDoS attacks during COVID-19 pandemic. Neural Comput. Appl. **4**, 1–14 (2021). https://doi.org/10.1007/s00521-021-06389-6

23. Benmoussa, K., Hamdadou, D., Roukh, Z.E.A.: GIS-based multi-criteria decision-support system and machine learning for hospital site selection: case study Oran, Algeria. Int. J. Softw. Sci. Comput. Intell. (IJSSCI) **14**(1), 1–19 (2022)

24. Masud, M., et al.: Pre-trained convolutional neural networks for breast cancer detection using ultrasound images. ACM Trans. Internet Technol. **21**, 4. Article 85, 17 pages (2021). https://doi.org/10.1145/3418355

25. Yu, H.Q., Reiff-Marganiec, S.: Learning disease causality knowledge from the web of health data. Int. J. Semant. Web Inf. Syst. (IJSWIS) **18**(1), 1–19 (2022)

26. Dan, S.: NIR spectroscopy oranges origin identification framework based on machine learning. Int. J. Semant. Web Inf. Syst. (IJSWIS) **18**(1), 1–16 (2022)

27. Singh, G., Malhotra, M., Sharma, A.: An adaptive mechanism for virtual machine migration in the cloud environment. Int. J. Cloud Appl. Comput. (IJCAC) **12**, 1–10 (2022). https://doi.org/10.4018/IJCAC.297095

28. Piper, J., Rodger, J.A.: Longitudinal study of a website for assessing American Presidential candidates and decision making of potential election irregularities detection. Int. J. Semant. Web Inf. Syst. (IJSWIS) **18**(1), 1–20 (2022)

Blockchains and Cross-Blockchains: Privacy-Preserving Techniques

Mojgan Rayenizadeh[1], Marjan Kuchaki Rafsanjani[1($\boxtimes$)],
Alicia García-Holgado[2], and Ali Azadi[2]

[1] Department of Computer Science, Faculty of Mathematics and Computer, Shahid
Bahonar University of Kerman, Kerman, Iran
`mojgan.rayeni@math.uk.ac.ir`, `kuchaki@uk.ac.ir`
[2] University of Salamanca, Salamanca, Spain
`aliciagh@usal.es`, `ali.azadi@usal.es`

Abstract. Blockchain is a new and revolutionary technology whose use is increasing in recent years. The blockchain network has recently been used in a wide variety of scenarios far beyond digital currencies, such as the Internet of Things (IoT), unmanned aerial vehicles (UAVs), and healthcare, where the need for collaboration between different blockchains is critical. The concept of blockchain interoperability, which means connecting multiple blockchain networks, has attracted the attention of many researchers. However, achieving interoperability between blockchains is a challenge for academics and the industry, as concerns about security and privacy are very evident in blockchain interoperability. Although there have been studies on blockchain privacy protection, cross-blockchain privacy protection is still in its infancy and more research is needed. In this article, we examine and compare the privacy protection provided for blockchain and cross-blockchain. We hope this work will provide future researchers with insight into blockchain and cross-blockchain privacy protection technologies.

Keywords: Blockchain · Cross-blockchain · Privacy-preserving · Security

1 Introduction

Blockchain is a ledger (a type of database) that provides the possibility of recording information permanently and without change. A series of specific features (decentralized, distributed, secure, and not requiring a trusted central authority) distinguish blockchain from other databases. This database is a chain of blocks [1]. The chain structure is a key feature that facilitates the reliability and immutability of the ledger [2]. The inherent features of blockchain provide features such as transparency and security [3,4]. There are different applications of this technology such as finance [5,6], cloud computing [7], and intelligent transportation systems.

© The Author(s), under exclusive license to Springer Nature Switzerland AG 2023
N. Nedjah et al. (Eds.): ICSPN 2021, LNNS 599, pp. 92–100, 2023.
https://doi.org/10.1007/978-3-031-22018-0_9

Due to the nature of blockchain technology (distributed and immutable features inherent in this technology), its use in various applications is very popular, such as used in UAVs [8], industry, healthcare, and transportation [9]. At present, most of the existing blockchains operate as an independent and isolated environment without communication with each other [10] while many applications require inter-blockchain interoperability [11]. So inter-blockchain interoperability becomes a challenge that attracts attention in both industry and academic research. For example, the full use of network-based applications of UAVs is still limited since each application runs on an isolated blockchain, requiring a cross-blockchain design for UAV networks [12]. According to our knowledge, security and privacy for inter-blockchain interoperability are the new topics in the blockchain. Creating trustless cross-blockchain trading protocols is challenging. So far, methods have been proposed to supply the security and privacy of the blockchain. There are several surveys on blockchain technology and security, such as [13–16], as well as some review studies on cross-blockchain methods, such as [10,17,18]. Recently, security and privacy-preserving methods have been introduced for cross-blockchain [19]. Presented a cross-blockchain-based privacy-preserving data sharing for electronic health records [20]. Designed an efficient blockchain interoperability scheme that achieves interoperability and trusted services at the same time. So far, there is no systematic research on the recent progress of privacy protection in cross-blockchain interactions [22]. To the best of our knowledge, no paper has comprehensively surveyed the methods that provide security and privacy for mutual blockchain and cross-blockchain (consensus mechanism, privacy preservation type, privacy strategy used, simulation environment, and application). This prompted us to present a survey on cross-blockchain security and privacy practices. In this article, the techniques that provide privacy for blockchain and cross-blockchain are explored.

1.1 Contribution

In this study, our focus is on privacy in blockchain and cross-blockchain. Our contribution includes reviewing the existing protocols and algorithms presented for privacy-preserving in blockchain and cross-blockchain and providing a comparison of the discussed methods.

The rest of the paper is organized as follows: Sect. 2 presents the blockchain technology, cross-blockchains, and privacy services. Section 3 describes a summary of each privacy-preserving technique and a comparison of them. Finally, in Sect. 4, we conclude this study.

2   Blockchain, Cross-Blockchain, and Privacy Security

This section introduces the main terms and concepts. We describe the blockchain technology; explain the cross-blockchain, and then privacy services.

2.1 Blockchain

Blockchain is a type of information recording and reporting system where the information stored in it is shared among all members of the network. Each system that connects to the network receives a copy of the entire blockchain. This technology allows us to record information permanently and without change.

The most important elements and contents of blockchain technology are [23]:

- Transaction: A record in the ledger that provides a piece of information or an operation before a transaction.
- Block: Contains several transactions and a cryptographic hash of the previous block, transaction data, timestamp, and other information such as a digital signature.
- Timestamp: Required to preserve the creation time of the block.
- Hash: It is an encryption function that takes data of any size from the input and produces an output whose length is always fixed.
- Genesis block: The first network block before which there is no other block.
- Chain: A linked list of blocks starting with the origin block.
- Consensus: An algorithm implemented by network nodes to agree on the state of the blockchain.

The description of the blockchain process by an example is shown in Fig. 1. If user A wants to make a transaction with user B using the blockchain: a new block is created that contains the transaction and then broadcasts to the network nodes for confirmation. If the new transaction is not verified, it is deleted, but if it is, the new block is added to the blockchain and distributed among the nodes of the network so that other nodes can update their blockchain. Finally, the transaction is received by user B [24].

2.2 Cross-Blockchain

A new concept in blockchain technology is cross-blockchain, which aims to enable interoperability between blockchain networks [25]. Interoperability in a blockchain network means that multiple blockchain networks can connect without any restrictions and transfer information between networks [10]. Interoperability between blockchains remains a research challenge [26]. Solutions proposed to connect blockchains can revolutionize blockchain technology by enabling the transfer of data and assets between different (homogeneous and heterogeneous) blockchains [27,28]. Presented a cross-blockchain protocol for asset transfer. Blockchain interoperability is much broader than digital currencies and on-chain asset transfers [29]. Researchers are trying to find the optimal communication solution between blockchains [10].

2.3 Privacy Services

Privacy refers to the user's ability to control the disclosure of their information and prevent unauthorized access and processing. Access to data can be controlled by "defining a set of rules about who can access what information and

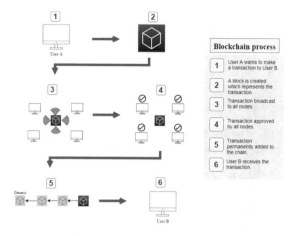

Fig. 1. Blockchain process

when". Mechanisms for providing privacy-preserving services are proposed. A traditional technique for protecting data privacy is to use encryption to prevent others from accessing the data. The homomorphic encryption method allows certain types of computations to be performed on the cipher text and produces an encrypted result [30]. Another aspect of privacy is hiding the user's identity [31,32]. Introduced a framework that provides stronger privacy guarantees. Traditional methods face challenges. For example, techniques that use complex cryptography have scalability and performance challenges. The question of who owns a piece of data is mostly a difficult one, data ownership is another challenge.

Some privacy protection strategies in blockchain networks include:

- Encryption: This strategy is almost used in every blockchain and uses public and private keys to preserve identity and privacy.
- Smart contract: In a smart contract-based blockchain, transaction details (including all information and data related to the transaction) are written on a smart contract. This contract runs after being installed on the network and no user can stop running it.
- Anonymization: In this strategy, after identifying the PII (personally identifiable information) contained in the data, these PIIs are removed from the data set before being published or transmitted on the network so that the information about the hidden data is not available to any adversary.
- Mixing: This strategy is designed to prevent user linking and facilitate anonymity in blockchain-based financial transactions by merging transactions to hide the original identifier.

3   Privacy-Preserving Techniques

Blockchain is a type of distributed and decentralized database that is updated after each transaction to maintain stability. Blockchain content is accessible to all members of the blockchain network. The leakage of this information is a threat to the blockchain; in addition, users' identities can be cracked using various strategies. Privacy is considered from two aspects: user identity privacy and transaction privacy.

3.1   Privacy-Preserving for Blockchain and Cross-Blockchains

Researchers have proposed various privacy preservation methods for blockchains and cross-blockchain. In the following, we describe some of these methods.

- Ruffing et al. [33] have presented the CoinShuffle protocol which is a decentralized bitcoin shuffling protocol. In this protocol, users can combine their coins with other users without the need for a third party. CoinShuffle only requires standard cryptography basics.
- Yang et al. [34] have presented a blockchain-based decentralized trust management system in vehicular networks. In this system, each vehicle is assigned a rank to the neighboring vehicles after confirming the message received from them. Then they upload the ranking and the RSUs (roadside units) calculate the amount of trust value offset and put this information in a block to add to the trust blockchain. As a result, all RSUs participate in a decentralized update of trust value.
- Aitzhan et al. [35] have proposed a system to develop the security and privacy transaction in decentralized smart grid energy trading without needing a trusted third party. This system uses blockchain technology, multi-signature, and an anonymous encrypted message propagation stream. The proposed system for consensus in the blockchain implemented a PoC (proof-of-concept) which is one of the alternative solutions to the problem of high energy consumption in PoW (Proof of Work) systems and cryptocurrency hoarding in PoS (Proof of Stake) systems.
- Li et al. [36] have presented a blockchain-based announcement network called CreditCoin to provide privacy via the proposed protocol (Echo-Announcement) in the vehicular network. CreditCoin allows users to generate signatures and send notices anonymously in untrusted surroundings. This uses the blockchain to motivate users to share traffic information.
- Mishra et al. [37] proposed a model for Defensive Approach using Blockchain Technology against Distributed Denial of Service attacks.
- Desai et al. [38] have presented an architecture that is a mixture of public and private blockchains to enable privacy-enhanced, accountable auctions. In the private blockchain, bids are opened and the winner of the auction is found securely using smart contracts founded on this blockchain. In the public blockchain, payments are made and smart contracts based on this type of blockchain are used to announce the winner of the auction. This architecture utilizes cryptographic tools on the public blockchain to prevent information disclosure.
- Cao et al. [39] have proposed a scheme to provide security-privacy for EHR (Electronic Health Record), called CEPS (Cross-Blockchain based EHR

Privacy-Preserving Scheme). The CEPS uses relay-chain which is a cross-blockchain technology to transfer EHR between different hospitals' private blockchains without privacy leakage problems. In addition, RaaS (relay-chain as a service) is presented that solves the trouble of deleting EHR information on the private blockchain per hospital.

- Wang and He [19] have presented a cross-blockchain-based privacy-preserving data sharing (CPDS). In this technique, EHRs are held securely in the cloud. The generated index is stored in the hospital's private blockchain. They presented Wanchain as a Service (WaaS), which is cross-blockchain technology and is utilized to solve the privacy leakage issue of EHR transmission on distinct blockchains.
- Mamata et al. [21] proposed Blockchain-Assisted Secure Fine-Grained Searchable Encryption for a Cloud-Based Healthcare Cyber-Physical System.
- Xiang et al. [40] have presented a blockchain-based hybrid medical image framework (BMIF) with privacy protection. In this method, consensus nodes are used to calculate the combined tasks of privacy-preserving medical images. BMIF proposed a new hybrid medical image model founded on a convolutional neural network (CNN) and Inception, which can acquire a better image presentation. In this framework, CKKS fully homomorphic encryption strategy is used to encrypt the data set to protect data privacy.

Table 1. Comparative view of privacy preservation techniques in blockchain and cross-blockchain.

Ref.	Year	Application	Type	Consensus mechanism	Privacy preservation type	Privacy strategy used	Simulation environment
[33]	2014	Finance	Blockchain	PoC	User identity privacy	Mixing	Python
[34]	2018	Vehicular network	Blockchain	Joint PoW and PoS	User identity privacy	Encryption	MATLAB
[35]	2018	Energy systems	Blockchain	PoC	User identity privacy	Anonymization	Python
[36]	2018	Vehicular network	Blockchain	PoW	User identity privacy	Anonymization	NS2 and JAVA
[38]	2019	Online auctions	Cross-blockchain	PoW	Transactional privacy	Smart contracts	Ethereum virtual machine (Ropsten as public blockchain)
[39]	2020	Electronic health record	Cross-blockchain	BFT (Byzantine fault tolerance)	Transactional privacy	Anonymization	Rust
[19]	2021	Electronic health record	Cross-blockchain	Galaxy	Transactional privacy	Anonymization	N/A
[40]	2022	Electronic health record	Blockchain	PoDL (proof of deep learning)	Transactional privacy	Encryption	Python

3.2 Comparison of Techniques

The comparison of discussed approaches is shown in Table 1. As shown in Table 1, cross-blockchain is an emerging technology and recently, many works have been introduced in the field of its security. Among consensus mechanisms, the PoW mechanism has been the most widely used. The type of privacy protection for blockchain is user identity privacy and for cross-blockchain is transaction privacy. Anonymization privacy strategy has also been more widely used than other strategies.

4 Conclusion

Today, with the widespread use of blockchain in various applications, it is expected that the connection of different blockchains will be necessary. Blockchains can be used to enhance security services in various applications. Due to the fact, that in such networks the content is accessible to all members of the network, as a result, information leakage is a threat. Therefore, one of the major challenges in this technology is improving security and privacy in blockchain and cross-block chain. In this paper, we reviewed and compared the privacy approaches available on blockchain and cross-blockchain.

Acknowledgement. This research was partially funded by the Spanish Government Ministry of Science and Innovation through the AVisSA project grant number (PID2020-118345RB-I00).

References

1. Gorkhali, A., Li, L., Shrestha, A.: Blockchain: a literature review. J. Manag. Anal. **7**(3), 321–343 (2020)
2. Gonczol, P., Katsikouli, P., Herskind, L., Dragoni, N.: Blockchain implementations and use cases for supply chains—a survey. IEEE Access **8**, 11856–11871 (2020)
3. Lu, J., et al.: Blockchain-based secure data storage protocol for sensors in the industrial internet of things. IEEE Trans. Ind. Inform. **18**(8), 5422–5431 (2021)
4. Casino, F., Dasaklis, T.K., Patsakis, C.: A systematic literature review of blockchain-based applications: current status, classification and open issues. Telemat. Inform. **36**, 55–81 (2019)
5. Vandezande, N.: Virtual currencies under EU anti-money laundering law. Comput. Law Secur. Rev. **33**(3), 341–353 (2017)
6. Treleaven, P., Brown, R.G., Yang, D.: Blockchain technology in finance. Computer **50**(9), 14–17 (2017)
7. Sharma, P.K., Chen, M.-Y., Park, J.H.: A software defined fog node based distributed blockchain cloud architecture for IoT. IEEE Access **6**, 115–124 (2017)
8. Alkadi, R., Alnuaimi, N., Shoufan, A., Yeun, C.: Blockchain interoperability in UAV networks: state-of-the-art and open issues, pp. 1–11 (2021). arXiv preprint arXiv:2111.09529

9. Maidamwar, P., Saraf, P., Chavhan, N.: Blockchain applications, challenges, and opportunities: a survey of a decade of research and future outlook. In: International Conference on Computational Intelligence and Computing Applications (ICCICA), pp. 1–5. IEEE, Nagpur, India (2021)

10. Qasse, I.A., Talib, M.A., Nasir, Q.: Inter blockchain communication: a survey. In: 6th Annual International Conference Research Track, pp. 1–6 (2019)

11. Tewari, A., et al.: Secure timestamp-based mutual authentication protocol for IoT devices using RFID tags. Int. J. Semant. Web Inf. Syst. (IJSWIS) 16(3), 20–34 (2020)

12. Alkadi, R., Alnuaimi, N., Yeun, C.Y., Shoufan, A.: Blockchain interoperability in unmanned aerial vehicles networks: state-of-the-art and open issues. IEEE Access 10, 14463–14479 (2022)

13. Ma, Y., Sun, Y., Lei, Y., Qin, N., Lu, J.: A survey of blockchain technology on security, privacy, and trust in crowdsourcing services. World Wide Web 23(1), 393–419 (2020)

14. Zheng, X., Zhu, Y., Si, X.: A survey on challenges and progresses in blockchain technologies: a performance and security perspective. Appl. Sci. 9(22), 4731 (2019)

15. Hathaliya, J.J., Tanwar, S.: An exhaustive survey on security and privacy issues in healthcare 4.0. Comput. Commun. 153, 311–335 (2020)

16. Liang, Y., et al.: PPRP: preserving-privacy route planning scheme in VANETs. ACM Trans. Internet Technol. (TOIT) (2022)

17. Vo, H.T., Wang, Z., Karunamoorthy, D., Wagner, J., Abebe, E., Mohania, M.: Internet of blockchains: techniques and challenges ahead. In: International Conference on Internet of Things (iThings) and IEEE Green Computing and Communications (GreenCom) and IEEE Cyber Physical and Social Computing (CPSCom) and IEEE Smart Data (SmartData), pp. 1574–1581. IEEE, Halifax, NS, Canada (2018)

18. Borkowski, M., Frauenthaler, P., Sigwart, M., Hukkinen, T., Hladký, O., Schulte, S.: Cross-blockchain technologies: review, state of the art, and outlook. White Paper, pp. 1–5 (2019)

19. Wang, Y., He, M.: CPDS: a cross-blockchain based privacy-preserving data sharing for electronic health records. In: 6th International Conference on Cloud Computing and Big Data Analytics (ICCCBDA), pp. 90–99. IEEE, Chengdu, China (2021)

20. Wang, G., Nixon, M.: InterTrust: towards an efficient blockchain interoperability architecture with trusted services. In: International Conference on Blockchain (Blockchain), pp. 150–159. IEEE, Melbourne, Australia (2021)

21. Mamta, et al.: Blockchain-assisted secure fine-grained searchable encryption for a cloud-based healthcare cyber-physical system. IEEE/CAA J. Autom. Sin. 8(12), 1877–1890 (2021). https://doi.org/10.1109/JAS.2021.1004003

22. Wang, J., Cheng, J., Yuan, Y., Li, H., Sheng, V.S.: A survey on privacy protection of cross-chain. In: Sun, X., Zhang, X., Xia, Z., Bertino, E. (eds.) ICAIS 2022, CCIS, vol. 1588, pp. 283–296. Springer, Cham (2022)

23. Bernabe, J., Canovas, J., Hernandez-Ramos, J.L., Moreno, R.T., Skarmeta, A.: Privacy-preserving solutions for blockchain: review and challenges. IEEE Access 7, 164908–164940 (2019)

24. Zubaydi, H.D., Chong, Y.-W., Ko, K., Hanshi, S.M., Karuppayah, S.: A review on the role of blockchain technology in the healthcare domain. Electronics 8(6), 679 (2019)

25. Pillai, B., Biswas, K., Hóu, Z., Muthukkumarasamy, V.: Cross-blockchain technology: integration framework and security assumptions. IEEE Access 10, 41239–41259 (2022)

26. Pillai, B., Biswas, K., Hóu, Z., Muthukkumarasamy, V.: Burn-to-claim: an asset transfer protocol for blockchain interoperability. Comput. Netw. **200**, 108495 (2021)

27. Ghaemi, S., Rouhani, S., Belchior, R., Cruz, R.S., Khazaei, H., Musilek, P.: A pub-sub architecture to promote blockchain interoperability, pp. 1–9 (2021). arXiv preprint arXiv:2101.12331

28. Sigwart, M., Frauenthaler, P., Spanring C., Sober, M, Schulte, S.: Decentralized cross-blockchain asset transfers. In: Third International Conference on Blockchain Computing and Applications (BCCA), pp. 1–12. IEEE, Tartu, Estonia (2020)

29. Belchior, R., Vasconcelos, A., Guerreiro, S., Correia, M.: A survey on blockchain interoperability: past, present, and future trends. ACM Comput. Surv. **54**(8), 1–41 (2021)

30. Byun, J.W., Kamra, A., Bertino, E., Li, N.: Efficient k-anonymization using clustering techniques. In: Kotagiri, R., Krishna, P.R., Mohania, M., Nantajeewarawat, E. (eds.) Advances in Databases: Concepts, Systems and Applications. DASFAA 2007. LNCS, vol. 4443, pp. 188–200. Springer, Heidelberg (2007)

31. Salman, T., Zolanvari, M., Erbad, A., Jain, R., Samaka, M.: Security services using blockchains: a state of the art survey. IEEE Commun. Surv. Tutor. **21**(1), 858–880 (2018)

32. Li, N., Li, T., Venkatasubramanian, S.: t-Closeness: privacy beyond k-anonymity and l-diversity. In: 23rd International Conference on Data Engineering, pp. 106–115. IEEE, Sanya, China (2007)

33. Ruffing, T., Moreno-Sanchez, P., Kate, A.: CoinShuffle: practical decentralized coin mixing for bitcoin. In: Kutyłowski, M., Vaidya, J. (eds.) Computer Security - ESORICS 2014. LNCS, vol. 8713, pp. 345–364. Springer, Cham (2014)

34. Yang, Z., Yang, K., Lei, L., Zheng, K., Leung, V.C.M.: Blockchain-based decentralized trust management in vehicular networks. IEEE Internet Things J. **6**(2), 1495–1505 (2018)

35. Aitzhan, N.Z., Svetinovic, D.: Security and privacy in decentralized energy trading through multi-signatures, blockchain and anonymous messaging streams. IEEE Trans. Dependable Secure Comput. **15**(5), 840–852 (2016)

36. Li, L., et al.: CreditCoin: a privacy-preserving blockchain-based incentive announcement network for communications of smart vehicles. IEEE Trans. Intell. Transp. Syst. **19**(7), 2204–2220 (2018)

37. Mishra, A., et al.: Defensive approach using blockchain technology against distributed denial of service attacks. In: International Conference on Smart Systems and Advanced Computing (Syscom-2021), Dec 2021

38. Desai, H., Kantarcioglu, M., Kagal, L.: A hybrid blockchain architecture for privacy-enabled and accountable auctions. In: International Conference on Blockchain (Blockchain), pp. 34–43. IEEE, Atlanta, GA, USA (2019)

39. Cao, S., Wang, J., Du, X., Zhang, X., Qin, X.: CEPS: a cross-blockchain based electronic health records privacy-preserving scheme. In: International Conference on Communications (ICC), pp. 1–6. IEEE, Dublin, Ireland (2020)

40. Xiang, T., Zeng, H., Chen, B., Guo, S.: BMIF: privacy-preserving blockchain-based medical image fusion. ACM Trans. Multimed. Comput. Commun. Appl. (TOMM) 1–23 (2022)

A Hybrid Approach for Protection Against Rumours in a IoT Enabled Smart City Environment

Anupama Mishra[1(✉)], Ching-Hsien Hsu[2], Varsha Arya[3,4(✉)],
Priyanka Chaurasia[5], and Pu Li[6]

[1] Swami Rama Himalayan University, Dehradun, India
anupamamishra@srhu.edu.in
[2] Asia University, Taichung, Taiwan
[3] Insights2Techinfo, New Delhi, India
varshaarya21@gmail.com
[4] Lebanese American University, Beirut 1102, Lebanon
[5] Ulster University, London, UK
p.chaurasia@ulster.ac.uk
[6] Software Engineering College, Zhengzhou University of Light Industry, Zhengzhou,
China
lipu@zzuli.edu.cn

Abstract. Smart cities have become a new ideal for maximising machine-to-machine, machine-to-human, and human-to-human communications to provide better amenities and quality of life. These ideas use diverse networks, sensor nodes, processing devices, and data repositories that collect, transfer, store, and process real-time information. Smart cities along with latest technology offer applications and services for health care, education, energy usage, and transportation. Smart cities offer potential, but security and privacy must be addressed. Our research aims to identify Intrusion Detection systems that detect cyber threats.

Keywords: Smart cities · Intrusion detection system · Cyber attacks

1 Introduction

The Internet of Things is a development in information communications that allows direct, permanent, and automatic device-to-device (including M2M and CPS) connection and has real-world applications. Drones and/or VR add another dimension. Industrial automation (Industrial IoT) and process control systems, including power grid administration, traffic tracking, smart cities, video surveillance, crowd sensing, and body area networks/e-health, have dominated IoT installations thus far. In 2018, 55.3% of the world's population lived in cities; by 2030, 60% will [1]. Rapid urbanisation and population growth have caused many social, technological, organisational, and economic difficulties that threaten the

N. Nedjah et al. (Eds.): ICSPN 2021, LNNS 599, pp. 101–109, 2023.
https://doi.org/10.1007/978-3-031-22018-0_10

economic and environmental sustainability of cities. Major nations are adopting 'smart' approaches to maximise transport, electricity distribution networks, natural resources, organisational and human capital. "Smart city" means using all available resources and technology in a coordinated and intelligent way to accomplish sustainable city development. IoT enables smart city applications [3]. Smart energy, for efficient generation and utilisation of various kinds of energy from multiple resources; smart buildings, which can control the temperature and manage the lightning and power consumptions without human intervention; and many other applications like smart mobility, smart networking, smart security, smart governance, and smart health care systems in which smart devices are connected to medical devices for monitoring. The smart city market might be worth $1.5 trillion by 2020 [4]. Thousands of sensor nodes can create a smart city. These nodes sense and collect data for big data management systems and apps. These systems give real-time information on traffic, parking lots, public transportation, air and water pollution, energy use, and catastrophe responses.

Generating, processing, analysing, distributing, and keeping sensitive data creates privacy and security concerns. Existing IoT devices collect data from numerous sources and transport it to data analysing or storage centres over existing networks. This increases the attack surface and provides an access point for malicious users or attackers. These malicious users or nodes use network or device flaws to launch cyberattacks like denial of service, SQL injection, eavesdropping, session hijacking, and others. These attacks harm smart city systems. Smart cities have active video surveillance and a GPS that can track inhabitants' movements. Attackers can obtain this data, endangering citizens' privacy. Smart cities need suitable security solutions [5,9,18]. In recent years, intrusion detection has evolved greatly, and several systems have been developed to fulfil diverse needs. Several occurrences have proved that authentication and encryption are insufficient for privacy and security. As systems become increasingly complicated, flaws grow, causing security difficulties. Intrusion detection is a second line of defence for networks and systems. An incursion might trigger a reaction to avoid or reduce damage [6,8]. After perimeter measures, authentication, access control, and firewall limit unexpected behaviours. These controls prevent malicious or anomalous activity. The next stage of defence is Intrusion detection systems (IDS). Intrusion detection systems only detect and alert the user or other systems. When IDS detects an intrusion, it logs the incident, stores the required data/traffic, notifies the administrator, and may intervene. The logged data can also be used for forensics and as proof against an attacker. Some IDS don't work in real-time because of how they analyse or because they're meant for forensics. The data is audited, reviewed, and compared with multiple detection systems to identify successful and failed intrusion attempts. With IDS and IPS, bad traffic or a cyber assault can be detected, protecting nodes in IoT networks of smart cities [10,12–14].

Section 2 presents an overview of IoT and smart cities' history and current security challenges. Section 3 provides a taxonomy of attacks and compares a few with smart cities. Section 4 covers IoT and smart city defences. Section 5 discusses IDS features and required datasets. Section 6 examines IDS classification based on parameters and IoT IDS models. Section 6 compares IDS with IPS. Section 8 discusses IDS and IPS design and modelling methodologies. Section 9 discusses open research difficulties.

2   Related Work

The physical aspects of the sensors [2,32], the wireless connection, and, to a certain point, analytics are vital in both areas. There is a third class of applications that are less concerned with the functional nature of the devices themselves and much more about data analysis: they tackle the fundamental transformation of business processes (BP) related to the specific commercial functions such as accounting, insurance, industry, and company (Including government functions) and medical care. In the light of (usually) restricted computing, storage, power, and control capacities of terminal nodes and physical access to those final nodes, protection in the IoT setting is seen to be crucial in the device space region. Although security is certainly critical for the applications, it is necessary for those corporate-based applications which deal almost exclusively with Personal Identifiable Information (PII). Apps such as ITSs, e-health, network controls, and drones will potentially have effects on life safety, and security problems are therefore main and predominant. Safety and security, Protection is a major end-to-end: beginning with the access/edge/fog network, the main network, and maybe, continuing with the cloud service. There has been substantial work and lobbying, but there is still an urgent need for technical and business needs. Author in [15] proposed spammer Detection Approaches in Online Social Network (OSNs): A Survey. Authors in [24] proposed a data mining approaches for sentiment analysis in online social networks (OSNs). In another work, authors [7] proposed Fake News Detection Using Multi-Channel Deep Neural Networks. Also, in [19] authors proposed a Hybrid approach for detection of malicious profiles in twitter.

3   Proposed Approach

In this section, we use several steps from collection of the data sets to classification. In this section discusses various techniques always had to gather data from social media allowing research into rumors, as well as approaches to gathering data annotations. APIs are effective for connecting, gathering, and storing social data. APIs provide well-defined methods for applications to request data. Before using an API, read its documentation to understand its methods and constraints. Each social media site has its own restrictions [17], which is important when using social media data to classify rumours. Twitter and Facebook are used to analyse rumours. Twitter provides comprehensive instructions on how to use its API, which allows access to a REST API to collect data from its servers and a streaming API to collect data in real time. After registering a Twitter application, the author can access all kinds of ways ("endpoints") for Twitter information collection through oath authentication. The largest disadvantage is meant to get actual or newest information, complicating data acquisition. Its API is research-friendly. Facebook offers a recorded API including collection of application design tools for the scripting languages and making it handy to use the data to build applications. Like Twitter's application program interface, WeChat too demands that an app be registered to obtain the API keys. Typically, the solution for accessing Facebook posts is to collect data from so-called Facebook pages that are public pages generated by organisations, authority. When using an API, it's important to grasp the platform's service restrictions, especially when releasing a data collection publicly. It is also not authorised to release raw data and is confined to particular content identifiers, such as tweeter identification.

4   Rumor-Gathering Strategies

The rumour classification system usually starts with the identification of unconfirmed details (rumour detection) and finishes with the determination of its veracity value. Complete cycle of rumors identification for determination of truthfulness is carried out mainly using 4 constituent - Detection, Tacking, Stance, Veracity.

1. **Rumour Detection** - Identifying whether or not a bit of data is a rumors. Collection of posts on social media could be an input to a rumour detection component, on which Binary classifier is used to identify rumour or non-rumour data stream. This component's output is collection of blogs and post in which each blog is classified as rumors or not.
2. **Rumour Tracking** - When rumour is detected using the component of rumour detection; this collects and filters the rumour after discussion. The output of such a portion would be a series of blogs addressing the speculation.
3. **Stance Classification** - Though this rumors tracking component scoops up articles related to a false story, the position classification component defines that each article orients the rumour's truthfulness and classifies the collected related post as predefined collection of stances (i.e., support, deny, query, and comment).
4. **Veracity Classification** - It gives the truth value of the rumour based on the stance calculation done in the stance classification. Truth value is the result of this classification, but it could provide contexts like Web addresses or some other sources of information records that assist consumer to determine the classification model's accuracy again by-checking the verifiable facts.

5   Performance and Evaluation

Process of detecting rumours is that, the application must detect rumours and information from series of posts on social sites that have yet to be checked. Explicitly, the task can take the timeline of the posts on social media TL = T[1], T[TL] as input, and the classification algorithm had to evaluate whether each of these posts, T[i], is a rumour or not by allocating a label from R = Y, N. To solve the problem binary classification will be used.

Datasets that differ based on the platforms from which the data are obtained, the forms of data used, whether the info on the propagation is registered, and so on. Table 1 lists the rumour detection datasets. Data set link of PHEME is avilable at https://figshare.com/articles/PHEME_dataset_of_rumours_and_non-rumours/4010619

Table 1. Rumour detection datasets

Dataset	Total rumours	Text	User info	Time stamp	Propagation info	Platform	Description
PHEME	6425	Yes	Yes	Yes	Yes	Twitter	Tweets from [Kochkina et al., 2018]
Facebook hoax	15.5k	Yes	Yes	Yes	–	Facebook	Data from [Tacchini et al., 2017]
Fake News net	23,196	Yes	Yes	Yes	Yes	Twitter	Data set from [Shu et al., 2019]
SemEval19	325	Yes	Yes	Yes	Yes	Reddit, Twitter	Task 7 dataset SemEval 2019
Ma twitter	992	Yes	Yes	Yes	–	Twitter	Tweets from [Ma et al., 2016]
PHEME-R	330	Yes	Yes	Yes	Yes	Twitter	Tweets from [Zubiaga et al., 2016]
Ma-Weibo	4664	Yes	Yes	Yes	–	Weibo	Weibo data from [Ma et al., 2016]

Most research was proposed in recent years to tackle the problem of detecting rumours in the world of social media. Almost all of the methods suggested in the study to identify false data that address the problem of classification - They seek to connect labels like whether it is rumour or non-rumour, true or false. Researchers are required to check various classification techniques to obtain the most appropriate for the data set and task available. Figure 1 contains various approaches that used to rumours identify in social media sites. Such approaches could be disproportionately divided into classified and unclassified approaches. Classified approaches can be categorised into deep learning and machine learning approaches.

Machine learning techniques are already demonstrated particularly effective for the resolution of various tasks in the field of information engineering. Machine learning approaches can be applied in a variety of research studies on the subject of fake information detection on the web. In general, plenty of the machine learning methods applied for rumour and fake new identification has implemented a supervised learning technique. SVM (Support Vector Machines) is the most commonly used classification techniques in a variety of research fields. In most support vector-based solutions to fake news and misinformation, content-based features have been abused and Afroz et al. [5] have achieved amazingly serious scores for the misleading assignment of a huge datasets by utilizing syntactic, lexical and content explicit features. Researcher Rubin et al. [11] instruct

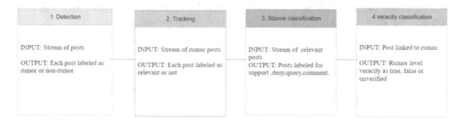

Rumour classification framework

Fig. 1. Classification of rumours: framework

the support vector machine to identify critical false news with the variety of content specific features, achieving a F1 score of 0.87. As far as rumour specific tasks includes identification, verification, etc. are concerned support vector machine-based approaches created the use of context-based features and also content-oriented attributes more popular and Qin et al. [13] introduced a series of design elements to detect uniqueness in tweets, allowing for a precision of 0.75 and the other popular Support vector machine-based approach was used in [14], where the researcher suggested the use of a graph kernel related support vector machine classifier to classify rumours by using content features and propagation structures. The accuracy 0.91 recorded on a limited dataset of Sina weibo. SVMs have also been used to detect clickbait in [6]. Authors [26,27], also consider a collection of content-oriented attributes to achieve F1 score 0.94. Random forests used in a variety of rumour research works and the random forest is a set of number of decision-making trees and the state of all observations of trees corresponds to final result. Qualitative research on machine learning techniques to detect false news and rumour have established random forests as more effective approach. More precisely Kwon et al. [21] have developed model based on random forest with collection of contextual and functional properties in a tweet graph with 0.90 precision for rumors classification and it can also use for stance detection that shown in [20,28,29]. Author Zeng et al. consider only content-based tweet features and had overall precision which lies in between 0.83 and 0.88 on a manual data-set. Random forest was also used to evaluate the credibility of users in [16]. Models based on logistic regression were used, for variety of studies, in general in the rumour stance classification problems and also regression (logistic) was utilized to arrange the position of news stories based on the cases and features in [22] the proposed approach had an accuracy of approximately 0.73 on the dataset which are emergent. An author Chua and Banerjee published research on semantic indicators of gossip veracity through using strategic relapse (regression) to characterize unmistakable ones with respect to valid and bogus bits of gossip and Hardalov et al. have utilized strategic relapse to assess the unwavering quality of genuine and bogus Bulgarian news, acquiring an exactness of 0.75, this accuracy is gotten on the hardest introduced datasets. Contingent Random Field Classifiers acquired less consideration so as to finish the assignment in this field, in spite of their relative ubiquity in NLP errands like PoS-Tagging. Zubiaga et al. contended that the CRF can utilize the past foundation so as to improve the recognizable proof of news-breaking bits of gossip. Creators detailed F1 of 0.6 on the dataset of PHEME. Tests have indicated that CRF additionally outperforms non-consecutive strategies. In additionally, Markov hidden models can also able to handle sequence-specific data and hidden Markov models have used for the classification veracity of rumour in paper [23, 25]. These papers suggested modelling of rumour functions are content-based as well as context-based and proposed method has achieved an average accuracy 0.75 for the proposed data set. Wang has introduced the use of content-based and context-based features that were used in many learning algorithms and it is capable of achieving 0.77 accuracy and data set is sued by author is RumourEval data set.

6   Evaluation Metrics

Every trained model's performance is determined by how close the observation is to actual events [22, 30, 31]. With labelled data, we can evaluate any model. Required some evaluation metrics to calculate the performance of any algorithms, the general metric of evaluation used in any algorithm is precision. Certain useful rumour identification metrics, apart from this, are accuracy, recall and F1-score. During trying to predict values toward labels, four bins were received that are FP, FN, TP, and TN. False Positive is a non-rumoured event that is estimated as a rumour, and False Negative is estimated as a non-rumoured event. True Positive is rumoured event is estimated as rumour, True Negative is non-rumoured event is estimated as non-rumour (Table 2).

Table 2. Evaluation metrics

Evaluation metrics	Formulae
Precision	TP/(FP + TP)
Accuracy	(TN + TP)/Total events
Recall	TP/(FN + TP)
F1-score	2 * (Recall * Precision)/ (Recall + Precision)

7   Conclusion

Paper consist specific literature of possibly fake info present in social media also, methodologies related to the rapid recognition of this kind of data. Rumours and fake information are becoming essential part of the social media lives. They already have demonstrated for being highly hazardous internally and outside the digital world.

References

1. Furnell, S.: Cybercrime: Vandalizing the Information Society, pp. 3–540. Addison-Wesley, London (2002)
2. Chawra, V.K., Gupta, G.P.: Optimization of the wake-up scheduling using a hybrid of memetic and Tabu search algorithms for 3D-wireless sensor networks. Int. J. Softw. Sci. Comput. Intell. (IJSSCI) **14**(1), 1–18 (2022)
3. Madhu, S., Padunnavalappil, S., Saajlal, P.P., Vasudevan, V.A., Mathew, J.: Powering up an IoT-enabled smart home: a solar powered smart inverter for sustainable development. Int. J. Softw. Sci. Comput. Intell. (IJSSCI) **14**(1), 1–21 (2022)
4. Gunjan, V.K., Kumar, A., Avdhanam, S.: A survey of cyber crime in India. In: 2013 15th International Conference on Advanced Computing Technologies (ICACT), Sept 2013, pp. 1–6. IEEE
5. Gordon, S., Ford, R.: On the definition and classification of cybercrime. J. Comput. Virol. **2**(1), 13–20 (2006)
6. Pandove, K., Jindal, A., Kumar, R.: Email spoofing. Int. J. Comput. Appl. **5**(1), 27–30 (2010)
7. Tembhurne, J.V., Almin, M.M., Diwan, T.: Mc-DNN: fake news detection using multi-channel deep neural networks. Int. J. Semant. Web Inf. Syst. (IJSWIS) **18**(1), 1–20 (2022)
8. Appati, J.K., Nartey, P.K., Yaokumah, W., Abdulai, J.D.: A systematic review of fingerprint recognition system development. Int. J. Softw. Sci. Comput. Intell. (IJSSCI) **14**(1), 1–17 (2022)
9. Chiang, T.A., Che, Z.H., Huang, Y.L., Tsai, C.Y.: Using an ontology-based neural network and DEA to discover deficiencies of hotel services. Int. J. Semant. Web Inf. Syst. (IJSWIS) **18**(1), 1–19 (2022)
10. Pittaro, M.L.: Cyber stalking: an analysis of online harassment and intimidation. Int. J. Cyber Criminol. **1**(2), 180–197 (2007)
11. Erickson, J.: Hacking: The Art of Exploitation. No Starch Press (2008)
12. Tewari, A., et al.: Secure timestamp-based mutual authentication protocol for IoT devices using RFID tags. Int. J. Semant. Web Inf. Syst. (IJSWIS) **16**(3), 20–34 (2020)
13. Lewis, J.A.: Assessing the Risks of Cyber Terrorism, Cyber War and Other Cyber Threats, p. 12. Center for Strategic & International Studies, Washington, DC (2002)
14. Jagatic, T.N., Johnson, N.A., Jakobsson, M., Menczer, F.: Social phishing. Commun. ACM **50**(10), 94–100 (2007)
15. Sahoo, S.R., et al.: Spammer detection approaches in online social network (OSNs): a survey. In: Sustainable Management of Manufacturing Systems in Industry 4.0, pp. 159–180. Springer, Cham
16. Bolton, R.J., Hand, D.J.: Statistical fraud detection: a review. Stat. Sci. **17**(3), 235–255 (2002)
17. Sahoo, S.R., et al.: Security issues and challenges in online social networks (OSNs) based on user perspective. Comput. Cyber Secur. 591–606 (2018)
18. Gaurav, A., et al.: A comprehensive survey on machine learning approaches for malware detection in IoT-based enterprise information system. Enterpr. Inf. Syst. 1–25 (2022)
19. Sahoo, S.R., et al.: Hybrid approach for detection of malicious profiles in twitter. Comput. Electr. Eng. **76**, 65–81 (2019). ISSN 0045-7906. https://doi.org/10.1016/j.compeleceng.2019.03.003

20. McMillen, J.: Understanding gambling. In: Gambling Cultures: Studies in History and Interpretation, pp. 6–42 (1996)
21. Fick, J.: Prevention is better than prosecution: deepening the defence against cyber crime. J. Digit. Forensics Secur. Law **4**(4), 3 (2009)
22. Buil-Gil, D., Miró-Llinares, F., Moneva, A., Kemp, S., Díaz-Castaño, N.: Cybercrime and shifts in opportunities during COVID-19: a preliminary analysis in the UK. Eur. Soc. 1–13 (2020)
23. Mishra, A., et al.: Security threats and recent countermeasures in cloud computing. In: Modern Principles, Practices, and Algorithms for Cloud Security, pp. 145–161. IGI Global (2020)
24. Gunti, P., et al.: Data mining approaches for sentiment analysis in online social networks (OSNs). In: Data Mining Approaches for Big Data and Sentiment Analysis in Social Media, pp. 116–141. IGI Global (2022)
25. Mishra, A., Gupta, N.: Analysis of cloud computing vulnerability against DDoS. In: 2019 International Conference on Innovative Sustainable Computational Technologies (CISCT), Oct 2019, pp. 1–6. IEEE
26. Bhushan, K., et al.: Distributed denial of service (DDoS) attack mitigation in software defined network (SDN)-based cloud computing environment. J. Ambient Intell. Humaniz. Comput. **10**(5), 1985–1997 (2019)
27. Alsmirat, M.A., et al.: Impact of digital fingerprint image quality on the fingerprint recognition accuracy. Multimed. Tools Appl. **78**(3), 3649–3688 (2019)
28. Dahiya, A., et al.: Multi attribute auction based incentivized solution against DDoS attacks. Comput. Secur. **92**, 101763 (2020)
29. Srinivasan, S., Babu, L.D.D.: A parallel neural network approach for faster rumor identification in online social networks. Int. J. Semant. Web Inf. Syst. (IJSWIS) **15**(4), 69–89 (2019)
30. Al-Qerem, A., et al.: IoT transaction processing through cooperative concurrency control on fog-cloud computing environment. Soft Comput. **24**(8), 5695–5711 (2020)
31. Gupta, S., et al.: PHP-sensor: a prototype method to discover workflow violation and XSS vulnerabilities in PHP web applications. In: Proceedings of the 12th ACM International Conference on Computing Frontiers, May 2015, pp. 1–8
32. Dahiya, A., et al.: A reputation score policy and Bayesian game theory based incentivised mechanism for DDoS attacks mitigation and cyber defense. Future Gener. Comput. Syst. (2020)

ImmuneGAN: Bio-inspired Artificial Immune System to Secure IoT Ecosystem

Vineeta Soni[(✉)], Siddhant Saxena, Devershi Pallavi Bhatt,
and Narendra Singh Yadav

Manipal University Jaipur, Jaipur-Ajmer Express Highway, Dehmi Kalan, Jaipur,
Rajasthan 303007, India
vineetasoni.mits@gmail.com

Abstract. The Internet of Things (IoT) has shown exponential growth
in the flow of data in-between numerous low-energy based devices, this
distributive and dynamic property of IoT network ecosystem possess a
huge threat to data security. This paper proposes a layered artificial
immune system approach, which is inspired by the natural immunity
mechanism and adapts an architecture called ImmuneGAN, the proposed approach solves the low-data availability issue to train the machine
learning model and adapts the implicit representation learning by using
the min-max optimization process, the ImmuneGAN architecture tends
to learn the underlying pattern distribution of the intrusion packets by
simulating antibody population and perform memory acquisition task in
order to provide enhanced response upon the subsequent interaction of
attack with the environment, also the ImmuneGAN outperforms existing
machine learning-based approaches and yields state of the art results to
classify the attack packets with an accuracy of 99.87% and F1-score of
0.99874 on MQTTsetdataset.

Keywords: Artificial immune systems · Generative adversarial
networks · Intrusion detection · IoT networks · Data security ·
Statistics · Neural network

1 Introduction

The Internet of things (IoT) has gained a lot of growth in the past few years,
but with the increase of devices used in daily life, there is exponential growth
in data collection, transmission, and exchange from these devices in order to
sustain an efficient and thriving IoT network environment. As this data involves
a lot of information which can be health-related tracking record of the individual or messaging information that can be sensitive for public domain any
type of malfunctioning can lead to directly affect the individual, and there is
a huge risk of safety for this continuous stream of data exchange, hence this is
a very challenging task to secure IoT environment [1]. However, the security of
the IoT domain has been a popular area of research in academia and there are

© The Author(s), under exclusive license to Springer Nature Switzerland AG 2023
N. Nedjah et al. (Eds.): ICSPN 2021, LNNS 599, pp. 110–121, 2023.
https://doi.org/10.1007/978-3-031-22018-0_11

several approaches developed to secure the network environment from malicious attacks. The IoT network eco-system operates on the basis of data exchange between small devices, which makes it vulnerable to advanced attacks and zero-day attack situations. Many existing security systems can handle regular and popular attacks but in the case of the IoT network environment, there are multimodal data streams and the devices used for the network workflow are very sophisticated in order to sustain the efficiency in terms of energy consumption.

1.1 Network Intrusion Detection

An Intrusion is defined as any type of suspicious activity which creates disruption in the regular flow of data in order to attack the network and steal the information from the data stream [4,12,14]. Due to the heavy flow of network packets in the IoT ecosystem, attackers target the network by sending malicious intrusions in the form of packets only. The detection of these intrusions comprises several different techniques such as graph-based network anomaly detection approaches, traditional machine learning-based classification of intrusion packets from the network, matrix manipulation-based approaches, etc.

1.2 Data Security in IoT Network Ecosystem

IoT network environments possess many different types of vulnerabilities due to their layered architecture and low energy-based device functionalities [2,3,23]. There are three basic building layers of the IoT ecosystem as described as follows:

1. Perception Layer: It is the fundamental layer of the architecture, consisting of physical hardware such as sensors, and transmitters that gather and transform various types of signals or data in the network.
2. Network Layer: It is responsible for message/Data passing in the Network ecosystem, it processes and transmits the data in-between the devices of the network.
3. Application Layer: This is the uppermost layer of the IoT network environment, application layer is responsible for providing the necessary services to the user and for the user-IoT interaction in the environment.

Each layer provides a different range of functionalities and can be exploited in various ways to attack the network [refer]. However, the exponential growth of data-centric approaches such as machine learning and deep learning domain provides a different approach to detecting intrusions from the normal flow of data in the real-time network. Another approach to secure IoT network data is Artificial Immune Systems (AIS), which is a broad research area in computer science and data security. AIS focuses on learning from the natural immune systems and building a robust system to secure platforms from advanced attacks as discussed in Sect. 2.1. This paper presents an artificial immune system based on generative adversarial networks in order to utilize the representation learning of the data and build an efficient and more robust security system for the IoT network ecosystem with the inspiration of the natural immune system.

2   Related Work on Artificial Immune Systems

The artificial immune system algorithms abstract out the elements of the natural immune system, their properties to adapt, memorizing the complex patterns and formulating these in a computationally efficient manner to secure data from attacks.

1. Negative Selection Algorithm (NS) [6]: It is based on the acquired immunity mechanism of self-non-self discrimination behavior. The family of NSs aims to achieve the fundamental objective of segregation between the self and non-self objects by producing detector objects that mitigate the T-cell working and binds with non-self objects in order to secure the system.
2. Clonal Selection Algorithm [8]: It is also based on acquired immunity theory but it does not replicate the behavior of cells instead it focuses on the generation of receptors that learn to respond to antigens over time, also called affinity maturation, and the algorithm work in the correlation between the mutation and cloning of these receptors, omitting those who degrade the autoimmunity of the environment.
3. Artificial Immune Networks [9]: This algorithm is based on the antibody theory from immunology and at the same time it also takes inspiration from the clonal selection algorithm. The antibodies are present in pairs and maintain the immune memory of the interaction-between cells even if there is no external antigen present in the ecosystem.
4. Danger Theory Algorithm [11]: This algorithm is based on an action-reaction mechanism from immunology, and proposes that immunity is controlled by the extent of damage caused by the foreign invaders instead of recognizing the structural properties of intruders.
5. Timestamp-based approach [5]: This algorithm provide secure timestamp-based mutual authentication protocol for IoT devices using rfid tags.
6. Secure routing model [7]: This algorithm supports an efficient and Secure Routing Protocol Based on Artificial Intelligence Algorithms With UAV-Assisted for Vehicular Ad Hoc Networks in Intelligent Transportation Systems.
7. Cryptosystem [10]: This algorithm supports an cryptosystem based on neural network.

3   Proposed Self-adaptive AIS Architecture

This paper proposes a data-centric approach toward a self-adaptive artificial immune system inspired by natural immune systems, that is robust to newly emerging attacks and efficiently detects anomalous packets in the normal flow of the IoT network environment. Figure 1 describes the overview of the data flow in the process of securing the IoT environment and monitoring attacks. The proposed artificial immune system is divided into two layers of immunity first is the innate layer and another is the adaptive immune layer, the basic workflow of the AIS to secure the IoT network ecosystem is as follows: The regular flow of data is captured using a tap connection between the nodes and can be stored As a pcap file, this raw pcap file is passed through a content feature extractor to get the feature-value mapping of the data. This processed data is passed through the innate layer module and adaptive layer module.

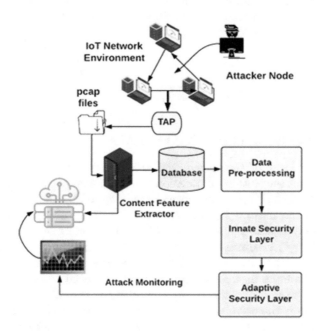

Fig. 1. Overview of the data flow from IoT network to AIS system.

3.1   Innate Immune Mechanism of AIS

The natural immune system provides a basic barrier layer as the first layer of defense to protect the ecosystem from foreign invaders, for the innate immune mechanism in the AIS, this paper proposes a statistical approach to distinguishing the packets as being normal or anomaly. The normal flow of data in the IoT network environment contains a lot of information and this information can be captured into a pcap file format, using a command-line tool called Wireshark [13]. These packet-captured files (pcap) contain different data points such as source address, destination address, protocol, length of the packet, etc. these data points can be converted into a data frame of feature-value mapping of the information related to each of the packets carrying information in the network. Therefore the innate immunity mechanism proposed is based on this feature-value extraction of the raw pcap files. The raw pcap file captured at the regular flow of the network contains the representation values of normal packets and forms a bell-shaped curve demonstrating the normal distribution of data and all the normal packets of the network will belong to a similar mean and distribution of the curve, but whenever the network is under attack from the malicious packets, and the captured pcap file is compared to the normal distribution of the raw pcap file, the feature-mapping of attack packets have a lot of variance from the normal distribution and they will lie far from the plotting of the normal distribution curve of the data as described in Fig. 3. The overall probability of a packet can be calculated using Eq. (1).

$$P(x) = \prod_{(j=1)}^{n} (\frac{1}{\sqrt{2 \prod \sigma_j}}) e^{-\frac{(x_j - v_j)^2}{2\sigma_j^2}} \tag{1}$$

3.2   Self-adaptive Immunity Layer of AIS

The secondary layer of security in the artificial immune system proposed is designed to adapt the feature information of packets that contain malicious intrusion to attack the IoT network environment, inspired by the adaptive immune mechanism of the natural immune system. The main objective of the adaptive immune layer is to memorize the effect of intruders on the ecosystem so that when there is a subsequent attack on the environment that matches the feature-value mapping learned by the adaptive layer from the previous attack, the acquired immune mechanism provides an enhanced response to eliminate the threat on the environment. Another motivation behind the secondary layer implementation is the correlation of the features of the packets in the network. The innate layer exploits the statistical distribution of the packets and calculates the joint probability of packets in order to calculate the overall probability of the packet. However, in the real-world data packet features are highly correlated with each other and hence the joint-probability based estimation of the normal and anomalous packets is biassed and can only be used to provide the first layer of defense against intrusion packets. Therefore we propose an architecture based on generative adversarial networks [15] as to be discussed in Sect. 3.3.

3.3   ImmuneGAN Architecture

The architecture of ImmuneGAN is based on min-max optimization between Antigen Detector and Antigen Generator as described in Fig. 4, the innate layer performs a statistical analysis of the packets in the network and generates an innate immunity response (IIR), it is the feature-value mapping of the detected intrusion samples, these are the ground truth values of the flagged anomaly in a tensor matrix format sampled to train the generator, Antigen generator initially takes values from a latent space matrix which contains the random values from normal distribution merged with noise to make generator robust from adversarial attacks. The Antigen Detector is the component designed to detect and discriminate the IIR from Antigen Generator and both the generator body and detector body are trained together, then there is a system immunity body that decides the accuracy of the discriminator and the tuning of generator and discriminator of antigens is done using backpropagation algorithm, to update the parameters of the model. This learning of feature representation values of the anomaly and generating new anomalies highly similar to existing ones resemble the antigen memory acquisition to train the model to detect anomalies containing highly correlated features as well.

$$min_\theta max_\varphi V(AG_\theta, AD_\varphi) = E_{\tilde{x} P_{IIR}}[\log AD_\phi(x)] + E_{\tilde{z} P_z}[\log(1 - AD_\phi(AG_\theta(z)))] \tag{2}$$

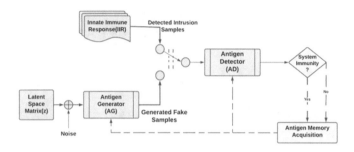

Fig. 2. Architecture of ImmuneGAN based on min-max optimization between generator and discriminator.

3.4 ImmuneGAN Training

The proposed ImmuneGAN architecture is based on simultaneously training the generator and discriminator in order to learn the probabilistic distribution of the attacks, through optimizing the objective function mentioned in Eq. (2).

Where the AG and AD denote the Antibody Generator and Antibody discriminator with respect to their trainable parameters θ and φ.

The learning of the implicit probability distribution and min-max optimization of the objective function is described in algorithm 1, which denotes the training of the ImmuneGAN, and further the results of the training process are described in the Experiments (Sect. 4).

Time Complexity Analysis:

If AG and AD have enough capacity, and at each step of Algorithm 1, the discriminator is allowed to reach its optimum given AG, and generator distribution is updated so as to improve the criterion then converges to actual data distribution, for a given batch m and latent representation z, the time complexity of Algorithm 1 can be observed as the upper bound on O(m * log(z)).

4 Experiments

The experimentation details of datasets used to train and test the proposed artificial immune system are mentioned in Table 1. Considering the IoT net-

---

Algorithm 1: Training algorithm of ImmuneGAN

Input: Prior distributions $P_\phi(z)$, $P_{IIR(x)}$, epochs = 200, k=1
Output:
1. Start
2. Read input
3. Initiate loop for epochs = 200

    1. Initiate loop for k=1 steps
    2. Sample minibatch of m from Prior distributions
    3. Update the generator parameters as

$$\nabla_\theta V(AG_\theta, AD_\varphi) = \frac{1}{m} \nabla_\theta \prod_{(j=1)}^{n} \log\left(1 - AD_\phi(AG_\theta(z))\right) \tag{3}$$

4. end loop
5. Update discriminator parameters as

$$\nabla_\theta V(AG_\theta, AD_\varphi) = \frac{1}{m} \nabla_\theta \prod_{(j=1)}^{n} [\log AD_\phi(x)] + E_{z^{\tilde{}} P_z}[\log\left(1 - AD_\phi(AG_\theta(z))\right)] \tag{4}$$

4. End of the loop

---

work environments are based on the MQTT protocol [16], training on large-scale datasets based on other protocols is irrelevant for IoT security. However to ensure the robustness of the intrusion detection the model is also tested on custom SlowITe attack packet captured data and the results are similar to the accuracy level achieved with the MQTTset dataset [17].

The implementation of the ImmuneGAN architecture is done using the PyTorch framework, and the architecture-specific details for the experimentation and results are as follows.

Table 1. Datasets used for the ImmuneGAN evaluation.

Dataset	Number of features (count)	Number of parameters (count)
MQTTset	Features: 33	Parameters: 3976
MQTT-IoT-IDS2020	Features: 28	Parameters: 2980
UFPI-NCAD-IoT-Attacks	Features: 29	Parameters: 2030
MQTT dataset	Features: 36	Parameters: 4312
KDD-CUP- 99	Features: 18	Parameters: 1875

The training flow of the ImmuneGAN model is described in Figs. 3 and 4, where Fig. 3 Indicates the performance of the ImmuneGAN in the classification of attacks and accuracy scores of the generator and discriminator models, similarly Fig. 4 explains the training loss optimization using min-max objective in a parallel loop.

- Batch size: 128
- Latent size: 128
- Discriminator activation function: LeakyReLU, Sigmoid
- Generator activation functions: ReLU, Tanh
- Loss function: Binary Cross Entropy
- Optimizer: Adam optimizer
- Epochs: 200
- Metrics: Accuracy, F1 Score.

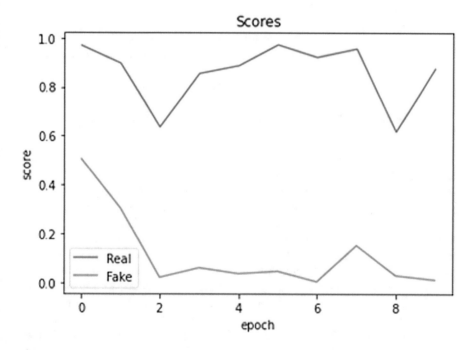

Fig. 3. Comparison of the real vs. fake attack packets in the optimization process of ImmuneGAN.

5   Results and Discussion

The traditional machine learning approaches for the anomaly detection [18] heavily rely on the given dataset, but in practice, there are very less amount of datasets available for the specific IoT data security task, and available options

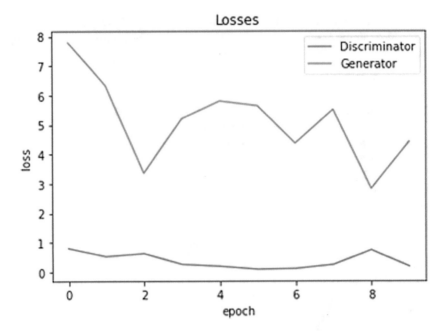

Fig. 4. Optimization of the ImmuneGAN by reducing the loss of generator and discriminator module.

are limited to certain types of attacks [19]. Therefore the proposed Immune-GAN architecture is designed to learn the underlying pattern of the attacks by simulating the attack packet features using a generator module. Hence the ImmuneGAN can classify the anomalous packets even when there is very less feature-value mapping available for the training. The analysis of the obtained results is described in Table 2, the proposed architecture represents the best results (bold solution) with an accuracy score of 0.99836 and F1 score of 0.9971, while the Random Forest and deep neural network model resulted in 0.97241 and 0.99352 of accuracy respectively, which is less than the proposed model. Moreover, on further analysis of Decision tree algorithms, particularly random forest performs best with an accuracy of 0.97241 and F1 score of 0.9890 in comparison with simple decision tree and gradient boost both having accuracy close to 0.9862. Finally, from the collection of supervised learning algorithms, the naive Bayes algorithm achieves an accuracy of 0.98836. This experimentation is performed using the CPU as a computation device and hence the training and testing time is high for the proposed model but the performance overpasses the other methodologies for intrusion detection in the IoT networks. As described in Table 3, the proposed ImmuneGAN architecture gives the best accuracy and F1 score for the MQTTset dataset, which provides a balanced distribution of attacks and feature-value mappings. However, the model surpasses prediction metrics for the traditional machine learning approaches for the KDD-CUP-99 dataset [20] and achieves 87.64% accuracy. The main reason behind achieving an accuracy of less than 90% is due to the imbalance distribution of feature values in the dataset, moreover, this dataset is not specific to attacks on MQTT protocol, and the dataset is processed before model training and testing.

The ImmuneGAN architecture is evaluated on the basis of accuracy, F1-score, training time, and testing time and the results of the test set are described in Table 3. Also, the training and testing procedure is iterated using the GPUs (Nvidia GeForce GTX 1650) therefore the training time and testing time is reduced in comparison to Table 2. In which the results are obtained using a CPU processor.

6   Conclusions

This paper proposed a self-adaptive artificial immune system solution to the security problem of the dynamic and distributed IoT network environment [21], this approach is highly efficient and reliable because the working of the IoT is analogous to the natural immune system [22]. This paper also discusses the problems in existing data-centric machine learning approaches for the anomaly detection task in the IoT networks, and hence the proposed solution does not directly rely on the datasets provided to train a neural network, but the Immune-GAN architecture trains on a min-max objective function which is inspired by generative adversarial networks to generate the antibody for the foreign anomalous packet in order to train the detector for the subsequent interaction of the attack packets with the environment, and as discussed in the results section the ImmuneGAN surpasses the accuracy of the existing methodologies and yields an accuracy of 99.87% and F1-score of 0.99874.

The proposed framework to approach a better security system for the IoT environment will benefit a huge spectrum of industrial segments such as Smart Homes, Smart Cities, Self-driven Cars, IoT Retail Shops, Farming, Wearables, Smart Grids, and Industrial Internet. Moreover, as the IoT has proven its need in IoT-Based Healthcare Data Preservation [24] the ImmuneGAN approach will boost the security of the patient data.

7   Future Perspective and Research Propagation

The ImmuneGAN architecture is highly compatible with the IoT network ecosystem because of the layered security mechanism and self-adaptive artificial immune system-based approach [25], this paper indicates that there is a lot of scope for research on the security of the IoT networks in this direction, also we are propagating the research to produce the hardware implementation of the ImmuneGAN. Moreover, this paper implements a different aspect of artificial immune systems in which instead of simulating the whole cells working, Immune-GAN takes inspiration from the working of the immune mechanism, therefore this paper opens another branch for the researchers in network security to work on the data-centric layered immunity paradigm for the dynamic and distributed environments.

References

1. Mahmoud, R., et al.: Internet of things (IoT) security: current status, challenges and prospective measures. In: 2015 10th International Conference for Internet Technology and Secured Transactions (ICITST), pp. 10250–10276. IEEE (2015). https://doi.org/10.1109/JIOT.2020.2997651
2. Gaurav, A., et al.: A comprehensive survey on machine learning approaches for malware detection in IoT-based enterprise information system. Enterp. Inf. Syst. 1–25 (2022)
3. Tewari, A., et al.: A lightweight mutual authentication approach for RFID tags in IoT devices. Int. J. Netw. Virtual Organ. **18**(2), 97–111 (2018)
4. Mukherjee, B., Todd Heberlein, L., Levitt, K.N.: Network intrusion detection. IEEE Netw. **83**, 26–41 (1994)
5. Tewari, A., et al.: Secure timestamp-based mutual authentication protocol for IoT devices using RFID tags. Int. J. Semant. Web Inf. Syst. (IJSWIS) **16**(3), 20–34 (2020)
6. Ji, Z., Dasgupta, D.: Revisiting negative selection algorithms. Evol. Comput. **15**(2), 223–251 (2007)
7. Fatemidokht, H., et al.: Efficient and secure routing protocol based on artificial intelligence algorithms with UAV-assisted for vehicular ad hoc networks in intelligent transportation systems. IEEE Trans. Intell. Transp. Syst. **22**(7), 4757–4769 (2021). https://doi.org/10.1109/TITS.2020.3041746
8. Haktanirlar Ulutas, B., Kulturel-Konak, S.: A review of clonal selection algorithm and its applications. Artif. Intell. Rev. **36**(2), 117–138 (2011)
9. Galeano, J.C., Veloza-Suan, A., González, F.A.: A comparative analysis of artificial immune network models. In: Proceedings of the 7th Annual Conference on Genetic and Evolutionary Computation (2005)
10. Peng, J., et al.: A biometric cryptosystem scheme based on random projection and neural network. Soft Comput. **25**(11), 7657–7670 (2021)
11. Zhang, C., Yi, Z.: A danger theory inspired artificial immune algorithm for on-line supervised two-class classification problem. Neurocomputing **73**(7–9), 1244–1255 (2010)
12. Sharma, K., Anand, D., Mishra, K.K., Harit, S.: Progressive study and investigation of machine learning techniques to enhance the efficiency and effectiveness of Industry 4.0. Int. J. Softw. Sci. Comput. Intell. (IJSSCI) **14**(1), 1–14 (2022)
13. Wang, S., Xu, D., Yan, S.L.: Analysis and application of Wireshark in TCP/IP protocol teaching. In: 2010 International Conference on E-Health Networking Digital Ecosystems and Technologies (EDT), vol. 2. IEEE (2010)
14. Wang, T., Pan, Z., Hu, G., Duan, Y., Pan, Y.: Understanding universal adversarial attack and defense on graph. Int. J. Semant. Web Inf. Syst. (IJSWIS) **18**(1), 1–21 (2022)
15. Wang, K., et al.: Generative adversarial networks: introduction and outlook. IEEE/CAA J. Autom. Sinica **4**(4), 588–598 (2017)
16. Soni, D., Makwana, A.: A survey on MQTT: a protocol of internet of things (IoT). In: International Conference on Telecommunication, Power Analysis and Computing Techniques (ICTPACT-2017), vol. 20 (2017)
17. Vaccari, I., et al.: MQTTset, a new dataset for machine learning techniques on MQTT. Sensors **20**(22), 6578 (2020)
18. Ahmed, T., Oreshkin, B., Coates, M.: Machine learning approaches to network anomaly detection. In: Proceedings of the 2nd USENIX Workshop on Tackling

Computer Systems Problems with Machine Learning Techniques. USENIX Association (2007)

19. Hodo, E., et al.: Threat analysis of IoT networks using artificial neural network intrusion detection system. In: International Symposium on Networks, p. 2016. Computers and Communications (ISNCC), IEEE (2016)

20. Tavallaee, M., et al.: A detailed analysis of the KDD CUP 99 data set. In: 2009 IEEE Symposium on Computational Intelligence for Security and Defense Applications. IEEE (2009)

21. Gandotra, V., Singhal, A.A., Bedi, P.: Layered security architecture for threat management using multi-agent system. ACM SIGSOFT Softw. Eng. Notes **36**(5), 1–11 (2011)

22. Ahammad, I., Khan, M.A., Salehin, Z.U., Uddin, M., Soheli, S.J.: Improvement of QoS in an IoT ecosystem by integrating fog computing and SDN. Int. J. Cloud Appl. Comput. (IJCAC) **11**(2), 48–66 (2021). https://doi.org/10.4018/IJCAC.2021040104

23. Bouarara, H.A.: N-gram-codon and recurrent neural network (RNN) to update Pfizer-BioNTech mRNA vaccine. Int. J. Softw. Sci. Comput. Intell. (IJSSCI) **14**(1), 1–24 (2022)

24. Sarrab, M., Alshohoumi, F.: Assisted-fog-based framework for IoT-based healthcare data preservation. Int. J. Cloud Appl. Comput. (IJCAC) **11**(2), 1–16 (2021). https://doi.org/10.4018/IJCAC.2021040101

25. Swarnakar, S., Bhattacharya, S., Banerjee, C.: A bio-inspired and heuristic-based hybrid algorithm for effective performance with load balancing in cloud environment. Int. J. Cloud Appl. Comput. (IJCAC) **11**(4), 59–79 (2021). https://doi.org/10.4018/IJCAC.2021100104

A Systematic Review of Recommendation System Based on Deep Learning Methods

Jingjing Wang[1(✉)], Lap-Kei Lee[1], and Nga-In Wu[2]

[1] School of Science and Technology, Hong Kong Metropolitan University,
Ho Man Tin, Hong Kong
{s1245831,lklee}@hkmu.edu.hk
[2] College of Professional and Continuing Education, Hong Kong Polytechnic University,
Hung Hom, Hong Kong
ngain.wu@cpce-polyu.edu.hk

Abstract. Recommender Systems (RSs) play an essential role in assisting online users in making decisions and finding relevant items of their potential preferences or tastes via recommendation algorithms or models. This study aims to provide a systematic literature review of deep learning-based RSs that can guide researchers and practitioners to better understand the new trends and challenges in the area. Several publications were gathered from the Web of Science digital library from 2012 to 2022. We systematically review the most commonly used models, datasets, and metrics in RSs. At last, we discuss the potential direction of the future work.

Keywords: Recommender system · Deep learning · Survey · Systematic literature review

1 Introduction

Recommender Systems (RSs) play an essential role in assisting online users in making decisions and finding relevant items of their potential preferences or tastes via recommendation algorithms or models. Due to the fast development of information technologies and the Internet, especially mobile applications, almost every daily aspect from working, studying, and entertaining to business relies on the Internet to complete. For example, e-commerce websites, e.g., Amazon, Alibaba, and eBay, satisfy users' shopping demands; social media platforms, e.g., WeChat, WhatsApp, and Facebook, immensely facilitates communication among individuals. When ones would like to share funny things in life, they will most likely upload videos to YouTube or TikTok, instead of meeting with friends physically. Especially, suffering from the COVID-19, students turn to online learning every day. On the one hand, these wide varieties of online services make life more convenient in various aspects. On the other hand, the data produced by these platforms is increasing at an unprecedented rate, which comes with the burden to choose interesting items from an immense number of possibilities. In 2022, TikTok has over 1 billion active users [1], where 55% of users are video makers [2], and 68% of TikTok active users watch others' videos; the world's largest video

N. Nedjah et al. (Eds.): ICSPN 2021, LNNS 599, pp. 122–133, 2023.
https://doi.org/10.1007/978-3-031-22018-0_12

platform YouTube has 14 billion monthly visits, and 694,000 h of videos are uploaded on YouTube each minute [3].

Despite the advantage of access to a significant amount of online services and products, users may be frustrated when they are looking for some desirable information. This phenomenon is referred to as "information overload". Compared to the search engine and traditional recommendation algorithms based on popularity, RSs explore individual users' interests or tastes based on interaction data and generate a set of recommended items via collaborative filtering and deep learning methods, which has attracted the interest of a broad scientific community. Various types of interaction data are provided in RSs, mainly including explicit feedback (e.g., rating of the movie, opinions of the restaurant) and implicit feedback (e.g., click, purchase and follow). Usually, an RS with fast and accurate recommendation results will not only stratify the demand of customers and improve the online service experience, but also bring benefits to service providers.

Therefore, exploring user preferences and needs from large-scale candidates based on user historical behavior sequences during a given period has become one of the most potential research issues in the field of RSs. The research of this study has significant value in academia and industry.

In this study, we aim to discuss the trends and developments of deep learning-based RS methods from 2012 to 2022 by reviewing various relevant publications.

2 Research Methods

This section presents how the dataset is obtained and discusses the relevant issues.

2.1 Data Collection

Our study aims to answer the following research questions (RQs):

RQ1: What are the challenges of deep learning-based RS methods?

RQ2: From 2012 to 2022, which deep learning methods are implemented to solve the above challenges in RS?

RQ3: From 2012 to 2022, what are the application domains of deep learning-based RS methods?

RQ4: From 2012 to 2022, what are the datasets and metrics used for evaluating these models' performance?

RQ5: From 2012 to 2022, what progress has been achieved in deep learning-based RSs? What are the future development trends for deep learning-based RSs?

2.2 Pre-processing of Papers

It is essential to collect relevant and high-quality articles for this review. In this study, we choose the Web of Science database[1] as the major digital library for its popularity and well reputation. This database contains more than 21,000 journals, 205,000 conferences, and 104,000 books. To cover these literatures abroad and accurately, we used

---

[1] https://clarivate.com/webofsciencegroup/solutions/webofscience-platform/.

multiple query words to retrieve articles: *recommendation system, personalized recommendation, session-based recommendation, sequential recommendation, diversified recommendation.* We choose these keywords based on the following reasons. At first, we choose recommendation system as it is, without doubt, the most related word to our review. Then, we further divide RSs to three categories: personalized recommendation, session-based recommendation, and sequential recommendation, based on the recommendation task. For example, session-based recommendation is used to solve the prediction problem based on the anonymous and chronological session sequences. Sequential recommendation aims to predict users' next action based on time-series interactions with users' profiles. At last, we included "diversified recommendation" as a supplementary keyword to RS, because diversity and accuracy are the two essential but orthogonal evaluation metrics to evaluate the model performance. Most existing works only focus on one of these two metrics.

We applied these searching keywords to the Web of Science and obtained 1276 articles as summarized in Fig. 1. The number of deep learning-based RS studies per year is increasing from 2012 to 2022. All papers have been carefully read and double-checked to further select the most related articles according to the following inclusion and exclusion criteria.

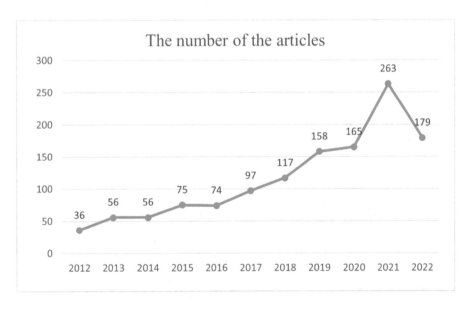

Fig. 1. Distribution of the number of articles on deep learning-based RS from 2012 to 2022.

First, each study must utilize deep learning methods to solve the RS issues and a described experiment analysis was provided. By adopting this criterion, review papers will be eliminated. Second, all studies must be published in journals or conference proceedings. As a result, the 97 papers that remain form the final dataset for our analysis. The procedure of the data collection and pre-processing is summarized and illustrated in Fig. 2.

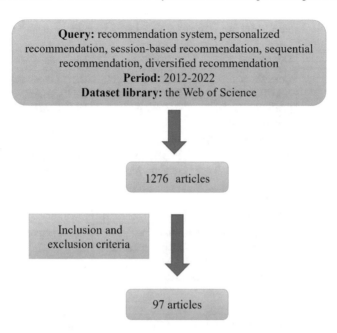

Fig. 2. The procedure of data collection and pre-processing.

We brief overviews all methods used in these studies as Table 1 shows. From Table 1, we observed that GNN is the most commonly used method, currently. Various studies attempts Transformer to solve the RS problem, a pure attention model, which has achieve encouragement results in natural language process. More works aim to the hybrid model to strength the advantage, for example RNN-CNN, Transformer-CNN.

3   Rating Prediction and Ranking Prediction

Evaluation is the essential component and ultimate goal of RSs to judge the effectiveness and compare the performance. Faced with different recommendation problems, diversity metrics are needed to estimate the recommendation quality. For example, in practical application, the time complexity is one of the important factors to ensure the real-time recommendation. In general, the metrics to evaluate the recommendation model can be divided into two categories according to the different usage scenarios: metrics about recommendation quality (such as accuracy, diversity, etc.) and metrics about method performance (scalability, recommendation efficiency, etc.). This study focuses on the study of the accuracy of the recommendation algorithm, we mainly describe the relevant metrics about recommendation quality. In specific, recommendation quality including rating and ranking prediction. In this section, we introduce these metrics used to evaluate recommendation quality in detail in above two categories.

Table 1. Brief overviews of different deep learning models.

Model	Description	Advantages	Disadvantages	Studies
CNN	Consists of convolution and pooling layer for feature extraction	Collective-level item decencies exploring	Can not memory the long-term relationship	[4–9]
RNN	A RNN cell consists of loops and memories unit. Two variants are commonly used LSTM and GRU	Memory the long-term relationship	Gradient vanishing and gradient explosion	[10–21]
GNN	Currently, GNN is the most commonly used method. It update the node features with its cross ponding neighbors	Modeling the Non-European-type data	High complexity	[22–38]
Transformer	A self-attention based model to encode the sequential signals	Parallel computation compared to RNN models	High complexity of the multi-head attention mechanism	[39–43]
AE	A generative deep neural network to minimize the difference between encoder and decoder modules	Suitable to the unsupervised and supervised learning	Sensitive to the noises	[44–46]
GAN	A generative deep neural network consisting discriminator and generator	Suitable to the unsupervised and supervised learning	Sensitive to the noises and difficult to training	[47–49]
RNN-CNN	Integrate CNN and RNN models	Better features learning	Complicated model	[50–52]
Transformer-CNN	Integrate transformer and CNN models	Robust model performance	A more complicated model	[53,54]
DBN	A generative graphical model multiple stacks of restricted Boltzmann machines layer	Suitable for the audio data	More parameters and difficult to train	[56]
RBM	A two-layer neural network with visible and hidden layer	Powerful ability to the low-rank features learning	Difficult to train	[57]

3.1   Rating Prediction Metrics

Rating prediction is a well-known recommendation task aiming to predict a user's rating on un-interact items. Metrics about the rating is used to measure the average difference between the ground truth and the prediction scores over the entire user set. The lower the value is, the closer the predicted score is to the real ratings and the fewer error of the system made. Commonly used metrics in this scenario include MSE, RMSE, and MAE.

3.2   Ranking Prediction Metrics

Another category of recommendation task is about the ranking prediction. Ranking prediction can deal with the implicit feedback which can be considered as a binary value, for example, click, purchase, or add to the cart, while rating prediction are only suitable to the explicit feedback. Compared to the explicit feedback, implicit interaction is easier to obtain and denser. The commonly used evaluation metrics in this scenario include recall, hit ratio, precision, F1 score, AUC (Area under the curve) and Mean Reciprocals Ranks (MRR), Mean average precision (MAP)s. The latter two metrics are sensitive to the ranking position while the former five are not [58]. We present each metric in the following section.

3.3   Prediction Diversity Metrics

Another type of recommendation quality is about the prediction diversity, which measures the novelty of the recommendation list. The original motivation of top-N recommender systems is to retrieve items based on users' similar tastes. It needs to be noted that it is easy to make items provided on the recommendation lists too similar to each other. Maximizing the diversity of the retrieved list is helpful to avoid monotonous recommendations as well as to promote enterprises' sales. We propose two metrics Intra-list distance and category coverage in this review to evaluate the recommendation diversity.

3.4   Datasets and Domain

Datasets and its domain are the essential factors to the model framework designing. In this section, we visualize the frequency of these datasets used in this review as shown in Fig. 3. Next, we analysis each dataset in detail as shown in Table 2.

4   Characteristics and Challenges

In this section, we systematically illustrate some research trends in RS which we believe are critical to the further development of the filed.

Table 2. Brief overviews of different datasets.

Dataset	Description	Domain
MovieLens[a]	MovieLens is released from the MovieLens platform[b], a movie watching and recommender system. It contains more than 25 million ratings and 1 million tags for about 62,000 movies by 162,000 users	Movie
Amazon	The Amazon dataset contains over 180 million relationships between products and 14 million ratings by 20 million users. This dataset is collected from the Amazon.com including 11 categories ranging from books, cell phone and accessories to toys and games	E-commerce
Yoochoose[c]	The Yoochoose dataset was released by the RecSys'15 Challenge, and it records six months of user click-streams collected from the e-commerce website Yoochoose.com. The raw session data including: Session ID, Time stamp, Item ID, and Item category	E-commerce
DIGINETICA[d]	The Diginetica dataset was obtained from CIKM Cup 2016, and it records the associated transaction data. The dataset consists of 1,235,380 view actions of 232,816 users over 184,047 items	E-commerce
Yelp[e]	Yelp is a review-sharing social platform. It records users' ratings ranging from 1 to 5 and comments on local business	E-commerce
Gowalla[f]	The Gowalla dataset collected from the Gowalla, a popular location-based social networking website. It contains more than 6 million check-in data of six hundred thousand users during February 2009 to October 2010	E-commerce
Foursquare[g]	The Foursquare dataset contains user's check-in data for different locations. It includes two subsets according to the location. One collects check-in data of NYC and Tokyo during 12 April 2012 to 16 February 2013. The other contains check-ins in global collected for about 18 month (from April 2012 to September 2013)	E-commerce
Lastfm[h]	Lastfm is a users' music listening dataset, collected from the Last.fm API. The raw data includes: <user, timestamp, artist, song>	Music
Tmall[i]	The Tmall dataset was released by the IJCAI-15 competition, and it records a series of anonymized shopping logs on the Tmall website in the past six months	E-commerce
Yahoo![j]	The Yahoo dataset collected from Yahoo. It contains a series subsets arranging from transactions to reviews.	E-commerce
Retailrocket[k]	The data has been collected from a real-world ecommerce website. The raw data includes item properties and behaviour data	E-commerce
Douban[l]	Douban is a popular social networking service in China	Review
XING[m]	The XING dataset released by ACM RecSys Challenge 2017, it is a job recommendation dataset by matching user's preference and recruiter demands	Job
Steam[n]	This dataset including player count history, price history	Game
MIND[o]	MIND is a real-world news recommendation dataset. It contains more than 160k English news and 15 million anonymized behavior logs by 1 million users	News
Epinions[p]	The Epinions dataset is trust network dataset. It contains user profile, ratings and trust relations	Review
Delicious[q]	This data set contains tagged web pages retrieved from the website delicious.com	Restaurant
Tiktok[r]	The dataset contains users trending tracks featured on TikTok	Video
Nowplaying[s]	The Nowplaying dataset was created by leveraging social media tweets to track the users' listening behaviors for one year	Music

[a] https://grouplens.org/datasets/movielens/
[b] https://movielens.org
[c] https://www.kaggle.com/chadgostopp/recsys-challenge-2015
[d] https://competitions.codalab.org/competitions/11161
[e] https://www.yelp.com/dataset
[f] https://snap.stanford.edu/data/loc-gowalla.html
[g] https://sites.google.com/site/yangdingqi/home/foursquare-dataset
[h] http://ocelma.net/MusicRecommendationDataset/lastfm-1K.html
[i] https://tianchi.aliyun.com/dataset/dataDetail?dataId=42
[j] http://webscope.sandbox.yahoo.com/catalog.php?datatype=r&did=75
[k] https://www.kaggle.com/retailrocket/ecommerce-dataset
[l] https://www.douban.com/
[m] http://2016.recsyschallenge.com/
[n] https://steam.internet.byu.edu/
[o] https://msnews.github.io/
[q] https://snap.stanford.edu/data/soc-Epinions1.html
[r] https://grouplens.org/datasets/hetrec-2011/
[s] https://www.kaggle.com/datasets/yamqwe/tiktok-trending-tracks
[t] https://www.kaggle.com/chelseapower/nowplayingrs

Incorporate more contextual factors into   RS

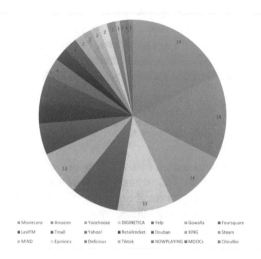

Fig. 3. The frequency of dataset usage.

Incorporate More Contextual Factors into RS

Contextual information is an essential factor in affecting the user's decision, such as weather, season and price. More contextual factors incorporated into RS might not only improve the recommended quality of RS, but also enhance the model's interpretability. Although several studies concentrate on this direction and provide some initial solutions, taking these contextual factors including external and internal into account is still the hot topic of RS.

Cross Domain Recommendation

In practice, users' tastes or preferences are consistent in different but relevant domains, e.g. movie domain and purchase domain. For example, customers prefer to buy the same cloth as the actor in a movie. In addition, taking multi domains into account might alleviate the data sparsity and cold start challenges. Exploit multi domains is interesting and quite challenging. It is hard to collect a user's consumed items from various domains together.

Heterogeneous Information Patterns

Heterogeneous information patterns include various multi types of users' behaviours, e.g. click, add to cart and purchase, multi kinds of auxiliary data, e.g. user social relation and item attributes. How to effectively discover and leverage more types of user behaviour and item information to improve SBR? On the one hand, different types of behaviours patterns contribute varying weight to the final decision, design a uniform architecture to encode this complex information is more difficult to identify the pattern dependencies in users' decision-making process. On the other hand, auxiliary data about items alleviate the risk of overfitting and bring encouragement potential to achieve a more accurate result at the same time introducing more noise and computation cost.

Personalized RS in Specific Scenario

In reality, it is essential to fine-tune the recommendation result over the time. For example, when user has already purchased the item, diverse items recommended will arouse users' attention instead of the similar items, which could be considered as a specific action scenario. A specific scenario is common in RS, for instance, repeat purchase, job hunting with gender bias. Existing works on SBRSs often ignore such kinds of constraints.

5   Conclusion

In this article, we present a systematic review of the state-of-the-art methods in RS based on deep learning. This study was conducted based on the web of science database from 2012 to 2022. We have thoroughly analyzed the commonly used models, datasets and metrics on these studies. In addition, we proposed some potential open issues and promising directions.

References

1. https://www.statista.com/statistics/272014/global-social-networks-ranked-by-number-of-users/
2. https://blog.gwi.com/trends/tiktok-music-social-media/
3. https://blog.hootsuite.com/youtube-stats-marketers/
4. Da'u, A., Salim, N., Idris, R.: An adaptive deep learning method for item recommendation system. Knowl.-Based Syst. **213**, 106681 (2021)
5. Huang, H., Luo, S., Tian, X., Yang, S., Zhang, X.: Neural explicit factor model based on item features for recommendation systems. IEEE Access **9**, 58448–58454 (2021)
6. Manoharan, S., Senthilkumar, R., Jayakumar, S.: Optimized multi-label convolutional neural network using modified genetic algorithm for popularity based personalized news recommendation system. Concurrency and Computation: Practice and Experience, p. e7033 (2022)
7. Srinivasan, S., Dhinesh Babu, L.D.: A parallel neural network approach for faster rumor identification in online social networks. Int. J. Semant. Web Inf. Syst. (IJSWIS) **15**(4), 69–89 (2019)
8. Safavi, S., Jalali, M.: Deepof: A hybrid approach of deep convolutional neural network and friendship to point-of-interest (poi) recommendation system in location-based social networks. Concurrency and Computation: Practice and Experience, p. e6981 (2022)
9. Xu, C., Feng, J., Zhao, P., Zhuang, F., Wang, D., Liu, Y., Sheng, V.S.: Long-and short-term self-attention network for sequential recommendation. Neurocomputing **423**, 580–589 (2021)
10. Dai, S., Yu, Y., Fan, H., Dong, J.: Spatio-temporal representation learning with social tie for personalized poi recommendation. Data Sci. Eng. **7**(1), 44–56 (2022)
11. Da'u, A., Salim, N., Idris, R.: Multi-level attentive deep user-item representation learning for recommendation system. Neurocomputing **433**, 119–130 (2021)
12. Sedik, A., et al.: Efficient deep learning approach for augmented detection of Coronavirus disease. Neural Comput. Appl. **34**(14), 11423–11440 (2022)

13. Du, W., Jiang, G., Xu, W., Ma, J.: Sequential patent trading recommendation using knowledge-aware attentional bidirectional long short-term memory network (KBiLSTM). J. Inf. Sci. (2021). https://doi.org/10.1177/01655515211023937

14. Duan, J., Zhang, P.F., Qiu, R., Huang, Z.: Long short-term enhanced memory for sequential recommendation. World Wide Web 1–23 (2022)

15. Habib, M., Faris, M., Qaddoura, R., Alomari, A., Faris, H.: A predictive text system for medical recommendations in telemedicine: a deep learning approach in the Arabic context. IEEE Access **9**, 85690–85708 (2021)

16. Hiriyannaiah, S., Siddesh, G.M., Srinivasa, K.: DeepLSGR: neural collaborative filtering for recommendation systems in smart community. Multimed. Tools Appl. 1–20 (2022)

17. Qiao, J., Wang, L., Duan, L.: Sequence and graph structure co-awareness via gating mechanism and self-attention for session-based recommendation. Int. J. Mach. Learn. Cybern. **12**(9), 2591–2605 (2021). https://doi.org/10.1007/s13042-021-01343-3

18. Qu, T., Wan, W., Wang, S.: Visual content-enhanced sequential recommendation with feature-level attention. Neurocomputing **443**, 262–271 (2021)

19. Rabiu, I., Salim, N., Da'u, A., Nasser, M.: Modeling sentimental bias and temporal dynamics for adaptive deep recommendation system. Expert Syst. Appl. **191**, 116262 (2022)

20. Sheng, X., Wang, F., Zhu, Y., Liu, T., Chen, H.: Personalized recommendation of location-based services using spatio-temporal-aware long and short term neural network. IEEE Access **10**, 39864–39874 (2022)

21. Wu, Y., Li, K., Zhao, G., Xueming, Q.: Personalized long-and short-term preference learning for next poi recommendation. IEEE Trans. Knowl. Data Eng. (2020)

22. Bhoi, S., Lee, M.L., Hsu, W., Fang, H.S.A., Tan, N.C.: Personalizing medication recommendation with a graph-based approach. ACM Trans. Inf. Syst. (TOIS) **40**(3), 1–23 (2021)

23. Cai, D., Qian, S., Fang, Q., Xu, C.: Heterogeneous hierarchical feature aggregation network for personalized micro-video recommendation. IEEE Trans. Multimed. **24**, 805–818 (2021)

24. Chou, Y.C., Chen, C.T., Huang, S.H.: Modeling behavior sequence for personalized fund recommendation with graphical deep collaborative filtering. Expert Syst. Appl. **192**, 116311 (2022)

25. Do, P., et al.: Developing a Vietnamese tourism question answering system using knowledge graph and deep learning. Trans. Asian Low-Resource Lang. Inf. Process. **20**(5), 1–18 (2021)

26. Cui, Y., Sun, H., Zhao, Y., Yin, H., Zheng, K.: Sequential-knowledge-aware next poi recommendation: a meta-learning approach. ACM Trans. Inf. Syst. (TOIS) **40**(2), 1–22 (2021)

27. Dai, Q., Wu, X.M., Fan, L., Li, Q., Liu, H., Zhang, X., Wang, D., Lin, G., Yang, K.: Personalized knowledge-aware recommendation with collaborative and attentive graph convolutional networks. Pattern Recognit. **128**, 108628 (2022)

28. Deng, Z.H., Wang, C.D., Huang, L., Lai, J.H., Philip, S.Y.: G3sr: global graph guided session-based recommendation. IEEE Trans. Neural Netw. Learn. Syst. (2022)

29. El Alaoui, D., Riffi, J., Sabri, A., Aghoutane, B., Yahyaouy, A., Tairi, H.: Deep graphSAGE-based recommendation system: jumping knowledge connections with ordinal aggregation network. Neural Comput. Appl. 1–12 (2022)

30. Feng, L., Cai, Y., Wei, E., Li, J.: Graph neural networks with global noise filtering for session-based recommendation. Neurocomputing **472**, 113–123 (2022)

31. Gu, P., Han, Y., Gao, W., Xu, G., Wu, J.: Enhancing session-based social recommendation through item graph embedding and contextual friendship modeling. Neurocomputing **419**, 190–202 (2021)

32. Gwadabe, T.R., Liu, Y.: Ic-gar: item co-occurrence graph augmented session-based recommendation. Neural Comput. Appl. **34**(10), 7581–7596 (2022)

33. Gwadabe, T.R., Liu, Y.: Improving graph neural network for session-based recommendation system via non-sequential interactions. Neurocomputing **468**, 111–122 (2022)

34. Jian, M., Zhang, C., Fu, X., Wu, L., Wang, Z.: Knowledge-aware multispace embedding learning for personalized recommendation. Sensors **22**(6), 2212 (2022)
35. Jiang, N., Gao, L., Duan, F., Wen, J., Wan, T., Chen, H.: San: attention-based social aggregation neural networks for recommendation system. Int. J. Intell. Syst. **37**(6), 3373–3393 (2022)
36. Li, C.T., Hsu, C., Zhang, Y.: FairSR: fairness-aware sequential recommendation through multi-task learning with preference graph embeddings. ACM Trans. Intell. Syst. Technol. (TIST) **13**(1), 1–21 (2022)
37. Li, Y.: A graph convolution network based on improved density clustering for recommendation system. Inf. Technol. Control **51**(1), 18–31 (2022)
38. Pan, Z., Cai, F., Chen, W., Chen, C., Chen, H.: Collaborative graph learning for session-based recommendation. ACM Trans. Inf. Syst. (TOIS) **40**(4), 1–26 (2022)
39. Chen, G., Zhao, G., Zhu, L., Zhuo, Z., Qian, X.: Combining non-sampling and self-attention for sequential recommendation. Inf. Process. Manag. **59**(2), 102814 (2022)
40. Huang, N., Hu, R., Xiong, M., Peng, X., Ding, H., Jia, X., Zhang, L.: Multi-scale interest dynamic hierarchical transformer for sequential recommendation. Neural Comput. Appl. 1–12 (2022)
41. Yang, N., Jo, J., Jeon, M., Kim, W., Kang, J.: Semantic and explainable research related recommendation system based on semi-supervised methodology using BERT and LDA models. Expert Syst. Appl. **190**, 116209 (2022)
42. Zang, T., Zhu, Y., Zhu, J., Xu, Y., Liu, H.: Mpan: multi-parallel attention network for session-based recommendation. Neurocomputing **471**, 230–241 (2022)
43. Zhang, Y., Liu, X.: Leveraging mixed distribution of multi-head attention for sequential recommendation. Appl. Intell. 1–16 (2022)
44. Duan, C., Sun, J., Li, K., Li, Q.: A dual-attention autoencoder network for efficient recommendation system. Electronics **10**(13), 1581 (2021)
45. Mahesh Selvi, T., Kavitha, V.: A privacy-aware deep learning framework for health recommendation system on analysis of big data. Visual Comput. **38**(2), 385–403 (2022)
46. Yang, Y., Zhu, Y., Li, Y.: Personalized recommendation with knowledge graph via dual-autoencoder. Appl. Intell. **52**(6), 6196–6207 (2022)
47. Cai, X., Han, J., Yang, L.: Generative adversarial network based heterogeneous bibliographic network representation for personalized citation recommendation. In: Proceedings of the AAAI Conference on Artificial Intelligence, vol. 32 (2018)
48. Elmisery, A.M., et al.: Cognitive privacy middleware for deep learning mashup in environmental IoT. IEEE Access **6**, 8029–8041 (2017)
49. Sasagawa, T., Kawai, S., Nobuhara, H.: Recommendation system based on generative adversarial network with graph convolutional layers. J. Adv. Comput. Intell. Intell. Inform. **25**(4), 389–396 (2021)
50. Daneshvar, H., Ravanmehr, R.: A social hybrid recommendation system using LSTM and CNN. Concurrency and Computation: Practice and Experience, p. e7015 (2022)
51. Fahad, S.A., Yahya, A.E.: Inflectional review of deep learning on natural language processing. In: 2018 International Conference on Smart Computing and Electronic Enterprise (ICSCEE), pp. 1–4. IEEE (2018)
52. Kang, W.C., McAuley, J.: Self-attentive sequential recommendation. In: 2018 IEEE International Conference on Data Mining (ICDM), pp. 197–206. IEEE (2018)
53. Jiang, J., Kim, J.B., Luo, Y., Zhang, K., Kim, S.: Adamct: adaptive mixture of CNN-transformer for sequential recommendation. arXiv preprint arXiv:2205.08776 (2022)
54. Kuo, R., Chen, J.: An application of differential evolution algorithm-based restricted Boltzmann machine to recommendation systems. J. Internet Technol. **21**(3), 701–712 (2020)
55. Afify, M., Loey, M., Elsawy, A.: A robust intelligent system for detecting tomato crop diseases using deep learning. Int. J. Softw. Sci. Comput. Intell. (IJSSCI) **14**(1), 1–21 (2022)

56. Venkatesh, M., Sathyalaksmi, S.: Memetic swarm clustering with deep belief network model for e-learning recommendation system to improve learning performance. Concurrency and Computation: Practice and Experience, p. e7010 (2022)
57. Wu, L., Li, S., Hsieh, C.J., Sharpnack, J.: Sse-pt: sequential recommendation via personalized transformer. In: Fourteenth ACM Conference on Recommender Systems, pp. 328–337 (2020)
58. Sahoo, S.R., et al.: Hybrid approach for detection of malicious profiles in twitter. Comput. Electr. Eng. **76**, 65–81 (2019). ISSN: 0045-7906. https://doi.org/10.1016/j.compeleceng.2019.03.003

COVID-19 Patient Recovery Prediction Using Efficient Logistic Regression Model

Shrawan Kumar Trivedi[1], Rajiv Kumar[2], Shubhamoy Dey[3(✉)],
Amit Kumar Chaudhary[3], and Justin Zuopeng Zhang[4]

[1] Rajiv Gandhi Institute of Petroleum Technology, Jais, Amethi 229305, India
[2] Indian Institute of Management Kashipur, Kashipur 244713, India
`rajiv.kumar@iimkashipur.ac.in`
[3] Indian Institute of Management Indore, Indore, India
`shubhamoy@iimidr.ac.in`, `f20amitc@iimidr.ac.in`
[4] University of North Florida, Jacksonville, USA
`justin.zhang@unf.edu`

Abstract. This research develops a COVID-19 patient recovery prediction model using machine learning. A publicly available data of infected patients is taken and pre-processed to prepare 450 patients' data for building a prediction model with 20.27% recovered cases and 79.73% not recovered/dead cases. An efficient logistic regression (ELR) model is built using the stacking of random forest (RF) and logistic regression (LR) classifiers. Further, the proposed model is compared with state-of-art models such as logistic regression (LR), support vector machine (SVM), decision tree (C5.0), and random forest (RF). All the models are evaluated with different metrics and statistical tests. The results show that the proposed ELR model is good in predicting not recovered/dead cases and handling imbalanced data.

Keywords: COVID-19 · Efficient logistic regression · Machine learning · Predicting patient recovery

1 Introduction

Coronavirus disease, also known as COVID-19, is an infectious disease caused by a newly discovered coronavirus that has impacted the entire world [1,2]. Coronavirus disease is surging rapidly and affecting the entire world so gravely that the outbreak disturbed worldwide the functioning of life as a whole. During the COVID-19 pandemic, hospitals are dealing with a variety of issues due to a lack of medical staff, life-support equipment, personal protective equipment, and other resources [10,16,17,21,24]. During the COVID-19 pandemic, to accommodate the high patients flow, hospitals are facing various issues due to limited medical staff, personal protective equipment, life-support equipment, and others [10,16,17,21]. In many countries like the United States, India, Brazil, Russia, Peru, etc., the pandemic is already overburdening the healthcare system [14].

N. Nedjah et al. (Eds.): ICSPN 2021, LNNS 599, pp. 134–149, 2023.
https://doi.org/10.1007/978-3-031-22018-0_13

To combat this pandemic, the whole world, including governments, individuals, and organizations from across industries and sectors, have come together. Researchers in various fields are working in this direction, from making medicines and vaccines to pandemic prediction. Prediction using different statistical models is a prominent area of research. Similarly, there are many other studies on COVID-19 predictions found in the recent literature. ML-based prediction models have been applied widely to understand patient recovery and can be used in the COVID-19 context. ML-based predictive models can support the health system in taking appropriate measures while treating an infected patient and can also guide clinical decision-making and improve operation and patient-centered outcomes. ML played a significant role in fighting the previous pandemics, particularly in drug/vaccine development, medication, disease diagnosis, and various types of predictions [18, 22]. Such methods may help fight the current COVID-19 pandemic.

Although much research has been done in developing the prediction models for COVID-19 prediction, a robust, rapid, and sensitive model is yet to be an area of research. As discussed earlier that several predictive models are proposed to fight this pandemic which gained major attention and shown powerful prediction capabilities. However, a major gap remains in the research on developing automated recovered and non-recovered/dead patient recovery models. A robust infected patient recovery prediction model using ML is a pressing need for resource planning and treatment therapy to improve patients' conditions. This study focuses on filling the gaps that arise from the above discussion, and the study tries to answer below three research questions.

RQ1: How well does the proposed efficient logistic regression (ELR) predict patients recovered from COVID-19 infection?

RQ2: How does the proposed efficient logistic regression (ELR) deal with the imbalanced COVID-19 patient data?

RQ3: Can the proposed efficient logistic regression (ELR) be recommended to develop a COVID-19 prediction model?

2   Related Work

This section primarily focused on the research related to COVID-19 prediction and patient recovery using ML methods. It also highlights some ML techniques used in building prediction models.

2.1   ML in COVID-19 Prediction

Different COVID-19 studies have used different types of ML techniques. For example, [4, 26] used time series analysis in their predictive model. Reference [12] adopted a seasonal ARIMA forecasting package with R statistical model to forecast registered and recovered patients. Reference [28] used iterative weighting to fit the Generalized Inverse Weibull distribution and develop a prediction framework for the pandemic's growth behavior. For real-time and more accurate prediction, they have also deployed the prediction model on a cloud computing platform. Among the studies on COVID-19, predicting patient recovery is also important to fight with COVID-19 pandemic, and therefore various researchers e.g., [17,22] have paid their attention. Author in [23] proposed a N-Gram-Codon and Recurrent Neural Network (RNN) to Update Pfizer-BioNTech mRNA Vaccine. However, research on COVID-19 patient recovery is nascent, and various aspects are still missing. The next subsection discusses such studies in detail and articulates the research gap in line with patient recovery prediction.

2.2   Works on COVID-19 Patient Recovery

Current ongoing research on COVID-19 patient recoveries has focused primarily on predicting the number of recovery cases. Very little attention has been paid to predicting a patient's recovery using some parameters like demographic, clinical, and epidemiological characteristics. For example, [22] focused on applying RF, LDA, and SVM as supervised ML techniques for COVID-19 patient survival prediction. Their predictive model is based on some demographic, clinical, and epidemiological characteristics as input variables. Yan et al. (2020) [32] selected three biomarkers, lactic dehydrogenase (LDH), lymphocyte, and high-sensitivity creative protein (hs-CRP). They used ML tools for the individual patient's survival prediction with more than 90% accuracy. Reference [17] tested three ML techniques, SVM, artificial neural networks (ANN), and regression model, to predict the patient's recovery. ML technique was used to predict the risk of positive COVID-19 diagnosis by [5]. They employed neural network, SVM, RF, LR, and gradient boosting trees, and found the number of lymphocytes, leukocytes, and eosinophils as the three most important variables among the fifteen identified variables for the predictive performance of the ML algorithms.

Summarizing most existing studies on patient recovery, we find they are mainly focused on the number of recovery cases prediction. There are limited studies on predicting the recovery of a COVID-19 patient. This research aims to address these gaps by proposing a novel efficient logistic regression model (ELR).

3   COVID-19 Patient Dataset

The COVID-19 data is prepared using publicly available COVID-19 patients' data retrieved from the Kaggle[1] source and created on January 30, 2020. This dataset is a modified version of the COVID-19 data created by John Hopkins University. At first, data cleaning is performed where the cases with maximum missing attributes are removed from the dataset. The final COVID-19 dataset of 450 patients with two categories, 20.27% recovered cases and 79.73% not recovered/dead cases, is presented in Table 1. Since a huge difference between the number of cases in the two categories is identified, data is considered imbalanced.

---

[1] https://www.kaggle.com/sudalairajkumar/novel-corona-virus-2019-dataset.

Such imbalanced data develops a strong possibility of the misclassification of the category with fewer cases. This research focuses on getting good specificity (true negative rate) and sensitivity (true positive rate) to assure the credibility of the prediction models in dealing with such imbalanced data.

Table 1. Description of the dataset

S. No.	Attributes	Description	% of categories	Mean	Std. Dev.
1	Age	Patient age	NA	51.80	17.33
2	Gender	Male/female	61% male, 39% female	NA	NA
3	Case in country	No. of Covid-19 cases on the date when infection discovered	NA	43.07	58.11
4	Country	Country where the person infected	NA	NA	NA
5	If_onset_ approximated	If the symptom detected	Onset detected = 4%; not detected=96%	NA	NA
6	Visiting Wuhan	If the infected person visited Wuhan	Visited = 25%; not visited = 75%	NA	NA
7	From Wuhan	If the infected person is from Wuhan	From Wuhan = 22%; other place = 78%	NA	NA
8	Symptoms	COVID-19 symptoms of infected person	NA	NA	NA
9	Gap_symptom_ hospitalized	Days between symptoms and admitted in hospital	3% present; 97% not present	3.73	3.13
10	Gap_reporting_ symptom	Days between case found and symptoms identified	NA	22.34	12.34
11	Response	Whether the person recovered from symptoms or not-recovered/dead	Recovered = 20.27%; not recovered/dead = 79.73%		

4   Machine Learning Algorithms

In this study, an efficient logistic regression (ELR) is proposed and compared with the state-of-art ML techniques which are built using different characteristics. The prediction model is made to predict the recovered and not recovered/dead patients from COVID-19. The description of the proposed classifier and state-of-art classifiers are given below.

4.1   Logistic Regression

Logistic regression (LR) is a statistical model that develops a model for a dependent variable using a logistic function [27]. LR is also referred as logit and has both binary and multinomial logit model [7]. The binary logit model is commonly used to identify how likely an observation belongs to a particular class. The binary logit model, also called LR, can be used to estimate the a binary event's probability Eq. 1 represents the binary logit model of an event with two outcomes, success and failure.

$$P = \frac{\exp\left(\sum_{i=0}^{k} a_i X_i\right)}{1 + \exp\left(\sum_{i=0}^{k} a_i X_i\right)} \tag{1}$$

where, P, X_i, and a_i are the probability of success of the event, independent variable i, and parameter to be estimated, respectively.

The maximum likelihood method is used to achieve a binary logit model, i.e., to estimate the parameters [11]. Cox & Snell R2 and Nagelkerke R2 are two commonly used measures of binary logit model fit that are based on the likelihood function. LR model is known for its simplicity and interpretability, and many studies [5,22] have found LR is a useful method for modeling patient recovery prediction.

4.2   C5.0 Decision Tree Classifier

C5.0 is an all-purpose classifier and one of the most well-known algorithms to generate decision trees, and it is applicable for many types of problems directly out of the box. C5.0 has become the industry standard and can be used to predict relevant or non-relevant attributes in classification; and can work well with noise and missing data as well as provides boosting [15]. C5.0 is useful, especially while working with high-dimensional datasets. The performance of the C5.0 algorithm is also as good as more advanced and sophisticated ML models (e.g., NN, SVM, etc.). However, because it allows you to view the large decision tree as a set of rules, it is simple to understand and deploy [15]. The C5.0 algorithm is an extension of the C4.5 and Id3 algorithms that employs an efficient tree punning technique to reduce model over-fitting. Where "C4.5 Decision tree learning is a method for approximating discrete-valued functions, in which a decision tree represents the learned function" [25] and "ID3 algorithm construct the decision tree by employing a top-down, greedy search through the given sets to test each attribute at every tree node" [29]. To handle numeric or nominal features, as well as missing data, C5.0 employs a highly automatic learning process. It eliminates less informative features and is effective for both small and large datasets. C5.0 model results can be interpreted without a mathematical background and are more efficient than other complex models.

4.3   Support Vector Machine-RBF

The SVM is a state-of-the-art classification method for both regression and classification [6,17,31]. Because of its high accuracy, ability to handle high-dimensional data, and flexibility in modelling diverse data sources, the SVM classifier is widely used in a variety of fields [6]. SVM works well for both linear and non-linear problems, and compared with other ML techniques, it has better generalization performance [17,33]. In various real-world applications, for instance image processing in medical applications, SVM has achieved a high level of performance [17]. For splitting the feature space in two distinct groups, SVM tries to identify hyperplane in a multidimensional space [22]. SVM tries to find an optimal hyperplane in multidimensional space to maximize the distance between the closest training sample data point and the separating hyperplane [20]. SVM tends to convert the features in higher dimensions which are separable by hyperplane [8,17]. The SVM's goal is then to maximize this hyperplane because the more features that are separated, the higher the classification accuracy [17]. SVM equations are expressed in Eq. (2).

Consider a series of data points $D = [x_i, d_i]$ (i to n) where n is the size of the data, $x_i d_i$ and d_i represents the sample's input space vector and target value, respectively. The SVM estimates the function as shown in the two equations below:

$$(f(x) = \omega\phi(x) + b), \tag{2}$$

$$(R_{SVMs}(c) = \frac{1}{2}\omega^2 + C(\frac{1}{n})\sum i = 1 \text{ to n})L(x_i, d_i)) \tag{3}$$

where ϕ is the high-dimensional space feature, b is a scalar, ω is a normal vector and $C(\frac{1}{n})\sum i = 1$ to n)$L(x_i, d_i)$ signifies the empirical error. Here Eq. (2) is used to assess b and ω.

4.4   Random Forest

Random forest (RF) is an ensemble learning method (also called Bootstrap Aggregation or bagging) for classification. Like a forest, a collection of many trees, an RF consists of many decision trees [3,30]. Derived from the decision trees, the RF technique overcomes the weaknesses revealed by the decision tree, such as the high sensitivity to small variations in data, improving the accuracy [19,30]. Different bootstrap samples are used to generate the decision trees in an RF by resampling the training data that limit each decision tree (Ado Osi et al., 2020). While creating a tree in RF, at each, a random subset of features is randomly selected to decide the optimal split [12,22]. RF trains different decision trees using different parts of the training dataset [13]. For the classification of new samples, the sample's input vector is required to be passed down through each decision tree for the RF. Classification outcome is then achieved by considering different parts of that input vector considered by each decision tree [30]. For discrete classification forest chooses the class having most votes, whereas for numeric classification average of all trees are selected by forest [22,30].

4.5   Efficient Logistic Regression (ELR)

This research proposes an efficient logistic regression classifier which is developed by stacking random forest classifiers and logistic regression classifier, which is an improvement on logistic regression for predicting Covid-19 patient's recovery. Here, COVID-19 training dataset is split into three subsets with equal size of data. Each random forest classifier was trained with a different training subset, resulting in three distinct trained random forest classifiers. Then classification decision by RF classifiers is taken as the input of Logistic Regression, which act as the meta classifier and takes the final classification decision. In order to ensure nearly equal distribution of classes, stratified sampling is done which creates random samples from training instances.

Figure 1 [31] depicts the operation of the proposed method. A class distribution vector formed by an individual RF classifier denoting a specific class of the Covid-19 case is used to test the model. Consider xth denoting the class distribution vector produced by the RF classifier. Here, for two classes, the class distribution vector is as follows:

$$(\Delta_x = [\partial_{1x} \ \partial_{2x}] \quad \text{for x} = 1, 2, ..., n) \tag{4}$$

where,

$$0 \leq \partial_{ix} \leq 1 \text{ for } i = 1, 2; \ x = 1, 2, \ldots, n,$$

$$\sum_{i=1}^{2} \partial_{ix} = 1.$$

The class vector for the 'n' RF classifiers can be represented as follows:

$$\Delta = [\Delta_1, \Delta_2, \ldots, \Delta_n]^T \tag{5}$$

Finally, the weight distribution vector is developed which assigns a weight to each individual classifier in order to create the meta-classifier. The weight distribution vector for the n classifier is as follows:

$$\varnothing = [\theta_1\theta_2\theta_3...\theta_n] \tag{6}$$

where,

$$0 \leq \theta_x \leq 1 \quad \text{for } x = 1, 2, ..., n,$$

$$\sum_{v=1}^{n} \theta_x = 1.$$

The meta-classifier creates a 1×2 distribution [see Eq. (4)] for classifying each test set instance using the class distribution matrix [see Eq. (2)] and the weight distribution matrix [see Eq. (3)].

$$\Delta' = \varnothing\Delta = [\partial^l_1 \ \partial^l_2] \tag{7}$$

where,

$$\partial^l_i = \sum_{x=1}^{n} \varnothing_i\partial_{ix} \text{ for } i=1,2.$$

The base classifier for making the stacking ensemble of the classifier is random forest. In the proposed IRF classifier, we have used n = 3 number of RF classifiers.

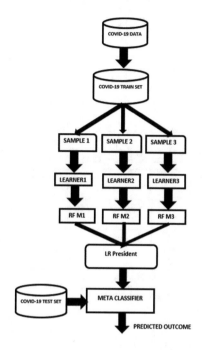

Fig. 1. Working on the proposed classifier

5   Results and Analysis

5.1   Experimental Design

We split the original dataset into a 'train' and 'test' dataset. We use the 'train' dataset to construct and tune the predictive (supervised) model and the 'test' dataset to assess the ML model's prediction performance. This study splits training and testing instances to balance the dataset with the train and test ratios of 50:50, 66:34, and 80:20% (please see Table 2 for details). 10-Fold cross-validation is also performed to validate the results received from splitting methods. In each split and 10-fold, the training is controlled with 15 iterations.

After testing, a confusion matrix of each classifier is developed using which accuracy, sensitivity, specificity, ROC, and AUC values are calculated for analysis. In addition, some of the statistical techniques such as Kappa statistics and Wilcoxon sign ranked test for pairwise data are also obtained to assess the credibility of the proposed ELR model with other state-of-the-art models to predict

Table 2. Training set instance for different partitions

Training-testing partition	Total training records	Positive records in training set	Negative records in training set
50-50	223	182	41
66-34	152	120	32
80-20			22
	89	67	
10-Fold cross-validation	410	345	65

Table 3. Confusion matrix and performance evaluation metrics and statistical tests

S/N	Metrics	Formula/Description		
1	Confusion Matrix		Actual	
			Not-Rec/Dead	Recovered
		Predicted — Not-Rec/Dead	True Positive (TP)	False Positive (FP)
		Predicted — Recovered	False Negative (FN)	True Negative (TN)
2	Accuracy	$\dfrac{TP+TN}{TP+TN+FP+FN}*100$		
3	Sensitivity	$\dfrac{TP}{TP+FN}*100$		
4	Specificity	$\dfrac{TN}{TN+FP}$		
5	Kappa statistics	$\dfrac{pa\text{-}pac}{(1\text{-}pac)}$, 'pa' stands for total agreement probability, and 'pac' stands for probability 'by chance.'		
6	Receiver operating characteristic (ROC)	The ROC curve is drawn between sensitivity and (1-Specificity). The area under the curve (AUC) quantifies how much the curve is up in the northwest corner.		
7	Wilcoxon's Rank-Sum (WRS) test	The WRS test method is used to determine whether two samples (prediction accuracy) differ significantly from one another.		

the COVID-19 case are performed on the models. Table 3 provides the confusion matrix and other performance evaluation and statistical techniques used in this study.

5.2   Results

This study proposes an efficient logistic regression (ELR) classifier, which is then compared with four Machine Learning state-of-art classifiers such as logistic regression (LR), decision tree (C5.0), support vector machine with radial basis kernel function (SVM-RBF), and random forest (RF). All the experiments are done on r-studios using different libraries such as "caret," "caretEnsemble," and "C50" for machine learning models. The proposed ELR model is built using a stacking ensemble of RF and LR with "caret" and "caretEnsemble" R libraries. One among these classifiers is parametric, namely LR; the rest of the three classifiers, DT, SVM, and RF, are non-parametric. There are various ways of evaluating the performance of machine learning models. Therefore, to compare the performance of popular classifiers (LR, DT, SVM, RF) with the proposed model, i.e., ELR, the performance measures such as accuracy, sensitivity, specificity, ROC, and AUC are used. In addition, a statistical test is done using Kappa statistics and Wilcoxon pairwise sign test, as specified in the previous section.

The result from Table 4 depicts that all different models have different accuracy levels. This table shows that the performance of non-parametric classifiers is better than the parametric classifier. The performance accuracy of famous classifiers is 80.90–89.27% for LR, 83.15–97.32% for DT, 77.53–90.73% for SVM-RBF 85.53–92.44% for RF. And when the accuracy of the proposed model ELR is compared against these other famous models, the accuracy value is found to be good for 50-50, 66-34, 80-20 split, and 10-fold cross-validation 87.44%, 85.53%, 87.64%, 98.78% respectively. Classification accuracy for 10-fold cross-validation is taken to validate the results from data splitting methods. In considerations of performance accuracy, overall, the proposed ELR is found accurate in detecting COVID-19 patients; however, for clearer evaluation, other metrics will justify our proposed model.

Table 4. Comparison of accuracy

Train-test partition	Accuracy				
	LR	C5.0	SVM	RF	ELR
50-50	83.86	84.75	87.44	85.65	87.44
66-34	83.55	82.92	82.24	85.53	85.53
80-20	80.90	83.15	77.53	86.52	87.64
10-Fold	89.27	97.32	90.73	92.44	98.78

The sensitivity and specificity of the proposed model ELR and four state-of-classifiers (LR, DT, SVM, RF) are reported in Table 5. The proposed model gives the highest sensitivity value, i.e., from 95% to 100%, compared to other classifiers with sensitivity values, 87–95% for LR, 89–100% for C5.0, 92–98% for SVM-RBF, and 90–100% for RF within all the splitting methods and 10-fold cross-validation method. The values of sensitivity showcase that although LR and SVM-RBF are comparable with ELR with 100% accuracy, the ELR model performs consistently well among all the training-testing experimental setups.

Since the data is imbalanced with 80% "Not Recovered-Dead" cases and only 20% "Recovered" cases, it imposes the necessity of specificity metric for evaluating the correctness of recovered cases of Covid-19 patients. Now considering the specificity, the proposed ELR model is found to be good, with sensitivity values 50.00–92.31% in comparison to the other classifiers with specificity values 40.91–65.85% for LR, 50.00–83.08% for C5.0, 13.64–63.41% for SVM-RBF and 52.31–65.85% for RF within all the splitting methods and 10-fold cross-validation method. The proposed model is shown to be good for imbalanced data, too, where it has given accuracy up to 92.31% for the recovered Covid-19 cases, which are less in numbers.

Table 5. Comparison of sensitivity and specificity

Training testing	LR		C5.0		SVM		RF		ELR	
	Sen	Spe	Sen	Spe	Sen	Spe	Sen	Spe	Sen	Spe
50-50	87.91	65.85	89.01	65.85	92.86	63.41	90.11	65.85	95.60	51.22
66-34	92.50	50.00	91.67	50.00	99.17	18.75	93.33	56.25	95.00	50.00
80-20	94.03	40.91	91.04	59.09	98.51	13.64	94.03	63.64	95.52	63.64
10-Fold	95.36	56.92	100	83.08	98.26	50.77	100	52.31	100	92.31

Sen Sensitivity, *Spe* Specificity

Figure 2 represents the ROC comparison chart of 10-fold cross-validation models for all models, i.e., four famous ML classifiers (LR, DT, SVM, RF) and the proposed model (ELR). The ROC curve of each model is plotted simultaneously in this figure, and the AUC-ROC value is provided. The area under an entire ROC curve is measured by Area Under Curve (AUC). If the AUC-ROC value is greater than 0.5, the model is considered better and more suitable for developing a prediction model. For this study, the AUC-ROC value of four well-known ML classifiers is 0.87 for LR, 0.89 for DT, 0.82 for SVM, and 0.87 for RF. And the AUC-ROC value of this proposed model (ELR) is 0.86, which is significantly greater than 0.5, implying that the proposed model, along with the other models, is suitable for developing a prediction model and does not fall under random guesser. However, the models' credibility can be tested using other different performance metrics.

If the comparison between ML classifiers' performance is made merely on performance accuracy metrics, it may produce ambiguous results. Cohen's Kappa Statistic (CKS) result is considered more realistic and considers correct prediction based on good training but not due to chance or random guess. The value of CKS lies in the range of −1 to +1, where −1 implies total disagreement, 0 implies random classification, and 1 means total agreement.

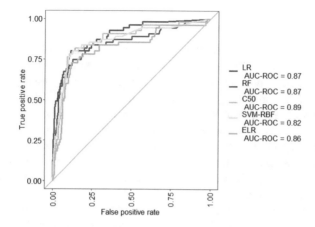

Fig. 2. ROC and AUC for 10-fold cross-validation

Table 6 represents the values of CKS for four popular ML classifiers and the proposed ELR model. The value of CKS for four popular machine-learning classifiers is in the range (40–56%) for LR, (44–89%) for DT, (16–58%) for SVM, and (53–64%) for RF, whereas the CKS value for the proposed model ELR is in range (50–95%). CKS value is observed to be the highest for the proposed ELR.

Table 6. Kappa statistic for each model

Training-testing (%)	LR	DT	SVM	RF	ELR
50-50	49.99	51.92	57.36	53.93	52.81
66-34	46.21	44.74	25.11	53.24	50.12
80-20	40.50	52.54	16.74	61.42	64.04
10-Fold	56.52	89.20	58.53	64.86	95.28

WRS (Wilcoxon Rank Sum), a non-parametric statistical test, is also performed to compare the statistical differences between four machine learning classifiers and the proposed model ELR. WRS is a hypothesis testing, where the null hypothesis (H0) is that two samples are taken from the same population, i.e., both samples have the same distribution with the same median. For this, the medians of the two samples are compared. The p-value from the WRS test was performed on the performance accuracies of the models. The p-value for all the models from the proposed ELR comes out to less than 0.05, which signifies the median difference between the performance accuracies of the two classifiers are different, and the null hypothesis (Ho) is rejected. This represents that the results produced from the proposed model, ELR, are comparatively better than all four existing famous models.

6   Conclusion

This research was started with the motivation to build a robust, sensitive, and accurate COVID-19 prediction model to help the doctors and hospitals predict a patient's recovery potential so that an early decision may be taken to save the life. We started by proposing efficient logistic regression (ELR) model to build a COVID-19 recovery prediction model by considering three research objectives (1) to build an accurate prediction model; (2) to build a prediction model which can deal with the imbalanced data; (3) to build a prediction model which may be recommended over the other state-of-art models. We have successfully achieved our objectives after doing an in-depth evaluation of the proposed ELR model with other popular models.

The comparative study suggested that the proposed ELR model was good in performance accuracy (85.53–98.78%) compared to other models for all the split methods as well as for 10-fold cross-validation. The sensitivity of a proposed ELR classifier was also found good (95–100%) compared to the other models, which confirmed that the proposed model is excellent in predicting not-recovered/dead cases (also known as true positive rate) up to 100%. As the percent of positive cases (not recovered/dead) in the dataset was high, the sensitivity of almost all the classifiers was satisfactory, but the proposed model was come out better than all the models. The critical challenge in building the COVID-19 prediction model was to tackle the imbalanced nature of the data and specifically the cases of the recovered patients, which resulted in poor training of the models. In such a situation, if a model gives good specificity (true negative rate), i.e., in this case, it is the proportion of patients that do not have the disease, and the test gives a negative result only. The specificity value of the ELR model was found to be good compared to the other models in all the split methods and 10-fold cross-validation, which is up to 92%. The AUC-ROC of the proposed model is 86%, which implies a good model. Cohen's Kappa statistic value of the proposed model ELR was also found to be good (Up to 95%), which showed the model's strength. Wilcoxon rank-sum (WRS) test showed a significant difference between all the classifiers and the proposed model ELR. After evaluating all the models, the proposed ELR model is strongly recommended for building the COVID-19 patient recovery prediction model.

This research also has some limitations. The data used in this study has been taken from various countries; hence some unique country attributes may be missing. The same proposed prediction model may be built separately for different countries with unique infrastructure, atmospheric and demographic conditions attributes, and symptomatic patients. This study tried to incorporate all the essential state-of-art classifiers for comparing from proposed classifiers; however, other classifiers may also be considered. In this research, stacking is done using LR and RF classifiers to build the proposed model; however, various other combinations of the stacking classifiers may be taken to further a comparative study. This study is limited to building only a prediction model; however, feature selection techniques may also be performed to check the importance of each individual attribute.

References

1. Zhou, Z., et al.: A statistical approach to secure health care services from DDoS attacks during COVID-19 pandemic. Neural Comput. Appl. (2021). https://doi.org/10.1007/s00521-021-06389-6
2. Gaurav, A., et al.: A novel approach for DDoS attacks detection in COVID-19 scenario for small entrepreneurs. Technol. Forecast. Soc. Change **177**, 121554 (2022). ISSN: 0040-1625. https://doi.org/10.1016/j.techfore.2022.121554
3. Adnan, M.N., Islam, M.Z.: Forest PA: constructing a decision forest by penalizing attributes used in previous trees. Expert Syst. Appl. **89**, 389–403 (2017)
4. Anastassopoulou, C., Russo, L., Tsakris, A., Siettos, C.: Data-based analysis, modelling and forecasting of the COVID-19 outbreak. PLoS ONE **15**(3), e0230405 (2020). https://doi.org/10.1371/journal.pone.0230405
5. Batista, A.D.M., Miraglia, J.L., Donato, T.H.R., Chiavegatto Filho, A.D.P., de Moraes Batista, A.F., Miraglia, J.L., Donato, T.H.R., Chiavegatto Filho, A.D.P.: COVID-19 diagnosis prediction in emergency care patients: a machine learning approach. MedRxiv (2020). https://doi.org/10.1101/2020.04.04.20052092
6. Ben-Hur, A., Weston, J.: A user's guide to support vector machines. In: Data Mining Techniques for the Life Sciences, pp. 223–239. Humana Press (2010). https://doi.org/10.1007/978-1-60327-241-4
7. Castaño, A., Fernández-Navarro, F., Gutiérrez, P.A., Hervás-Martínez, C.: Permanent disability classification by combining evolutionary Generalized Radial Basis Function and logistic regression methods. Expert Syst. Appl. **39**(9), 8350–8355 (2012)
8. Carrasco, M., López, J., Maldonado, S.: A second-order cone programming formulation for nonparallel hyperplane support vector machine. Expert Syst. Appl. **54**, 95–104 (2016)
9. Chintalapudi, N., Battineni, G., Amenta, F.: COVID-19 virus outbreak forecasting of registered and recovered cases after sixty day lockdown in Italy: a data driven model approach. J. Microbiol. Immunol. Infect. **53**(3), 396–403 (2020). https://doi.org/10.1016/j.jmii.2020.04.004
10. Chowdhury, M.E., Rahman, T., Khandakar, A., Al-Madeed, S., Zughaier, S.M., Doi, S.A
11. De Menezes, F.S., Liska, G.R., Cirillo, M.A., Vivanco, M.J.: Data classification with binary response through the Boosting algorithm and logistic regression. Expert Syst. Appl. **69**, 62–73 (2017)
12. De, A., Chowdhury, A.S.: DTI based Alzheimer's disease classification with rank modulated fusion of CNNs and random forest. Expert Syst. Appl. **169**, 114338 (2021)
13. El-Askary, N.S., Salem, M.A.M., Roushdy, M.I.: Features processing for Random Forest optimization in lung nodule localization. Expert Syst. Appl. 116489 (2022)
14. Firstpost: www.firstpost.com/health/coronavirus-outbreak-this-pandemic-threatens-to-overwhelm-indias-health-care-system-8296101.html (2020)
15. Galathiya, A.S., Ganatra, A.P., Bhensdadia, C.K.: Improved decision tree induction algorithm with feature selection, cross validation, model complexity and reduced error pruning. Int. J. Comput. Sci. Inf. Technol. **3**(2), 3427–3431 (2012). https://ijcsit.com/docs/Volume3/Vol3Issue2/ijcsit2012030227.pdf
16. Grasselli, G., Pesenti, A., Cecconi, M.: Critical care utilization for the COVID-19 outbreak in Lombardy, Italy: early experience and forecast during an emergency response. JAMA **323**(16), 1545–1546 (2020). https://doi.org/10.1056/NEJMoa2002032

17. Hassanien, A.E., Salam, A., Darwish, A.: Artificial intelligence approach to predict the covid-19 patient's recovery. EasyChair Preprint (3223) (2020). www.egyptscience.net

18. Lalmuanawma, S., Hussain, J. and Chhakchhuak, L.: Applications of machine learning and artificial intelligence for Covid-19 (SARS-CoV-2) pandemic: a review. Chaos Solitons Fractals **139**, 110059 (2020). https://doi.org/10.1016/j.chaos.2020.110059

19. Loureiro, A.L., Miguéis, V.L., da Silva, L.F.: Exploring the use of deep neural networks for sales forecasting in fashion retail. Decis. Support Syst. **114**, 81–93 (2018)

20. Melgani, F., Bruzzone, L.: Classification of hyperspectral remote sensing images with support vector machines. IEEE Trans. Geosci. Remote Sens. **42**(8), 1778–1790 (2004)

21. Moghadas, S.M., Shoukat, A., Fitzpatrick, M.C., Wells, C.R., Sah, P., Pandey, A., Sachs, J.D., Wang, Z., Meyers, L.A., Singer, B.H. and Galvani, A.P.: Projecting hospital utilization during the COVID-19 outbreaks in the United States. Proc. Natl. Acad. Sci. **117**(16), 9122–9126 (2020). https://doi.org/10.1073/pnas.2004064117

22. Osi, A.A., Dikko, H.G., Abdu, M., Ibrahim, A., Isma'il, L.A., Sarki, H., Muhammad, U., Suleiman, A.A., Sani, S.S., Ringim, M.Z.: A Classification Approach for Predicting COVID-19 Patient Survival Outcome with Machine Learning Techniques. medRxiv (2020). https://doi.org/10.1101/2020.08.02.20129767

23. Bouarara, H.A.: N-gram-codon and recurrent neural network (RNN) to update Pfizer-BioNTech mRNA vaccine. Int. J. Softw. Sci. Comput. Intell. (IJSSCI) **14**(1), 1–24 (2022)

24. Mohammed, S.S., Menaouer, B., Zohra, A.F.F., Nada, M.: Sentiment analysis of COVID-19 tweets using adaptive neuro-fuzzy inference system models. Int. J. Softw. Sci. Comput. Intell. (IJSSCI) **14**(1), 1–20 (2022)

25. Polat, K., Güneş, S.: A novel hybrid intelligent method based on C4.5 decision tree classifier and one-against-all approach for multi-class classification problems. Expert Syst. Appl. **36**(2), 1587–1592 (2009)

26. Mishra, A., et al.: A comparative study of distributed denial of service attacks, intrusion tolerance and mitigation techniques. In: Proceedings of the 2011 European Intelligence and Security Informatics Conference (EISIC '11). IEEE Computer Society, USA, pp. 286–289 (2011). https://doi.org/10.1109/EISIC.2011.15

27. Surówka, G., Ogorzalek, M.: Wavelet-based logistic discriminator of dermoscopy images. Expert Syst. Appl. **167**, 113760 (2021)

28. Tuli, S., Tuli, S., Tuli, R., Gill, S.S.: Predicting the growth and trend of COVID-19 pandemic using machine learning and cloud computing. Internet Things **11**, 100222 (2020). https://doi.org/10.1016/j.iot.2020.100222

29. Ture, M., Tokatli, F., Kurt, I.: Using Kaplan-Meier analysis together with decision tree methods (C&RT, CHAID, QUEST, C4. 5 and ID3) in determining recurrence free survival of breast cancer patients. Expert Syst. Appl. **36**(2), 2017–2026 (2009)

30. Uddin, S., Khan, A., Hossain, M.E., Moni, M.A.: Comparing different supervised machine learning algorithms for disease prediction. BMC Med. Inform. Decis. Making **19**(1), 1–16 (2019). https://doi.org/10.1186/s12911-019-1004-8

31. Wang, T., Huang, H., Tian, S., Xu, J.: Feature selection for SVM via optimization of kernel polarization with Gaussian ARD kernels. Expert Syst. Appl. **37**(9), 6663–6668 (2010)

32. Yan, L., Zhang, H., Goncalves, J., Xiao, Y., Wang, M., Guo, Y., Sun, C., Tang, X., Jin, L., Zhang, M., Huang, X.: A machine learning-based model for survival prediction in patients with severe COVID-19 infection (2020). https://doi.org/10.1101/2020.02.27.20028027

33. Zouhri, W., Homri, L., Dantan, J.Y.: Handling the impact of feature un-certainties on SVM: a robust approach based on Sobol sensitivity analysis. Expert Syst. Appl. **189**, 115691 (2022)

Ensemble Feature Selection for Multi-label Classification: A Rank Aggregation Method

Amin Hashemi[1], Mohammad Bagher Dowlatshahi[1], Marjan Kuchaki Rafsanjani[2(✉)], and Ching-Hsien Hsu[3]

[1] Department of Computer Engineering, Faculty of Engineering, Lorestan University, Khorramabad, Iran
hashemi.am@fe.lu.ac.ir, dowlatshahi.mb@lu.ac.ir
[2] Department of Computer Science, Shahid Bahonar University of Kerman, Kerman, Iran
kuchaki@uk.ac.ir
[3] College of Information and Electrical Engineering, Asia University, Taichung, Taiwan

Abstract. Multi-label classification is an important task in machine learning applications. However, today, these methods have more challenges due to high-dimensional data. Existing methods in the literature have not yet reached an acceptable performance, and as the number of labels in the dataset increases, their performance becomes weaker. Thus providing effective feature selection methods is necessary and is suitable for all data. Ensemble feature selection is also a new approach based on combining the results of several feature selection methods. Ensemble approaches have been used in many single-label applications and have provided adequate results. However, this technique is less considered in multi-label feature selection. For this purpose, in this article, we have presented a method based on an ensemble of feature selection methods. This paper presents an ensemble multi-label feature method using rank aggregation algorithms. Three filter rankers with different structures are utilized to obtain the final ranking. This rank aggregation process is treated like an election where the rankers are assumed to be the voters, and the features are the alternatives. The Weighted Borda Count (WBC) method conducts the election process. Finally, the performance and results obtained by the proposed method are compared with the nine filter-based algorithms used based on six multi-label datasets. Most of the selected datasets have a high number of class labels, and the proposed method has recorded the best classification performance in all these datasets. Also, the results show that this method is not well affected by the adverse effects of the algorithms used.

Keywords: High-dimensional data · Multi-label feature selection · Rank aggregation · Weighted Borda Count

1 Introduction

Over time and with the emergence of new technologies, many challenges have been posed in data science. We are beholding this data overgrowing every moment. In addition to the volume of data, they also increase their dimensions. Today, there is super-high-dimensional data, i.e., each instance has many features. Datasets typically have

© The Author(s), under exclusive license to Springer Nature Switzerland AG 2023
N. Nedjah et al. (Eds.): ICSPN 2021, LNNS 599, pp. 150–165, 2023.
https://doi.org/10.1007/978-3-031-22018-0_14

unrelated and redundant features that do not affect or adversely affect the achievement of the output [3]. Feature Selection (FS) can be used to remove irrelevant and redundant features as a dimensionality reduction method that selects the significant subset of features [33]. Increasing the classification accuracy, reduction of data dimensionality, computational cost, training time, and storage complexity can be achieved by using FS [7, 8, 13, 14].

Based on the search strategy, there are three types of feature selection methods: filters, wrappers, and embedded. Statistical measures such as Mutual Information (MI), Fisher Score, Information Interest, Symmetric Uncertainty (SU), and ReliefF are utilized to assess the features before the learning stage in filter methods. Wrapper strategies assess the features within the learning method. Different feature subsets are considered to discover the best features for the learning stage. Embedded strategies address the weaknesses of the previous two methods and take advantage of both [1, 9, 12].

The features are assessed based on their power to predict the class label in supervised feature selection methods. These methods are useful for performing feature selection in single-label (SL) [4, 10] and multi-label (ML) data. ML data consist of more than one label for each sample. Thus, to estimate the predictive power of each feature, its correlation is computed against the set of labels [18, 19, 21]. Unsupervised methods consider the relationship between the features due to the absence of class label information in data [26, 27]. Semi-supervised methods deal with datasets containing a fraction of labeled samples among unlabeled ones [2, 24, 25, 31]. Figure 1 shows two different categorizations of feature selection methods.

Since each instance in the real world can belong to multiple data classes simultaneously, multi-label classification is used in many applications, and the curse of dimensionality challenge exists in this type of data. Therefore, providing an effective feature selection method is seen as a fundamental challenge. Many multi-label feature selection algorithms have been proposed so far. However, these methods have not led to the elimination of existing challenges. The low accuracy of classification in datasets with a high number of features or labels and the dependence of algorithms on the number of labels are the reasons for the need to provide newer methods. On the other hand, ensemble feature selection is also a new approach based on combining the results of several feature selection methods. The logic of using this strategy is based on the fact that the variety of different methods can lead to better decisions in selecting features. Ensemble approaches have been used in many single-label applications and have provided adequate results. However, this technique is less considered in multi-label feature selection. For this purpose, in this article, we have presented a method based on an ensemble of feature selection methods.

Ratings gained by different feature selection algorithms may have different classification performances on particular data [15]. For example, if three feature selection methods apply to the same data set, each algorithm will capture a ranking with a different classification performance since algorithms have their strengths and weaknesses. In this regard, ensemble feature selection is a powerful strategy to take advantage of the strength of feature evaluation metrics and decrease their weaknesses. For this matter, we have considered three filter-based rankers in the proposed method. These methods consider the correlation between features and class labels and their redundancy.

Thus, we aim to construct a council of feature evaluation methods that can maximize the relevancy of selected features against the class labels and minimize their similarity. Thus, each ranker is applied to the dataset to achieve a feature ranking vector. We have assumed these rankings as the judgments of several voters in an election operation where the features are the alternatives. These rank vectors are assumed to be the preferences of each voter in the election. Then the results obtained from the FS algorithms are embedded in a matrix in which each column corresponds to the ranks of an FS algorithm. The WBC method is used to perform the rank aggregation process and achieve a final ranking. To the best of our knowledge, until now, no election-based method for multi-label feature selection has been proposed. The proposed method tries to ignore algorithms' weaknesses and order the features based on the strengths of various feature rankings.

Nine multi-label feature selection algorithms are used to assess the proposed method's performance on different real-world datasets. The results demonstrate the superiority of the proposed method based on classification accuracy and hamming loss. Also, the run-time of all algorithms is compared, and the proposed method achieves a good performance in this field, too.

The article's organization is as follows: In Sect. 2, some related works in the literature are reviewed. Section 3 presents the main strategy of the proposed method, WBC algorithm is presented. Section 4 describes the proposed algorithm in detail, and the experiment results are displayed in Sect. 5. At last, the paper's conclusion is presented in Sect. 6.

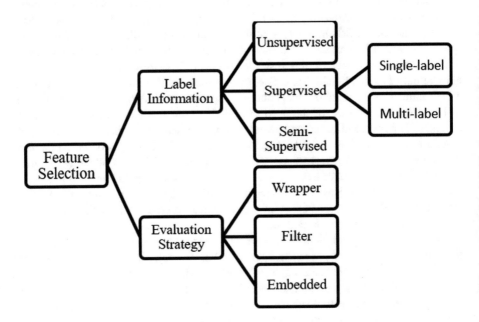

Fig. 1. The categorization of feature selection methods

2   Related Works

In this section, we discuss the multi-label feature selection methods used to evaluate the performance of the proposed algorithm. MLACO was presented by Paniri et al. in 2020 [20]. MLACO algorithm combines correlation with class labels and redundancy of features based on Ant Colony Optimization algorithm (ACO) for multi-label feature selection. This method uses a combination of supervised and unsupervised functions to evaluate the features. The cosine similarity between the features and labels sets the initial values of the pheromones. Also, two heuristic functions are used to update values in the MLACO algorithm.

The LRFS method [32] is a new method for selecting a multi-label feature that uses a new correlation criterion using the redundancy of labels in the feature selection process. In this method, labels are first divided into two categories: independent and dependent labels. In this new criterion, the conditional mutual information between the candidate features and each label is computed under the condition of other labels. Finally, this new criterion aggregated with the redundancy of features.

The Cluster-Pareto method [17] uses the concept of Pareto dominance and clustering to solve the problem of multi-label feature selection. First, the symmetric uncertainty of each feature with each label is computed. Then the features are placed in a multi-objective space based on the symmetric uncertainty values and are evaluated based on the Pareto dominance concept. The optimal feature set is the Pareto optimal front in this method. The hierarchical clustering method and the full link communication mode are applied to rank the features in Pareto optimal front.

The MCLS method [16] is an algorithm based on the Manifold structure and the Laplacian score, consisting of three basic steps. The first step is to convert discrete (binary) labels to continuous values based on Manifold learning. Binary labels are converted to constant values. In the second step, three similarity criteria were used based on the correlation of labels and local data. The first criterion is the constraint score method, which uses similarities between continuous labels as constraints for two instances. Finally, a combination of the three criteria states that a final score is assigned to each feature.

This method presented by Zhang et al. [29] uses an embedded approach for multi-label feature selection that calculates the local correlation between features and labels. The MDFS algorithm first converts the main feature space into a low-dimensional embedded space based on the Manifold structure. In this new space, if two features are very similar, it can be concluded that the features are also close to each other in the main space. The correlation of features in the embedded space is used locally in the feature selection process. This criterion considers redundancy between features. Next, the correlation between the features and the labels is obtained using the Manifold setting. Finally, these two criteria are used in the feature selection process.

One of the approaches used in binary transformation methods to reduce the problem's computational complexity and single-label problems is the Pruned Problem Transformation (PPT) method, which converts only labels that have occurred above a threshold. For example, a label that occurs only once in a data set with a thousand instructional samples does not change. One of the most popular multi-label feature selection methods that use this approach is the PPT-Relief method [22]. In this method, multi-label data is first subdivided into multiple single-label data by the number of labels. Then, labels that have occurred less than six times are pruned or deleted.

The BMFS method was proposed by Hashemi et al. [11]. For the first time, they have modeled the multi-label feature selection problem into a bipartite graph. The first part is the features, and the other is class labels. Generally, a bipartite graph is a model in which each node on the left side of the graph can connect to all nodes on the right side. There is no connection between the nodes in each section. BMFS method uses a two-step approach. In the first step, based on the bipartite graph model, one attribute is selected for each label to maximize the properties associated with all labels. This operation is performed in bipartite graphs by graph matching algorithms. They used the Hungarian algorithm, a famous graph algorithm method. The second criterion, called weighted correlation distance, calculates the distance between features in the label space and is based on the weight of the labels.

The MGFS method [5] was presented by Hashemi et al. in 2020, which models the multi-label feature selection problem into a complete graph. In this graph, the nodes represent the features, and each edge refers to the Euclidean distance between two features. This distance is based on the correlation between the features and the labels. To calculate these values, first, the correlation distance matrix is calculated. Each layer of this matrix equals the correlation distance between the two features in the row and column. Next, the Euclidean distance between features is calculated based on the previously obtained correlation distance, representing the edges between nodes. Finally, executing the PageRank algorithm on the obtained graph ranks the features.

The MFS-MCDM [6] method is a multi-label feature selection method that performs the feature selection process as a multi-criteria decision-making (MCDM) process. In general, MCDM means making decisions based on several different criteria from several candidates. In feature selection, features are our candidates for selection, which are evaluated based on several criteria: the labels.

3   Weighted Borda Count Method

Voting is a famous collective decision-making procedure and preference fusion commonly used in multi-agent systems, political elections, and more [33]. It is also used in machine learning applications when dealing with rank aggregation problems. WBC is a common voting method that evaluates alternatives based on a weighted-counting strategy. Based on WBC, the votes are getting more weight (importance) from the first vote to the last. This weight is multiplied by the number of votes in each stage to achieve a score for each alternative [19].

To better understand the algorithm, we use a numerical example in the field of feature selection. Suppose that we have five features in a dataset evaluated based on five feature selection methods. Then the features are ranked for each method and placed in Table 1, where the rows of the matrix represent the priorities.

Table 1 shows the priority matrix based on five voters. Each row represents the stage of priority of each voter. For example, in the first row of this table that shows the voter's first priority, F1 is repeated three times, which means that feature 1 is the first priority

Table 1. The example matrix for election.

Weights	FS-1	FS-2	FS-3	FS-4	FS-5
5	F1	F5	F1	F4	F1
4	F5	F3	F5	F3	F3
3	F3	F1	F4	F1	F5
2	F4	F2	F3	F5	F4
1	F2	F4	F2	F2	F2

by three voters. Based on WBC, the priorities are assigned by a weight equal to the number of alternatives. The second priority alternatives are getting a weight equal to the number of alternatives minus one. Thus, each row is assigned a weight one less than the weight of the previous row. In WBC, the number of occurrences of each alternative in each row is multiplied by the weight of that row. Finally, these values are added together to represent the importance of each alternative (feature). Thus, the calculation for this example is as follows:

$$F1 = (5 \times 3) + (2 \times 3) = 21, \quad F2 = (1 \times 2) + (4 \times 1) = 6$$
$$F3 = (3 \times 4) + (1 \times 3) + (1 \times 2) = 17$$
$$F4 = (1 \times 5) + (1 \times 3) + (2 \times 2) + (1 \times 1) = 13$$
$$F5 = (1 \times 5) + (2 \times 4) + (1 \times 3) + (1 \times 2) = 18$$

The final ranking of the features will be $\{F1, F5, F3, F4, F2\}$ where the features are ordered in descending order based on the obtained values. The WBC algorithm steps are shown in Algorithm 1.

4   Proposed Method

This section describes the proposed algorithm in full detail. An ensemble strategy is utilized in the proposed method by using three multi-label feature selection rankers. The resulting ranks by each feature selection ranker are stored in a matrix. This rank matrix is delivered to the WBC algorithm to deliver the final ranking of the features. Figure 2 demonstrates a graphical summary of the proposed method. Algorithm 2 also shows the step-by-step algorithm.

Single-label data and multi-label data are known as two machine learning categories, and their structures are shown in Figs. 3 and 4. Each data sample belongs to more than one class label in multi-label datasets. A multi-label data sample is presented by a feature vector as $X_i = (x_{i1}, x_{i2}, \ldots\ldots, x_{iM})$ and a binary label vector as $Y_i = (y_{i1}, y_{i2}, \ldots\ldots, y_{iL})$. The number of features and labels are shown by M and L parameters, respectively.

As can be seen from Fig. 2, the ensemble algorithm can be viewed as a voting system. Feature selection methods (Ridge Regression, Maximum Correlation with class

---

Algorithm 1: Weighted Borda Count (WBC) method

Input: d × h rank matrix R
Output: Alternatives ranking
Begin
R* = ∅
for *i=1 to d* **do**
    for *j=1 to h* **do**
        temp = R (i, j)
        R* (temp) + = d − i
    end
end
Rank the alternatives based on **R*** in descend order
End

---

labels, Minimum correlation among features) are assumed to be voters, and the features are the candidates. Each voter submits its vote (feature ranking) to the system (WBC algorithm), and votes are aggregated according to multiple strategies.

We will walk through this method step by step according to Algorithm 1. In the first step, we introduce an empty matrix A_R in which each column corresponds to the final ranking based on one of the rankers. We execute the three multi-label feature selection rankers in the next three steps. We have used the Pearson correlation coefficient as our first ranker. The following equation is conducted to capture the correlation between each feature (X_i) and the class labels (Y_i):

$$Corr(X_i, Y_i) = \frac{\sum_{i=1}^{n}(X_i - \overline{X})(Y_i - \overline{Y})}{\left(\sqrt{(X_i - \overline{X})^2}\right)\left(\sqrt{(Y_i - \overline{Y})^2}\right)} \tag{1}$$

Fig. 2. Summary of the proposed method

X_1	X_1	$\cdots$	X_M	Y
X_{11}	X_{12}	$\cdots$	X_{1M}	0
X_{21}	X_{22}	$\cdots$	X_{2M}	1
$\vdots$	$\vdots$	$\ddots$	$\vdots$	$\vdots$
X_{N1}	X_{N2}	$\cdots$	X_{NM}	1

Fig. 3. Single-label data structure

X_1	X_1	$\cdots$	X_M	Y_1	Y_2	$\cdots$	Y_L
X_{11}	X_{12}	$\cdots$	X_{1M}	0	1	$\cdots$	0
X_{21}	X_{22}	$\cdots$	X_{2M}	1	0	$\cdots$	0
$\vdots$	$\vdots$	$\ddots$	$\vdots$	$\vdots$	$\vdots$	$\ddots$	$\vdots$
X_{N1}	X_{N2}	$\cdots$	X_{NM}	0	1	$\cdots$	1

Fig. 4. Multi-label data structure

By calculating the correlation between all features and labels, the correlation matrix is formed as follows:

$$Correlation = \begin{bmatrix} Corr(X_1, Y_i) & Corr(X_1, Y_2) & \ldots & Corr(X_1, Y_L) \\ Corr(X_2, Y_i) & Corr(X_2, Y_2) & \ldots & Corr(X_2, Y_L) \\ \vdots & \vdots & \ddots & \vdots \\ Corr(X_M, Y_i) & Corr(X_M, Y_2) & \ldots & Corr(X_M, Y_L) \end{bmatrix} \quad (2)$$

Since we are dealing with multi-label data, the maximum value of each column of the *Correlation* matrix is considered a representative value as follows:

$$P1 = \begin{bmatrix} max(Correlation(1, :)) \\ max(Correlation(2, :)) \\ \vdots \\ max(Correlation(M, :)) \end{bmatrix} \quad (3)$$

Since in feature selection, the goal is to remove irrelevant and redundant features, we have considered a redundancy metric too. The Pearson correlation coefficient is utilized to capture the redundancy between features. Thus, based on Eq. 1, the redundancy matrix is constructed as follows:

$$redundancy = \begin{bmatrix} Corr(X_1, X_1) & Corr(X_1, X_2) & \dots & Corr(X_1, X_M) \\ Corr(X_1, X_1) & Corr(X_1, X_2) & \dots & Corr(X_1, X_M) \\ \vdots & \vdots & \ddots & \vdots \\ Corr(X_M, X_1) & Corr(X_M, X_2) & \dots & Corr(X_M, X_M) \end{bmatrix} \tag{4}$$

Algorithm 2: Proposed method

Input: N × M Feature matrix X, N × 1 label vector Y
Output: feature ranking
Begin
A_R=∅
R1=Feature ranking based on maximum Pearson correlation coefficient (relevancy)
R2=Feature ranking based on maximum Ridge regression
R3=Feature ranking based on minimum Pearson correlation coefficient (redundancy)
A_R=[R,R2,R3]
Perform algorithm (1) to conduct the rank aggregation
The final feature ranking is the output of algorithm 1, and the user can select the desired
   number
End

The minimum value of each column of the redundancy matrix is considered as the redundancy value as follows:

$$P2 = \begin{bmatrix} min((redundancy(1,:)) \\ min(redundancy(2,:)) \\ \vdots \\ min(redundancy(M,:)) \end{bmatrix} \tag{5}$$

Due to the importance of relevancy in feature selection, the Ridge Regression is used to capture the relevancy between the features and class labels to evaluate the feature's relevance in another aspect. The RR matrix is computed as follows:

$$RR = \left(XX^T + \lambda I\right)^{-1} = \begin{bmatrix} W(X_1, Y_1) & W(X_1, Y_2) & \dots & W(X_1, Y_L) \\ W(X_2, Y_1) & W(X_2, Y_2) & \dots & W(X_2, Y_L) \\ \vdots & \vdots & \ddots & \vdots \\ W(X_M, Y_1) & W(X_M, Y_2) & \dots & W(X_M, Y_L) \end{bmatrix} \tag{6}$$

where $W(X_i, Y_j)$ shows the relevancy degree of feature i and label j. The regularization parameter (λ) is set to 10 in this algorithm based on experimenting with different values. Also, the maximum value of each column of the RR matrix is considered as the relevancy value as follows:

$$P3 = \begin{bmatrix} max((RR(1,:)) \\ max(RR(2,:)) \\ \vdots \\ max(RR(M,:)) \end{bmatrix} \tag{7}$$

By sorting the features based on vectors P1 and P3 in descending order and P2 in ascending order, vectors R1, R2, and R3 are obtained. In step 5, the A_R matrix is constructed based on three ranking vectors. A_R matrix is delivered to the algorithm (2) to perform the aggregation. The output of algorithm (2) is the final ranking of the features. Here the user can select as many features as they want.

5   Experimental Studies

In this section, we will introduce the datasets and experimental conditions. Also, the results of experimenting with the proposed algorithm against competitive methods will be presented.

5.1   Datasets

In the experiments of this empirical study, six multi-label text datasets are used. A brief description of the datasets, including the number of samples, features, and labels, is presented in Table 2.

Table 2. Datasets characteristics

Num	Dataset	Samples	Features	Labels
1	Chess	1675	585	227
2	Philosophy	3971	842	233
3	Chemistry	6961	715	175
5	Arts	7484	23,150	26
5	Bibtex	7395	1836	159
6	Cs	9270	635	274

5.2   Multi-label Evaluation Measures

This paper used two classification-based criteria: accuracy and hamming loss (HL) [18].
    Suppose h is a classifier, $T = \{(x_i, Y_i) : i = 1, \ldots, n\}$ is a set of test data and $L = \{l_j : j = 1, \ldots, q\}$ is a set of all labels in the dataset, while Y_i is a subset of L $(Y_i \subseteq L)$. The necessary explanations for each of the evaluation criteria are given below.

– Accuracy

One of the most common evaluation methods is classification algorithms [21]. Accuracy is the ratio of documents that the classifier has correctly predicted. The higher the accuracy, the better because it is a measure of profit.

$$Accuracy(h, T) = \frac{1}{n} \sum_{i=1}^{n} \frac{|Y_i \Delta Z_i|}{|L|} \tag{8}$$

The labels are predicted by the classifier shown as Z_i based on sample x_i.

– Hamming Loss

In multi-label classification, the HL is related to the Hamming distance between y_true and y$_{predicted}$. If y$_{predicted}$ does not precisely match the actual set of labels, HL does not deem the entire tags of a given sample to be false but only penalizes the individual labels [28].

$$Hamming\_loss(h, T) = \frac{1}{n} \sum_{i=1}^{n} \frac{|Y_i \Delta Z_i|}{|L|} \tag{9}$$

where Δ is the symmetric difference between two sets.

5.3   Experimental Results

Nine multi-label feature selection algorithms are provided for comparison with the proposed algorithm in these experiments. These algorithms are MGFS [5], BMFS [11], Cluster-Pareto [17], MDFS [29], LRFS [32], MFS-MCDM [6], MCLS [16], MLACO [20], and PPT-ReliefF [22]. For all simulations, the parameter's value for all algorithms is set according to the recommendation of the following article. We used the Multi-Label K-Nearest Neighbor classifier (ML-KNN) to assess the categorization of our proposed algorithm [30]. The k parameter utilized in the KNN classifier is set to 10. To assess the performance of the multi-label feature selection algorithms, different number of features are selected in the range {10, 20, 30, 40, 50, 60, 70, 80, 90, 100}. All the experiments are conducted using MATLAB R2018a and a windows server 2013-64 bit machine with 64 GB Ram and Intel (R) Xeon (R) Gold 6254 CPU with 16 3.10 GHz processors. The dataset is randomly partitioned into two training and test sets with no subscription for all experiments. For this matter, 70% of the samples are randomly chosen for the training set and the other 30% for the test set. It should also be noted that the announced results are based on an average of 30 independent runs for each algorithm. Figures 5 and 6 show the obtained results based on two mentioned multi-label evaluation metrics in Sect. 5.2. The number of features selected in mentioned figures is shown by the horizontal axis, and the vertical axis shows the comparison criteria. Algorithms are first evaluated based on the top 10 features of the ranking system. This process then continues for the 20, 30, ..., and 100 top features.

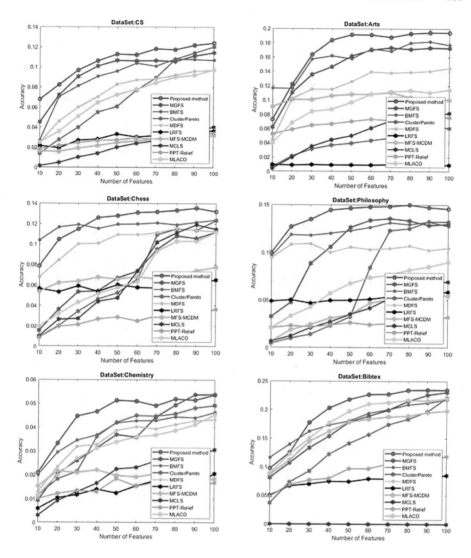

Fig. 5. The results based on classification accuracy

5.4 Discussion

In this paper, we have proposed a new multi-label feature selection based on an ensemble strategy. In fact, the ensemble of feature evaluation metrics is utilized to consider different behaviors of features and achieve an optimal feature ranking system. For the first time, an election-based method is proposed for multi-label feature selection. One of the advantages of this method over other algorithms is that it is not affected by the weaknesses of the algorithms used in this method.

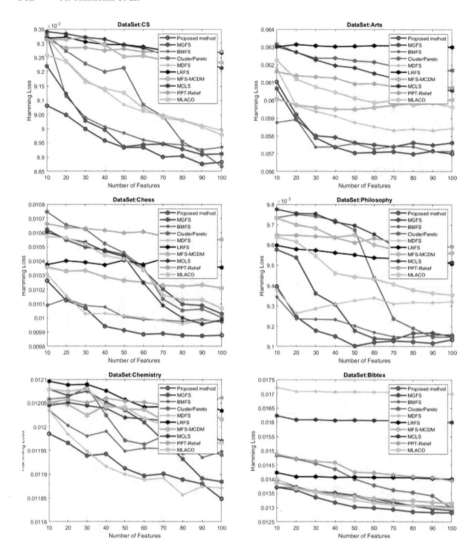

Fig. 6. The results based on hamming loss

Nine new and powerful methods in multi-label feature selection are utilized for this test. These methods use various strategies for feature selection. MFS-MCDM uses a multi-criteria decision-making approach, MDFS and MCLS are manifold-based methods, MGFS and BMFS are two graph-based methods, Cluster-Pareto is based on the Pareto dominance concept and clustering approach, LRFS uses the mutual information, PPT-ReliefF is based on pruned problem transformation and ReliefF ranker, and MLACO is an ACO-based algorithm.

Based on the obtained results and statistical tests, It can be seen that the proposed method is superior to the other methods in terms of multi-label evaluation criteria.

Table 3. The run-time of algorithms

Dataset	Bibtex	CS	Chess	Philosophy	Arts	Chemistry
Proposed methods	2.66	0.33	0.09	0.26	0.11	0.19
MGFS	1.86	0.26	0.13	0.30	0.10	0.17
BMFS	22.62	1.92	1.05	2.53	0.49	0.84
Cluster Pareto	47.04	25.92	5.35	14.74	1.92	11.31
MDFS	511.6	643.8	32.58	210.86	8.39	247.43
LRFS	55.19	37.11	5.51	17.74	2.89	16.41
MFS-MCDM	1.28	0.17	0.12	0.23	0.08	0.12
MCLS	259.4	638.5	51.01	504.13	11.10	250.50
PPT-ReliefF	73.94	28.91	0.85	8.02	14.81	15.11
MLACO	46.67	4.27	3.49	7.62	1.53	3.18

Based on Figs. 5 and 6, we can observe that the proposed method is superior to this all algorithms based on all evaluation measures.

The results indicate that the proposed method has performed well in the multi-label feature selection problem on all datasets.

Also, in Table 3, the execution time of all methods is reported in seconds. Based on these values, it can be seen that the proposed method is not the fastest. However, it is one of the fastest methods since there is no complex computation in the proposed method, all the steps are simple, and no iterative procedure exists.

6   Conclusion

This paper presents an ensemble multi-label feature selection method using rank aggregation. This method aims to reduce the dependency of the algorithm on the number of class labels to be an effective method for all data dimensions. Also, a combination of redundancy-based and relevancy-based is used to achieve the optimal ranking of the features. In this approach, we first implement multi-label rankers on the data so that each algorithm presents its ranking. We put the obtained ranks in a whole matrix. Multi-label ranking methods are first applied to the data to obtain a rank vector for each method. Then the vector ranks are entered across a matrix.

The rank obtained for each feature is shown in the row of the matrix. In addition, the columns display the ranking of features for each method. Then we combine the rankings obtained by different methods using the WBC rank aggregation algorithm. At last, we compare the experimental results of the proposed method and the nine competitive algorithms. All chosen datasets are high-dimensional, including many samples, features, and especially class labels, the superiority of this method is clearly shown. The proposed method can be used in several real-world applications like image classification, bioinformatics, and video and text classification.

References

1. Bayati, H., Dowlatshahi, M.B., Hashemi, A.: MSSL: a memetic-based sparse subspace learning algorithm for multi-label classification. Int. J. Mach. Learn. Cybern. (2022). https://doi.org/10.1007/s13042-022-01616-5

2. Dalvand, A., Dowlatshahi, M.B., Hashemi, A.: SGFS: a semi-supervised graph-based feature selection algorithm based on the PageRank algorithm. In: 2022 27th International Computer Conference, Computer Society of Iran (CSICC), pp. 1–6. IEEE (2022)

3. Hancer, E., Xue, B., Zhang, M.: A survey on feature selection approaches for clustering. Artif. Intell. Rev. **53**(6), 4519–4545 (2020). https://doi.org/10.1007/s10462-019-09800-w

4. Hashemi, A., Dowlatshahi, M.B.: An ensemble of feature selection algorithms using OWA operator. In: 2022 9th Iranian Joint Congress on Fuzzy and Intelligent Systems (CFIS), pp 1–6. IEEE (2022)

5. Hashemi, A., Dowlatshahi, M.B., Nezamabadi-pour, H.: MGFS: a multi-label graph-based feature selection algorithm via PageRank centrality. Expert Syst. Appl. **142**, 113024 (2020). https://doi.org/10.1016/j.eswa.2019.113024

6. Hashemi, A., Dowlatshahi, M.B., Nezamabadi-pour, H.: MFS-MCDM: multi-label feature selection using multi-criteria decision making. Knowl. Based Syst. **206**, 106365 (2020). https://doi.org/10.1016/j.knosys.2020.106365

7. Hashemi, A., Dowlatshahi, M.B., Nezamabadi-pour, H.: An efficient Pareto-based feature selection algorithm for multi-label classification. Inf. Sci. (NY) **51**, 428–447 (2021). https://doi.org/10.1016/j.ins.2021.09.052

8. Hashemi, A., Dowlatshahi, M.B., Nezamabadi-Pour, H.: Minimum redundancy maximum relevance ensemble feature selection: a bi-objective Pareto-based approach. J. Soft Comput. Inf. Technol. (2021)

9. Hashemi, A., Dowlatshahi, M.B., Nezamabadi-pour, H.: VMFS: a VIKOR-based multi-target feature selection. Expert Syst. Appl. **182**, 115224 (2021). https://doi.org/10.1016/j.eswa.2021.115224

10. Hashemi, A., Dowlatshahi, M.B., Nezamabadi-pour, H.: A Pareto-based ensemble of feature selection algorithms. Expert Syst. Appl. **180**, 115130 (2021). https://doi.org/10.1016/j.eswa.2021.115130

11. Hashemi, A., Dowlatshahi, M.B., Nezamabadi-Pour, H.: A bipartite matching-based feature selection for multi-label learning. Int. J. Mach. Learn. Cybern. **12**(2), 459–475 (2020). https://doi.org/10.1007/s13042-020-01180-w

12. Hashemi, A., Dowlatshahi, M.B., Nezamabadi-pour, H.: Ensemble of feature selection algorithms: a multi-criteria decision-making approach. Int. J. Mach. Learn. Cybern. **13**(1), 49–69 (2021). https://doi.org/10.1007/s13042-021-01347-z

13. Hashemi, A., Joodaki, M., Joodaki, N.Z., Dowlatshahi, M.B.: Ant Colony Optimization equipped with an ensemble of heuristics through multi-criteria decision making: a case study in ensemble feature selection. Appl. Soft. Comput. 109046 (2022). https://doi.org/10.1016/j.asoc.2022.109046

14. Almomani, A., et al.: Phishing website detection with semantic features based on machine learning classifiers: a comparative study. Int. J. Semant. Web Inf. Syst. (IJSWIS) **18**(1), 1–24 (2022)

15. Hashemi, A., Pajoohan, M.-R., Dowlatshahi, M.B.: Online streaming feature selection based on Sugeno fuzzy integral. In: 2022 9th Iranian Joint Congress on Fuzzy and Intelligent Systems (CFIS), pp. 1–6. IEEE (2022)

16. Huang, R., Jiang, W., Sun, G.: Manifold-based constraint Laplacian score for multi-label feature selection. Pattern Recognit. Lett. **112**, 346–352 (2018). https://doi.org/10.1016/j.patrec.2018.08.021

17. Kashef, S., Nezamabadi-pour, H.: A label-specific multi-label feature selection algorithm based on the Pareto dominance concept. Pattern Recognit. **88**, 654–667 (2019). https://doi.org/10.1016/j.patcog.2018.12.020
18. Kashef, S., Nezamabadi-pour, H., Nikpour, B.: Multilabel feature selection: a comprehensive review and guiding experiments. Wiley Interdiscip. Rev. Data Min. Knowl. Discov. **8**, e1240 (2018). 10.1002/widm.1240
19. Miri, M., Dowlatshahi, M.B., Hashemi, A.: Evaluation multi label feature selection for text classification using weighted Borda count approach. In: 2022 9th Iranian Joint Congress on Fuzzy and Intelligent Systems (CFIS), pp. 1–6. IEEE (2022)
20. Paniri, M., Dowlatshahi, M.B., Nezamabadi-pour, H.: MLACO: a multi-label feature selection algorithm based on ant colony optimization. Knowl. Based Syst. **192**, 105285 (2020). https://doi.org/10.1016/j.knosys.2019.105285
21. Pan, X., Yamaguchi, S., Kageyama, T., Kamilin, M.H.B.: Machine-learning-based white-hat worm launcher in botnet defense system. Int. J. Softw. Sci. Comput. Intell. (IJSSCI) **14**(1), 1–14 (2022)
22. Reyes, O., Morell, C., Ventura, S.: Scalable extensions of the ReliefF algorithm for weighting and selecting features on the multi-label learning context. Neurocomputing **161** (2015). https://doi.org/10.1016/j.neucom.2015.02.045
23. Sebastiani, F.: Machine learning in automated text categorization. ACM Comput. Surv. (CSUR) **34**, 1–47 (2002)
24. Sheikhpour, R., Sarram, M.A., Gharaghani, S., Chahooki, M.A.Z.: A Survey on semi-supervised feature selection methods. Pattern Recognit. **64**, 141–158 (2017). https://doi.org/10.1016/j.patcog.2016.11.003
25. Gaurav, A., et al.: A comprehensive survey on machine learning approaches for malware detection in IoT-based enterprise information system. Enterp. Inf. Syst. 1–25 (2022)
26. Solorio-Fernández, S., Carrasco-Ochoa, J.A., Martínez-Trinidad, J.F.: A review of unsupervised feature selection methods. Artif. Intell. Rev. **53**(2), 907–948 (2019). https://doi.org/10.1007/s10462-019-09682-y
27. Cvitić, I., Peraković, D., Periša, M., Gupta, B.: Ensemble machine learning approach for classification of IoT devices in smart home. Int. J. Mach. Learn. Cybern. **12**(11), 3179–3202 (2021). https://doi.org/10.1007/s13042-020-01241-0
28. Tsoumakas, G., Katakis, I.: Multi-label classification: an overview. Int. J. Data Warehous. Min. (IJDWM) **3**, 1–13 (2007)
29. Zhang, J., Luo, Z., Li, C., Zhou, C., Li, S.: Manifold regularized discriminative feature selection for multi-label learning. Pattern Recognit. **95**, 136–150 (2019). https://doi.org/10.1016/j.patcog.2019.06.003
30. Zhang, M.L., Zhou, Z.H.: ML-KNN: a lazy learning approach to multi-label learning. Pattern Recognit. **40**, 2038–2048 (2007). https://doi.org/10.1016/j.patcog.2006.12.019
31. Brdesee, H.S., Alsaggaf, W., Aljohani, N., Hassan, S.U.: Predictive model using a machine learning approach for enhancing the retention rate of students at-risk. Int. J. Semant. Web Inf. Syst. (IJSWIS) **18**(1), 1–21 (2022)
32. Zhang, P., Liu, G., Gao, W.: Distinguishing two types of labels for multi-label feature selection. Pattern Recognit. **95**, 72–82 (2019). https://doi.org/10.1016/j.patcog.2019.06.004
33. Zhang, R., Nie, F., Li, X., Wei, X.: Feature selection with multi-view data: a survey. Inf. Fusion **50**, 158–167 (2019). https://doi.org/10.1016/j.inffus.2018.11.019

Fire Neutralizing ROBOT with Night Vision Camera Under IoT Framework

Neelam Yadav[1(✉)], Dinesh Sharma[1], Shashi Jawla[2], and Gaurav Pratap Singh[3]

[1] Chandigarh College of Engineering and Technology, Chandigarh, India
{neelam_cse,dsharma}@ccet.ac.in
[2] Swami Vivekanand Institute of Engineering and Technology, Banur, IKGPTU, Kapurthala, India
[3] EEE, Bharati Vidyapeeth College of Engineering, New Delhi, India

Abstract. Organizations and individuals are increasingly accepting and using the smart home or smart building applications features. Real-time tracking, remote control, intruder protection, gas/fire alarm, and as well as other features are common in smart houses and buildings. In our proposed research work, the major concern is Safety and Security. Firefighting has always been a risky profession, and there have been numerous tragic casualties due to a lack of technological innovation. Furthermore, firefighting methods used today are inadequate and inefficient, as they heavily depend on humans who are susceptible to errors despite their extensive training. A fire extinguishing robot is a technology that may be used for firefighting. In the firefighting domain, the current technology focuses on extinguishing any fire with carbon dioxide if it comes across and then navigating the area to avoid any obstacles on its way. The goal of using IoT to create the robot is to use a camera to see what is going on around it and a Raspberry Pi to monitor the robot's status.

Keywords: Internet of Things · Security · ROBOT · Raspberry-Pi · Firefighting

1 Introduction

Robotics is now used in almost every sector of science and has become increasingly important in everyday life. Recently, there has been a rise in research into robots that can assist humans in their regular lifestyle, such as service robots, office robots, security robots, and so on [2,3,27]. Humans think that robots, particularly security robots, will take a significant part of everyday lives in the future. Robots have primarily been used in the automation of mass production industries, where the same definable activities must be performed repeatedly in the same manner [16–18]. Domestic robots that handle simple activities like vacuuming and grass cutting are also now accessible.

The FIRE FIGHTING robot can move in both forward and backward directions, as well as turn, left and right [7,28,29]. As a result, we can control a robot over a vast distance without having to send a human anywhere near a fire. For fire detection, we employed the flame detector. It is a highly sensitive device that can detect even the

© The Author(s), under exclusive license to Springer Nature Switzerland AG 2023
N. Nedjah et al. (Eds.): ICSPN 2021, LNNS 599, pp. 166–176, 2023.
https://doi.org/10.1007/978-3-031-22018-0_15

smallest of fires. To put out fires, the robot has a water tank and sprinkler built inside it. The major goal of this research is to create an RF-controlled firefighting robot toolkit that can replace traditional firefighting methods [8, 24, 30, 31]. The Raspberry Pi 3 is used as the controller in this paper. This robot can also be used as a fire extinguisher, spraying water on fires when they occur [13, 19]. If a fire is detected, the motor is activated, which subsequently activates the water sprinkler (pump). The RF Encoder connects the switches to the RF transmitter. Each encoder constantly reads the condition of the switching devices, transmitting the data to the Transceiver, which then sends the data. A 12V battery is sent to power this paper. The detection and tracking of mines will be greatly aided by this paper [21–23].

A machine that can replicate human actions and replace humans is developed using these technologies [15, 25, 27, 32]. It is possible to use robots in many situations and for many different purposes, but in today's world, they are most commonly used in dangerous environments (such as bomb detection and deactivation), manufacturing, and places where humans cannot survive (such as fire, space, underwater, and radioactive and hazardous material cleanup and containment). Research into robots' functionality and uses has only steadily grown since the 20th century, despite the concept of autonomous machines dating back to ancient times [14, 20, 26].

2 Literature Survey

See Table 1.

Table 1. Literature survey

S. No.	Author name	Year of publication	Definition	Objective/advantage
1.	Megha Kanwar and Agilandeeswari L [1]	2018	Introduced this concept, based on which the robot can detect a fire, display the location of the fire, and interact with humans in need of assistance. When they need assistance, the robot will follow its orders and start with its water or carbon dioxide pump, due to the type of fire	It has sensors built in that can detect the amount of carbon monoxide, as well as a graph may be examined as well. This will allow it to act in the impacted region in the proper manner
2.	Jayanth Suresh [4]	2017	Proposed this robot with the technology of replacing robots in the place of humans in life-threatening situations	In this, it can detect the fire with help of some sensors and it will extinguish them with carbon dioxide
3.	Ahmed Hassanein, Mohanad Elhawary, Nour Jaber, and Mohammed El-Abd [5]	2015	Stated that the absence of technological improvement made risky and devastating losses in the field of firefighting. They created the robot in this way so that it can find and put out fires in the environment	During its journey, it will maneuver the robot and mitigate the risks

(*continued*)

Table 1. (*continued*)

S. No.	Author name	Year of publication	Definition	Objective/advantage
4.	C. Sevanthi, R. Kaaviya Priya and V. Suganthi [6]	2018	Proposed this model for communicating to various devices or things with each other by sharing usable and relevant data	To maintain security, they are keeping an eye on both public and private spaces. They are using a program to link the Raspberry Pi to Google Drive and then use the image sensor to observe it
5.	Kadam, K., Bidkar, A., Pimpale, V., Doke, D., and Patil, R. [9]	2018	Proposed it finally approaches the fire at a fixed distance and extinguishes it with the help of water or carbon dioxide	The objective of this research is to create a robot firefighter that helps society prevent fire accidents. The robot encourages individuals to seek immediate medical attention for anyone who has been hurt in a fire mishap, assisting in liberation and helping the general public
6.	Wakade, A. [10]	2019	The goal of presenting this firefighter is to safeguard people's lives, property, and natural assets on the planet against fire as well as other emergencies	In this, they designed the Arduino microcontroller robot for monitoring systems for fire detection. The circuit for this contained a camera, sensors, and buzzers. With the help of all of these, the surveillance system can be remotely displayed and all of the information from the fire detection and camera will be taken
7.	Perumal, K. A. P., Ali, M. A., and Yahya, Z. H. [11]	2019	Developed this paper in order to prevent the longer duration of fires initially, the robot will send a message when it is triggered by a smoke detector then it will navigate the robot to the fired area and extinguish the fire with the help of water.	When a fire breaks out, the fire detection system acts as a real-time surveillance system that identifies smoke in the air caused by the fire and takes pictures using a camera mounted within the room
8.	Saravanan P1, Soni Ishwarya V2 and Padmini K M [12]	2015	In this study, we looked at a technique for controlling a mobile firefighting robot's entire kinematics via an Android-powered smartphone and a built-in Bluetooth module	Its objective is to enable successful human-machine interaction by making it simple for the user to manage different aspects of the machine and get the desired results. These may be activated with membrane switches and pens, but in today's environment gestures have become common and seamless

3 Methodology

In every domain of science and technology, robots are used. Robotics has become increasingly important in everyday life. The more study that is done on robots, the more they will be able to assist humans in their daily lives, such as business robots, service robots, security robots, and many others. We assume that, in the future, robots will play a vital role in everyday life, mostly for security reasons. Robots have primarily been used in the automation of mass production businesses, where the same tasks must be completed repeatedly in the same manner. Domestic robots can perform duties like mowing the lawn and vacuuming the house. This paper firefighting robot can move forward and backward in addition to turning right and left. So that we may control the robot from far without the need for a human. The camera was used to observe the fire. We're putting a water tank in this robot to feed water and a sprinkler on it to put out the fire. To preserve human lives, the typical fire extinguisher has been replaced by this firefighting robot.

As seen in Fig. 1 the main component used is Raspberry pi 3 which is a microcontroller from this all the process is done. The raspberry pi 3 is activated by the power supply. Which is given from the power band through the USB cable. The raspberry pi is connected to the 2 motor drivers which are L293D. The L293D 1 is connected to motor1 and motor2. Where motor1 is the left motor of robot and the motor2 is the right motor. These motor drivers are used to control the robot in different directions based on the input given by the user.

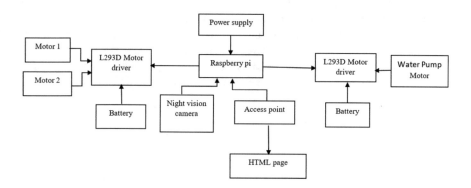

Fig. 1. Block diagram

In which both motors 1 and 2 move in the clockwise motion, the robot moves FORWARD; when both motors move in an anti-clockwise manner, the robot moves BACK; whenever the left motor moves in the circular motion and the right motor is off, the robot moves to the right; whenever the right motor only moves in the circular motion and the left motor is off, the robot shifts to the left. The motor driver 2 is connected to the water pump motor. When the water pump motor is ON state the motor pump the water which is used to extinguish the fire. When the water pump motor is OFF then the motor cannot pump the motor. The 12V power supply is given to both the motor drivers ON the motor drivers. Raspberry pi 3 contains USB ports so we can directly connect the USB camera to it. We can control the robot on the HTML page using the buttons which are present on the page. The access point acts as the interface between the HTML page and the robot.

3.1 Software Requirement

This paper uses languages such as Python, HTML, Putty Software, and Flask Library. Python is an interpreter, a general-purpose language of programming that is translated. The use of considerable indentation in its design philosophy emphasizes code readability. Its language elements and object-oriented approach are aimed at assisting programmers in writing clear, logical code for both small and large-scale projects. Python emphasizes program strength and power and code reusability by supporting modules and packages.

HTML is the abbreviation for Hyper Text Markup Language. This is the language that explains the structure of a Web page and is used to create them. It is made up of a number of elements, each of which instructs the browser on how to display the material. Web applications translate HTML documents received from a web application or personally saved files into multimedia web pages.

Putty is an open device emulator, generated by the computer, and internet file transfer program. It provides SCP, Secure shell, Telnet, rlogin, and raw socket sessions, along with other web services. Putty was originally designed for Microsoft Windows, but it has since been translated to a variety of other operating systems.

Flask is a simple Java-based software development kit. It's designed to be easy to learn and use, with the potential to expand to more sophisticated applications. As among the most extensively used Python custom application platforms, it has received increasing attention. Flask can assist you in organizing your job. The programmer can use any hardware and software he or she wants. A variety of modifications have been made by the organization to create introduce new features straightforwardly.

3.2 Hardware Requirement

Fig. 2. Raspberry pi 3

The Raspberry pi 3 is having Chipset of Broadcom BCM2837, it has a CPU-1.2 GHz quad-core and 64-bit ARM cortex A53, The Raspberry pi 3 uses a USB device having Four USB 2.0 with 480 Mbps data transfer Storage of Micro SD card or via USB-attached storage. It uses wireless LAN -802.11n (Peak transmit/receive throughput of 150 Mbps), Bluetooth 4.1. The Raspberry pi 3 has multimedia and a Memory of 1GB LPDDR2-900 SDRAM (Fig. 2).

- **Motor driver**
  The L293D is a common motor driver or microcontroller Chip that can control DC motors in either way.
- **DC Motor**
  A DC motor is a mechanism that transfers direct current (electrical energy) into mechanical energy in a straightforward way. It is essential for the industry today, as well as for engineers, to understand the working concept of DC motors.
- **USB Camera**
  The QHM495LM is a USB camera the features of this camera made use of this device.
- **Water Pump Motor**
  A pump is a mechanical device that uses a mechanical device to transfer fluids (fluids, gases, and sometimes liquid). Pumps can be classified into three types based on how they transfer fluid: straight lift, displacing, and gravitational pumps. Pumps are powered by a mechanism and utilize energy to perform useful work through moving fluid (usually revolving or rotary working principle, electricity, motors, and wind and solar are all examples of renewable resources used by pumps. They are available in a variety of forms, ranging from small health applications to big industrial pumps.

4   Implementation and Results

As in this paper for implementation above work to capturing images using commands as shown in Fig. 3. This is done by the *fswebcam* command. This command will work whenever we install those particular libraries through which the command will execute.

Using *fswebcam* commands not only capturing of images and also capture the videos for a particular set of frame. It executes the commands in the console window only.

Fig. 3. Capturing a picture using commands

For the live streaming using an IP address. In this, we used a putty application to run our particular server so that no other person going to use it. And also we will view this live streaming if and only if we are connected to the same Wi-Fi to which our raspberry pi is connected.

Fig. 4. Video streaming using IP address

In Fig. 4 we will view live streaming and controlling of the robot can be done by using HTML pages. In this, we are controlling our robot by using left, right, backward and forward keys. And if this observes any fire near to robot then we will move our robot to that place and we pump the water by pressing the key pump which is on the HTML page.

Figure 5 shows a Robot approaching fire by using the commands which are controlled by an HTML page which is having a particular IP address. The buttons which are present on the HTML page are used to control the robot in either direction based on the requirements. We will move our robot to the fired area some distance away from the fire and by pumping water from the tank, it may put out the fire.

Fig. 5. Fire occurred area

When the fire is extinguished by the robot using the commanded pump which is present on the HTML page this is shown in Fig. 6. The water will get pumped continuously until we release the pump key. After putting off the fire we will release the pump key and the fire gets exhausted.

Fig. 6. Putting off the fire by controlling the robot

5   Future Scope

The purpose of this research is to develop a system that can identify fires and respond appropriately without requiring human intervention. It may be able to use mobile agents for activities that require observation of external stimuli and reaction to that stimulus, even when the reaction needs a huge number of mechanical actions, thanks to the growth of sensor networks and the maturity of robotics. This enables us to assign tasks to robots that have been previously performed by humans but were dangerous to their health.

However, many of these components have been studied in many situations, such as mobile agent coordination, obstacle detection, and avoidance approaches, on-the-fly communication between humans and mobile agents, and so on. Putting all of this together into a practical autonomous fire-fighting service will be both exciting and demanding.

6   Conclusion

This research has offered a new perspective on the principles that are employed in this domain, with the goal of promoting technological innovation in order to produce a dependable and efficient result from the numerous instruments. In addition to enabling greater flexibility in control, operation, and expansion, these latest instruments will also be able to incorporate embedded intelligence, thereby increasing the instruments' resilience, and ultimately improving convenience and reliability for customers. Interface design made enormous strides for better human-machine interactions in the 1990s. The implementation of mechatronics provides a practical method of improving life by supplying more flexible and user-friendly features in computing equipment. The major features and functions of the many concepts that could be applied in this sector are presented in detail in this article through various categories. Because this initial study will not be able to address everything within the suggested framework and vision, more research and development activities will be required to completely implement the proposed framework through a collaborative effort of multiple entities. In a simulated house fire, an autonomous robot successfully executes the duties of a firefighter. Benefits from this technology since the cost of triggering other types of fire extinguishers may outweigh the cost of a robot in situations when precise fire control approaches could damage product stock.

References

1. Kanwar, M., Agilandeeswari, L.: IOT based fire fighting robot. In: 2018 7th International Conference on Reliability, Infocom Technologies and Optimization (Trends and Future Directions) (ICRITO) (2018). https://doi.org/10.1109/icrito.2018.8748619
2. Tewari, A., et al.: Secure timestamp-based mutual authentication protocol for IoT devices using RFID tags. Int. J. Semant. Web Inf. Syst. (IJSWIS) **16**(3), 20–34 (2020)
3. Elmisery, A.M., et al.: Cognitive privacy middleware for deep learning mashup in environmental IoT. IEEE Access **6**, 8029–8041 (2017)
4. Suresh, J.: Fire-fighting robot. In: 2017 International Conference on Computational Intelligence in Data Science (ICCIDS) (2017). https://doi.org/10.1109/iccids.2017.8272649
5. Hassanein, A., Elhawary, M., Jaber, N., El-Abd, M.: An autonomous firefighting robot. In: 2015 International Conference on Advanced Robotics (ICAR) (2015). https://doi.org/10.1109/icar.2015.7251507
6. Sevanthi, C., Kaaviya Priya, R.P., Suganthi, V.: A smart fire extinguisher using fire fighting robot. In: 2018 Asian J. Appl. Sci. Technol. (AJAST) (2018). Online ISSN: 2456-883X. Website: https://www.ajast.net
7. Wang, H., et al.: Visual saliency guided complex image retrieval. Pattern Recognit. Lett. **130**, 64–72 (2020)
8. Afify, M., Loey, M., Elsawy, A.: A robust intelligent system for detecting tomato crop diseases using deep learning. Int. J. Softw. Sci. Comput. Intell. (IJSSCI) **14**(1), 1–21 (2022)

9. Kadam, K., Bidkar, A., Pimpale, V., Doke, D., Patil, R.: Fire fighting robot. Int. J. Eng. Comput. Sci. **7**(01), 23383–23485 (2018). Retrieved from: https://www.ijecs.in/index.php/ ijecs/article/view/3939

10. Wakade, A.: Fire fighting robot using Arduino. Int. J. Res. Appl. Sci. Eng. Technol. **7**(5), 4100–4103. https://doi.org/10.22214/ijraset.2019.5693

11. Perumal, K.A.P., Ali, M.A., Yahya, Z.H.: Fire fighter robot with night vision camera. In: 2019 IEEE 15th International Colloquium on Signal Processing & Its Applications (CSPA) (2019). https://doi.org/10.1109/cspa.2019.8696077

12. Saravanan, P., Soni Ishwarya, V., Padmini, K.M.: Android controlled integrated semi-autonomous fire fighting mobile robot. Int. J. Innov. Sci. Eng. Technol. (IJISET) **2**(8) (2015). ISSN: 2348-7968

13. Zhang, Z., Jing, J., Wang, X., Choo, K.-K.R., Gupta, B.B.: A crowdsourcing method for online social networks security assessment based on human-centric computing. Hum.-centric Comput. Inf. Sci. **10**(1), 1–19 (2020). https://doi.org/10.1186/s13673-020-00230-0

14. Alsulami, M.H.: Government Services Bus (GSB): opportunity to improve the quality of data entry. Int. J. Semant. Web Inf. Syst. (IJSWIS) **17**(3), 35–50 (2021)

15. Srivastava, A.M., Rotte, P.A., Jain, A., Prakash, S.: Handling data scarcity through data aug-mentation in training of deep neural networks for 3D data processing. Int. J. Semant. Web Inf. Syst. (IJSWIS) **18**(1), 1–16 (2022)

16. Sundar, S.G.F.A.S., Shanmugasundaram, M.: Manually controlled enhanced wireless intel-ligent fire fighting robot. ARPN J. Eng. Appl. Sci. (2017). ISSN: 1819-6608

17. Rambabu, K., Siriki, S., Chupernechitha, D., Pooja, Ch.: Monitoring and controlling of fire fighting robot using IoT. Int. J. Eng. Technol. Sci. Res. (IJETSR) (2018). ISSN: 2394-3386

18. Nagesh, M.S., Deepika, T.V., Michahial, S., Sivakumar, M.: Fire extinguishing robot. IJAR-CCE **5**(12), 200–202 (2016). https://doi.org/10.17148/ijarcce.2016.51244

19. Din, S., et al.: Service orchestration of optimizing continuous features in industrial surveil-lance using big data based fog-enabled internet of things. IEEE Access **6**, 21582–21591 (2018)

20. Gupta, B.B., Yadav, K., Razzak, I., Psannis, K., Castiglione, A., Chang, X.: A novel app-roach for phishing URLs detection using lexical based machine learning in a real-time envi-ronment. Comput. Commun. **175**, 47–57 (2021)

21. Nikhare, A., Shrikhande, S., Selote, C., Khalale, G.: Development of Autonomous Fire Fighting Robot (2022). https://doi.org/10.31224/2375

22. Rane, T., Gupta, N., Shinde, J., Kharate, S., Mhadse, P.: Design & implementation of fire fighting robot using wireless camera. Int. J. Sci. Res. Publ. (IJSRP) **10**(3) (2020). https://doi. org/10.29322/ijsrp.10.03.2020.p99110

23. John, A.J., Ashik, K., Vishnu, K.A., Fahmi, P., Henna, P.: Automated fire extinguishing robotic vehicle. J. Emerg. Technol. Innov. Res. (JETIR) (2016). https://www.jetir.org. ISSN: 2349-5162

24. Zou, L., Sun, J., Gao, M., Wan, W., Gupta, B.B.: A novel coverless information hiding method based on the average pixel value of the sub-images. Multimed. Tools Appl. **78**(7), 7965–7980 (2018). https://doi.org/10.1007/s11042-018-6444-0

25. Sharma, K., Anand, D., Mishra, K.K., Harit, S.: Progressive study and investigation of machine learning techniques to enhance the efficiency and effectiveness of Industry 4.0. Int. J. Softw. Sci. Comput. Intell. (IJSSCI) **14**(1), 1–14 (2022)

26. Patil, R.M., Srinivas, R., Rohith, Y., Vinay, N.R., Pratiba, D.: IOT enabled video surveillance system using Raspberry Pi. In: 2017 2nd International Conference on Computational Sys-tems and Information Technology for Sustainable Solution (CSITSS) (2017). https://doi.org/ 10.1109/csitss.2017.8447877

27. Jaiswal, D.A.: IOT and AI enabled fire detecting moving bot. Int. J. Res. Appl. Sci. Eng. Technol. **6**(4), 134–141 (2018). https://doi.org/10.22214/ijraset.2018.4028

28. Bahrudin, M.S.B., Kassim, R.A., Buniyamin, N.: Development of fire alarm system using Raspberry Pi and Arduino Uno. In: 2013 International Conference on Electrical, Electronics and System Engineering (ICET) (2013)

29. Undug, J., Arabiran, M. P., Frades, J. R., Mazo, J., Teogangco, M.: Fire locator, detector and extinguisher robot with SMS capability. In: 2015 International Conference on Humanoid, Nanotechnology, Information Technology, Communication and Control, Environment and Management (HNICEM) (2015). https://doi.org/10.1109/hnicem.2015.7393197

30. Sivasankari, G.G., Joshi, P.G.: Live video streaming using Raspberry Pi in IOT devices. Int. J. Eng. Res. Technol. (IJERT) (2017). https://www.ijert.org. ISSN: 2278-0181

31. Scavix, Instructables: Raspberry Pi as low-cost HD surveillance camera. Instructables (2017). Retrieved from: https://www.instructables.com/Raspberry-Pi-as-low-cost-HD-surveillance-camera/. 8 Aug 2022

32. Singla, D., Singh, S.K., Dubey, H., Kumar, T.: Evolving requirements of smart healthcare in cloud computing and MIoT. In: International Conference on Smart Systems and Advanced Computing (Syscom-2021) (2021)

Multi-dimensional Hybrid Bayesian Belief Network Based Approach for APT Malware Detection in Various Systems

Amit Sharma[1], Brij B. Gupta[2,3]($\boxtimes$), Awadhesh Kumar Singh[1], and V. K. Saraswat[4]

[1] Department of Computer Engineering, National Institute of Technology Kurukshetra, Kurukshetra 136119, Haryana, India
amitsharma@gov.in

[2] International Center for AI and Cyber Security Research and Innovations, & Department of Computer Science and Information Engineering, Asia University, Taichung 413, Taiwan

[3] Lebanese American University, Beirut 1102, Lebanon
bbgupta@asia.edu.tw

[4] NITI Aayog, New Delhi 110001, India

Abstract. We are living in a digital world where information flows in the form of bits. The world is witnessing rapid transformation in easy of living by utilizing smart devices ranging from Inter of Things (IoT), personal computing to advanced super computers for solving complex problems in domains like medical, transportation, banking and day-to-day living etc. So protecting the digital assets is an important aspect of modern world, which is susceptible to highly advanced Cyber attacks which have wide range of ramifications like financial loss, intellectual property loss and privacy violations etc. The modern Cyber adversaries are evolving with new tactics, techniques and procedures (TTP) for long-time strategic information gathering by neutralizing the victim's security defenses at the perimeter and system level. The sophistication and complexity of the Advanced Persistent Threats (APT) necessitate developing a new range of security solutions with unconventional defense technologies. The APT payloads are equipped with highly evasive manoeuvers to evade modern security solutions which rely on signature and rule-based methodologies. In these scenarios the malware behavioral analysis assists in detecting APT payloads. However it generates huge amount of behavioral logs which require domain experts review to decide the samples nature which is time consuming, resource intensive and less scalable. We need an automation process to mimic human intelligence by analyzing the internal dependencies of the behavioral patterns to solve the classification problem. So we are proposing an unconventional Bayesian Belief Network-based approach to address the APT malware detection problem by extracting unique features over the malware sample's static, dynamic, and event analysis. The proposed Threat Detection Bayesian Belief Network Model (TDBBNM) is a combination of three Bayesian models named Static Analysis Bayesian Belief Network

N. Nedjah et al. (Eds.): ICSPN 2021, LNNS 599, pp. 177–190, 2023.
https://doi.org/10.1007/978-3-031-22018-0_16

(SABBN), Dynamic Analysis Bayesian Belief Network (DABBN) and Event Analysis Bayesian Belief Network (EABBN) for better accuracy and fewer false-positives in malware detection. The system is evaluated over 10,413 (4733 APT payloads + 5680 Benign) samples by extracting unique features and constructing the BBN models using expert knowledge and empirical observations for conditional dependencies and probabilities. The proposed system exhibited 92.62% accuracy with a 0.0538% false-positive rate in detecting the APT malware, which is suitable for APT detection, a domain where reliable detection mechanisms are limited. Our approach is unique in nature and first of it's kind to introduce BBN Networks in APT detection. The BBN Networks are very good at mimicking human reasoning with respect to internal dependent patterns analysis which is very important aspect in malware detection. So we are expecting the BBN approach opens up a new direction in solving complex problem in Cyber security domain.

1   Introduction

The APT adversaries are states, nations or syndicates sponsored and equipped with ample resources in terms of people, money and infrastructure. So the APT attacks are highly targeted with strategic goals like long-run information stealing, disruption or denial of services, and financial attacks for money. The payloads used in these attacks are unique and contain highly evasive [8], stealth and complex functionalities with encrypted and evasive communication methods. The malware analysis plays an essential role in detecting APT payloads. However, the manual analysis is time-consuming and requires domain experts to perform it. The security industry has been trying to automate the process to reduce time and cost in recent days. However, it is hindered by a large number of analysis logs which require domain expert review for malware artefacts identification and behavioural correlation. The Bayesian Belief Network is a probabilistic graph model capable of solving the automated log analysis problem by exploring the conditional dependence of malware behavioural artefacts and events. Modern security companies should adopt new methodologies like BBNs to solve the APT payload detection problem.

1.1   Advanced Persistent Threat

The APT attacks use continuous, clandestine and sophisticated hacking techniques to gain access to a system and remain inside for a prolonged period, with potentially destructive consequences. The APT attack is performed in five stages:

1. The adversary uses various delivery mechanisms like phishing, spam mail with the infected document or exploitable vulnerabilities to gain initial access.
2. The attacker uses sophisticated malware to create a network backdoor with footprint hiding.

3. The adversary tries to get better control and access to the system resources through various privilege escalation techniques.
4. The attacker compromises other networked digital assets by lateral movement.
5. The attacker understands the victim's digital infrastructure and harvests the required information with long endurance.

1.2   Malware Analysis

Malware Analysis is a process of understanding the behaviour and purpose of a suspicious file or URL. It is performed in three ways:

1. `Static Analysis`: The sample is examined for malicious indicators without execution by performing code analysis and metadata extraction methods. It helps in identifying malicious infrastructure, libraries and packed files. The advanced attackers incorporate techniques like obfuscation and packing to delay or minimize the analysis results. Some of the example static analysis tools are *strings utility, ExifTool, pefile and IDA Pro disassembler* etc.
2. `Dynamic Analysis`: The suspected files are executed in controlled environments called sandboxes to observe the runtime behaviour and log the various IoCs like a filesystem, process, network and persistent mechanisms. The dynamic analysis provides quick intelligence on deciding the sample nature as malware or benign. However sometimes the modern adversaries are becoming smart to detect sandbox environments to deceive dynamic analysis. Some of the example tools are *Sysinternals Suite, WireShark and Ollydbg debugger* etc.
3. `Hybrid Analysis`: This method combines static and dynamic analysis techniques to overcome the adversary evasion techniques. A scenario in the hybrid analysis is patching a binary to make it run in a sandbox using static code analysis. This analysis provides more IoCs by simultaneously performing the analysis by sharing information and knowledge.

The manual malware analysis process is resource intensive, time consuming and repetitive. The analysis process is automated to some extent with automated sandboxes like Cuckoo, Joe Sandbox and Hybrid Analysis etc. which provide behavioral insight into the malware in quick time. The automation process has a bottleneck at deciding the sample nature as malware or benign by analyzing the huge amount of logs, where modern Cyber security industry is searching for reliable solutions. The Bayesian Belief Network is a Probabilistic Graph Model to represent a casual probabilistic relationship among random variables [6,9]. The Bayesian networks explore conditional dependence, which indicates the sequence of interrelated events that provides deeper insight into distinguishing malware from a benign sample where independent events are less effective. The Bayesian models are widely accepted in various expert system domains like medical diagnosis [10], solving financial problems [3,7] and machine diagnosis problems [4,13]. So we are planning to use BBN as an unconventional technique to detect APT malware. The significant contributions of the research work are summarized as follows:

- `APT Malware Collection:` The samples are collected by extracting the hashes by parsing major APT attacks related publications over the past decade, searching the hashes at various online malware repositories like Virus-Total, MalwareBazar and VirusShare etc. and downloading them.
- `Feature Identification and Dataset Creation:` The features are identified by extensive empirical research by manual malware analysis and observations of APT payloads and benign samples (Operating System and Application Software files). The Dataset is created by transforming the logs into a feature vector of binary attributes where 1 indicates feature presence and 0 indicates feature absence.
- `TDBBNM Creation:` The model is a majority aggregation hybrid BBN over static, dynamic and event-based BBN's created with respective analysis features, identifying the internal dependencies and assigning the probabilities based on domain expert knowledge and empirical observations.

The rest of this paper is organized as follows. Section 2 describes the related work, Sect. 3 describes the proposed hybrid BBN model, Sect. 4 provides experimental results and Sect. 5 discusses conclusion and future work in the domain.

2   Related Work

The APT malware detection mechanisms are gradually increasing in recent times, with many researchers proposing various methodologies ranging from network traffic analysis, security event association and modern malware analysis combined with various statistical and machine learning methods.

Han et al. [5] proposed ontology-based APT malware detection and clustering using dynamic analysis Application Programming Interface (API) sequences. An APT detection mechanism by creating attack sequences using intrusion kill chain mapping and fuzzing clustering-based correlation on IDS (Intrusion Detection System) alarm logs is proposed by Zhang et al. [12]. The authors used a probability transfer matrix to construct attack scenarios by mining the attack sequence set, useful for APT attack detection.

Baksi et al. [1] proposed a Deception-based hidden Markov model to counter the APT attacks, especially ransomware. The authors used deception methodology combined with the Kidemonas framework to detect and counter ransomware attacks by traffic analysis with HMM models in an automated way at large corporate networks. Niu et al. [11] used web request graph-based HTTP traffic analysis combined with URL similarity and redirection to detect APT malware.

Filiol et al. [2] provided theoretical characterization of Bayesian Network-based malware detection models using soundness, completeness, robustness, complexity properties to assess the detection engine capabilities and limitations. Zimba et al. [14] proposed Bayesian network-based weighted APT attack path modelling in cloud computing. In this research, the authors evaluated each attack path for the occurrence of the APT attack and used an optimized algorithm to find out the shortest attack path. The author evaluates the model based on the wannaCry ransomware attack [15] and characterized the attack time expense as High, Medium and Low of the APT attack.

We are proposing a new direction in APT malware detection by leveraging the conditional dependence of the malware behavioural patterns to improve the detection rate and reduce the false-positive rate. We are constructing a practical Bayesian Belief Network model over important static, dynamic and event-based feature nodes for the APT detection problem. The research is the first practical approach for APT malware detection using BBN models to the best of the author's knowledge. Various researchers used BBN models in the theoretical perspective of APT malware detection, medical diagnosis domains. Some authors relied on traditional machine learning and deep learning mechanisms for APT detection. The majority of the mechanisms ignored the feature dependence characteristics of the malware.

3   Proposed System

The proposed system architecture is depicted in Fig. 1, and it is named as Threat Detection Bayesian Belief Network Model (TDBBNM). It combines different models named Static Analysis BBN, Dynamic Analysis BBN and Event Analysis BBN. The TDBBNM system utilizes the hybrid approach of malware analysis and mimics the human intelligence reasoning in the form of conditional and prior probabilities to reason about the sample class as malware or benign. In the first step, the system analyses the samples in sandboxed automated analysis environment to collect static, dynamic and event analysis based logs. In the next stage, the system creates a dataset by identifying important features and transforming the logs based on features. In a later stage, the system constructs the relevant BBNs of Static, Dynamic and Event Analysis by exploring the relationship between feature nodes and assigning the conditional and prior probabilities based on the human expert knowledge of various domain experts and empirical malware analysis observation. In the last stage, the model will be evaluated

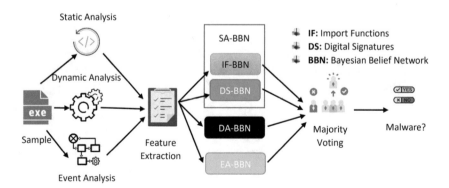

Fig. 1. System architecture

over the Dataset to check the system's efficacy. The TDBBNM is an aggregate model based on majority voting where it decides the sample nature as malware or benign by majority decision of 4 individual BBN networks.

3.1   Static Analysis BBN

This BBN is constructed by parsing the Import Address Table (IAT) of the binary along with the digital signature information. The import functions are grouped based on the activity, and each activity is considered as a feature. The nature of the digital certificate indicates the trustworthiness and genuinity of the sample. So the nature of the digital signature is considered as a feature vector. Two BBN models IFBBN and DSBBN are constructed using static analysis information.

1) IFBBN: The windows Portable Executable (PE) file contains a list of Import Function/API references from system or application-specific Dynamic Link Libraries (DLL). The import functions will provide an overview of sample behaviour. The import functions are mapped to malware activity like process injection, key logging and network connection etc. Each activity from the mapping is considered as one feature. We have identified total 20 activity mappings so the BBN constructed using import functions consists of 21 nodes, of which 20 are feature nodes, and 1 is a decision node. The IFBBN model, along with example conditional probabilities, is depicted in Fig. 2. The node description, along with example API mapping are provided in Sect. 3.1. The full list of API mapping with feature nodes is large in size, so it will be provided as supplementary material.

2) DSBBN: In this BBN, the binary integrity check characteristics will be examined to determine the binary nature as malware or benign. In general, the malware authors won't sign the binaries or provide spoofed/fake signatures for anonymity. So digital signatures are one of the important features in deciding whether the sample is benign or malicious. The DSBBN is illustrated in Fig. 3 along with sample conditional probabilities and contains seven nodes

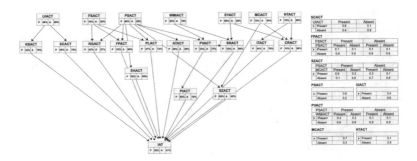

Fig. 2. Import functions BBN and sample conditional probabilities

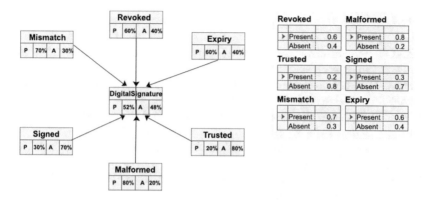

Fig. 3. Digital signatures BBN and sample conditional probabilities

as Mismatch, Revoked, Expiry, Trusted, Malformed, Signed, and Decision Node named DigitalSignature. The respective node descriptions are provided in Sect. 3.1.

IFBBN Nodes Description

- **FSACT** File System Manipulation Behavior. *Example API: DeleteFileW, WriteFile etc.*
- **PSACT** Process Manipulation Behavior. *Example API: CreateProcessW, ShellExecuteW etc.*
- **NTACT** Windows Network Management Activity. *Example API: NetUser-Add, NetUserDel etc.*
- **ITACT** Internet Related Activity. *Example API: HttpOpenRequestW, FtpGetFileW etc.*
- **RGACT** Registry Manipulation Behavior. *Example API: RegSetValueExW, RegCreateKeyW etc.*
- **SRACT** Service Manipulation Behavior. *Example API: CreateServiceW, OpenSCManagerW etc.*
- **SYACT** Collecting & Manipulating System Parameters. *Example. API: Get-SystemInfo, GetSystemDEPPolicy etc.*
- **CRACT** Cryptographic Activity. *Example API: CryptDecryptMessage, CryptGenKey etc.*
- **KBACT** Credential Theft and Key Stroke Capture Activity. *Example API: SetWindowsHookExW, CallNextHookEx etc.*
- **SCACT** Screen Capture Activity. *Example API: GetWindowDC, CreateDCW etc.*
- **ATACT** Sandbox Detection. *Example API: IsDebuggerPresent, OutputDebugStringW etc.*
- **WMIACT** System Manipulation Via Windows Management Instrumentation and Component Object Model interlace. *Example. API: CoCreateInstance, ExecQuery etc.*

- **UIACT** User Interface Manipulation Behavior. *Example API: DrawTextExW, FindWindowW* etc.
- **SZACT** Synchronization Related Activity. *Example API: CreateMutexW, GetTickCount* etc.
- **EHACT** Exception Handling Activity. *Example API: SetUnhandledExceptionFilter, RtlDispatchException* etc.
- **PVACT** Privilege Escalation Activity. *Example API: AdjustTokenPrivileges, LookupPrivilegeValueW* etc.
- **PIACT** Process Injection Behavior. *Example API: CreateRemoteThread, WriteProcessMemory* etc.
- **FPACT** Foot-Print Hiding Related Behavior. *Example. API: ClearEventLogW, OpenEventLogW* etc.
- **PLACT** Payload Dropping Behavior. *Example API: FindResourceExW, SizeofResource* etc.
- **MCACT** Miscellaneous Behavior. *Example API: WriteConsoleW, GetUserNameExW* etc.
- **IAT** Import Functions Decision Variable.

DSBBN Nodes Description

- **Signed** Checks whether binary is digital signed or not
- **Trusted** Checks whether the signature is trustworthy or not
- **Mismatch** Checks for signature mismatching
- **Revoked** Checks whether the signature is revoked or not
- **Expiry** Checks whether the signature is expired or not
- **Malformed** Checks whether the signature is malformed or not
- **Digital Signature** The Decision Node.

3.2   Dynamic Analysis BBN

The dynamic malware analysis provides a wide range of behavioural characteristics ranging from process activity, file system activity, network activity and persistence activity etc. The activity is captured by executing the samples in a sandboxed environment with selective API hooking and logging the parameters values. The log information is processed and API sequences like *OpenProcess, VirtualAllocEx, WriteProcessMemory and CreateRemoteThread* etc. which indicates process injection activity are identified to map to feature nodes. The internal nodes are created by analyzing the internal dependencies of similar category like process related activity, file-system related activity and network related activity etc. This BBN consists of 35 nodes where 29 are the feature nodes, 5 are internal child nodes, and 1 is a decision node named DynamicBehaviourDecisionNode. Figure 4 provides the DABBN structure, and the conditional, prior probabilities for the nodes of the BBN are provided by the correlation of various domain experts' knowledge and manual empirical analysis of advanced APT malware by the authors. Figure 4 also provides DABBN sample conditional probabilities for the nodes. Section 3.2 provides the node descriptions of the DABBN.

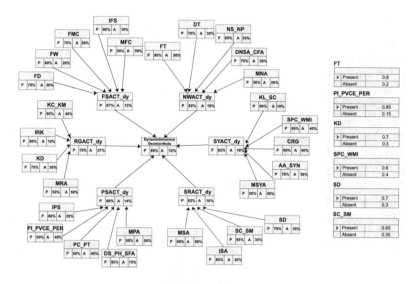

Fig. 4. Dynamic analysis BBN and sample conditional probabilities

DABBN Nodes Description

1. **IPS** Important Process Activity. **Conditions**:
   (a) *payload drop and process creation*
   (b) *Spawn a child process and exit*
   (c) *process spawning with image moves/copy*
2. **PI_PVCE_PER** Process Injection, Enumeration and Privilege Escalation
3. **PC_PT** Process Creation and Termination Behavior
4. **DS_PH_SFA** DLL Side Loading, Process Hooking and System File Abuse
5. **MPA** Miscellaneous Process Activity (Other than 1, 2, 3, 4 nodes)
6. **PSACT_dy** Dynamic Process Related Activity Child Node
7. **KC_KM** Registry Keys Creation and Modification Behavior
8. **IRK** Addition of Prominent Persistence Registry Entries. **Examples**:
   (a) *HKCU Software Microsoft Windows CurrentVersion Run*
   (b) *HKLM SOFTWARE Microsoft WindowsNT CurrentVersion Winlogon*
9. **KD** Registry Keys Deletion Activity
10. **MRA** aMiscellaneous Registry Activity (Other than 7, 8, 9 nodes)
11. **RGACT_dy** Dynamic Registry Related Activity Child Node
12. **FD** File or Folder Deletion Behavior
13. **FW** File Write activity
14. **FMC** File Move or Copy Behavior
15. **IFS** Important File System Activity. **Conditions**:
    (a) *File drop with process spanning*
    (b) *File Drop/Move/Copy with registry entry*
    (c) *File Drop/Move/Copy with service creation*
    (d) *File Drop/Move/Copy with process injection*

16. **MFS** Miscellaneous File System Behavior (Other than the 12, 13, 14, 15 nodes)
17. **FSACT_dy** Dynamic File System Activity Child Node
18. **FT** Network/Internet File Transfer Behavior
19. **DT** Network/Internet Data Transfer Activity
20. **NS_NP** Network Scanning or Profiling for Resource Mapping
21. **DNSA_CFA** DNS and Cloud Feed Abuse for Command and Control
22. **MNA** Miscellaneous Network Activity (Other than 18, 19, 20, 21)
23. **NWACT_dy** Dynamic Network Activity Child Node
24. **KL_SC** Key Logging or Screen Capture Behavior
25. **SPC_WMI** System Information Collection or Windows Management Instrumentation Activity
26. **CRG** Usage of Cryptographic Modules
27. **AA_SYN** Anti-Analysis and Process Synchronization Behavior
28. **MSYA** Miscellaneous System Activity (Other than 24, 25, 26, 27 nodes)
29. **SYACT_dy** Dynamic System Activity child node
30. **SC_SM** Dynamic System Activity child node
31. **SD** Service Deletion Behavior
32. **ISA** Important Service Activity. **Conditions**:
    (a) *Payload drop and Creates a Service*
    (b) *service spawns a process with image deletion/move/copy*
33. **MSA** Miscellaneous Service Activity (Other than 30, 31, 32 nodes)
34. **SRACT_dy** Dynamic Service Activity child node
35. **DecisionNode** Dynamic Analysis Decision Node.

3.3 Event Analysis BBN

An APT group usually utilizes numerous advanced techniques during its campaign for facilitating initial compromise, maintaining command and control centres within adversary target infrastructure, exfiltrating data, and hiding footprints within victim's systems. The analysts understand various attack stages and created reasoning and standards for detecting such attacks. The event-based analysis BBN mainly focuses on the APT attack vector sequencing and the dependencies among the stages of the attack vector. The APT payloads exhibit specific behaviour, so TTP plays a pivotal role in distinguishing between binary as malware or benign. Figure 5 describes the BBN structure along with sample conditional probabilities, consisting 10 nodes where 9 are feature nodes, and 1 is a decision node named TTPDecisionNode. The node descriptions are provided in Sect. 3.3.

EABBN Nodes Description

- **Execution** The adversary is attempting to run malicious code on a local or a remote system. The node observes the events related to similar behavior.
- **Command and Control** This observes the activity of payload communication with adversarial systems.

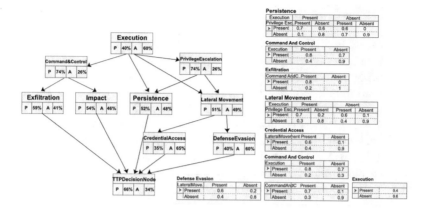

Fig. 5. Event analysis BBN and sample conditional probabilities

- **Exfiltration** The malware after gaining access to target systems, exfiltrates sensitive data from the network using different protocols and channels for bypassing defensive infrastructure. The node observes the events related to exfiltration.
- **Impact** The events in which adversaries are trying to manipulate, interrupt and destroy systems and data is observed on this node.
- **Privilege Escalation** The event related to gain higher-level permissions on a system or network is presented through this node.
- **Persistence** This observes the malware persistence events like Auto-runs entries, account manipulation, event trigger execution, and hijacking of legitimate code etc.
- **Lateral Movement** This node checks for network level spreading via lateral movement techniques like proxy creation and active directory comprise etc.
- **Credential Access** The node observes the events related to gather login credentials such as user names and passwords using keylogging and credential dumping.
- **Defense Evasion** The node observes the techniques utilized by payload to counter perimeter and system defenses.
- **Decision Node** The TTP Decision Node.

The features for all BBN Graphs (IFBBN, DSBBN, DABBN and EABBN) are binary in nature, where 1 represents the activity present in the sample and 0 represents the activity absence in the sample. The conditional probability tables at each node are calculated based on the domain expert knowledge combined with empirical analysis of various binaries and correlating of the results. The full conditional probability distribution for all BBN Graphs is large in size, so it will be provided as supplementary material.

4   Model Evaluation and Result Analysis

The model is evaluated over 10,413 samples, of which 4733 are advanced persistent threat malware of the past decade, and 5680 are benign samples collected from Operating systems and application software. The samples are analyzed and processed, and four datasets are created according to the identified features of individual Bayesian Networks named IF-BBN, DSBBN, DABBN, and EABBN. The four BBN models are evaluated with respect to their individual Dataset. The TDBBNM detection system prediction for a sample is calculated based on the majority voting of the four Individual BBN predictions for the sample. Table 1 provides the performance results of the Individual BBNS and TDBBNM with respect to accuracy and false-positive rates.

Table 1. Performance metrics

S. No.	BBN type	Accuracy	FPR (%)
1	IF-BBN	88.38	1.38
2	DS-BBN	88.39	1.14
3	DA-BBN	91.45	5.09
4	EA-BBN	88.38	1.38
5	TDBBNM	92.62	0.058

Figure 6 provides the performance comparison visualizations of the BBN models with respect to accuracy and false-positive rates. The IF-BBN produced 88.38% accuracy with 1.38% false positive rate over the sample set, DSBBN produced 88.39% accuracy with 1.14% false positive rate over the sample set, DABBN produced 91.45% accuracy with 5.09% false positive rate over the sample set, and EABBN produced 88.38% accuracy with 1.38% false positive rate over the sample set. Finally, the TDBBNM produced 92.62% accuracy with a 0.058% false positive rate over the sample set.

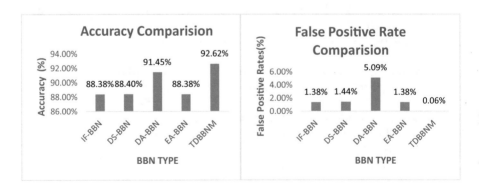

Fig. 6. Performance comparison with respect to accuracy and false positive rates.

In performance comparison, The Dynamic Analysis BBN produced the highest accuracy compared to individual BBNs, as expected because of the wide range of behavioural pattern nodes incorporated in it. However, it has the highest false-positive rate also because of the more overlapping feature nodes between benign and malware behavioural patterns. The IFBBN, DSBBN and EABBN produced almost similar accuracy with small deviations in false-positive rates. The TDBBNM produced the highest accuracy with the lowest false positive rate because of the majority voting combination of individual BBN networks, representing different aspects of human reasoning for malware detection. The TDBBNM model produced 92.68% accuracy in APT malware detection domain, which is acceptable because of the lack of reliable mechanisms in existing security solutions to detect APT malware. We plan to tune the model with respect to feature nodes and dependencies over larger data sets to improve the accuracy. The BBN approach is unique in detecting APT malware as it incorporates human-based reasoning, which is very important to mimic the APT adversary tactics, techniques and procedures. So we are of the opinion that this research will open a new direction in APT malware detection.

5   Conclusion and Future Work

APT attacks are complicated, and the payloads involved in such attacks are complex in nature and embedded with highly evasive techniques. So for the detection of APT attacks, there is a need for advanced mechanisms capable of detecting such malware payloads. This research carries out a novel BBN-based mechanism to detect APT malware. The proposed system TDBBNM consists of the four BBN models IFBBN, DSBBN, DABBN, and EABBN. Each BBN model utilizes the domain expert knowledge to mimic the human analyst-based reasoning with the features extracted from the sample analysis. The TDBBNM predicts the payload class based on the majority voting on the obtained result of each individual BBN model. The TDBBNM has evaluated over 10,413 payload samples (4733 Malware and 5680 Benign), producing 92.62% accuracy with a 0.058 false-positive rate. The authors are of the opinion that the model performance is good in APT detection because of the lack of reliable mechanisms in the domain for the same. In future, we are planning to improve the BBN performance by additions of the feature nodes, conditional dependency tweaking and using a large Dataset. We are also planning to extend the BBN model for solving APT attribution.

References

1. Baksi, R., Upadhyaya, S.: Decepticon: A Hidden Markov Model Approach to Counter Advanced Persistent Threats, pp. 38–54 (2020). https://doi.org/10.1007/978-981-15-3817-9_3
2. Filiol, E., Josse, S.: New trends in security evaluation of Bayesian network-based malware detection models. In: 2012 45th Hawaii International Conference on System Sciences, pp. 5574–5583 (2012). https://doi.org/10.1109/HICSS.2012.450

3. Gupta, B.B., Misra, M., Joshi, R.C.: An ISP level solution to combat DDOS attacks using combined statistical based approach. arXiv preprint arXiv:1203.2400 (2012)
4. Gupta, S., et al.: Detection, avoidance, and attack pattern mechanisms in modern web application vulnerabilities: present and future challenges. Int. J. Cloud Appl. Comput. (IJCAC) **7**(3), 1–43 (2017)
5. Han, W., Xue, J., Wang, Y., Zhang, F., Gao, X.: Aptmalinsight: identify and cognize apt malware based on system call information and ontology knowledge framework. Inf. Sci. **546**, 633–664 (2021). https://www.sciencedirect.com/science/article/pii/S0020025520308628
6. Korb, K.B., Nicholson, A.E.: Bayesian Artificial Intelligence. CRC Press, USA (2011)
7. Parnell, G.S., Bresnick, T.A.: Introduction to Decision Analysis, chap. 1, pp. 1–21. Wiley (2013). https://doi.org/10.1002/9781118515853
8. Sharma, A., et al.: Orchestration of apt malware evasive manoeuvers employed for eluding anti-virus and sandbox defense. Comput. Secur. **115**, 102627 (2022). www.sciencedirect.com/science/article/pii/S0167404822000268
9. Tewari, A., et al.: Secure timestamp-based mutual authentication protocol for IoT devices using RFID tags. Int. J. Semant. Web Inf. Syst. (IJSWIS) **16**(3), 20–34 (2020)
10. Wasyluk, H., Oni'sko, A., Druzdzel, M.: Support of diagnosis of liver disorders based on a causal Bayesian network model. Med. Sci. Monit. **7**(Suppl 1) (2001). https://pubmed.ncbi.nlm.nih.gov/12211748/
11. Niu, W.-N., Xie, J., Zhang, X.-S., Wang, C., Li, X.-Q., Chen, R.-D., Liu, X.-L.: Http-based apt malware infection detection using URL correlation analysis. Secur. Commun. Netw. **2021** (2021). https://doi.org/10.1155/2021/6653386
12. Zhang, R., Huo, Y., Liu, J., Weng, F., Qian, Z.: Constructing apt attack scenarios based on intrusion kill chain and fuzzy clustering. Secur. Commun. Netw. **2017** (2017). https://doi.org/10.1155/2017/7536381
13. Zhao, Y., Xiao, F., Wang, S.: An intelligent chiller fault detection and diagnosis methodology using Bayesian belief network. Energy Build. **57**, 278–288 (2013). https://www.sciencedirect.com/science/article/pii/S0378778812005968
14. Zimba, A., Chen, H., Wang, Z.: Bayesian network based weighted apt attack paths modeling in cloud computing. Future Gener. Comput. Syst. **96**, 525–537 (2019). www.sciencedirect.com/science/article/pii/S0167739X18308768
15. Zimba, A., Wang, Z., Chen, H.: Multi-stage crypto ransomware attacks: a new emerging cyber threat to critical infrastructure and industrial control systems. ICT Express **4**(1), 14–18 (2018). www.sciencedirect.com/science/article/pii/S2405959517303302

Software Quality Attributes Assessment and Prioritization Using Evidential Reasoning (ER) Approach[*]

Ahmad Nabot[1]([✉]), Hamzeh Aljawawdeh[1], Ahmad Al-Qerem[1], and Mohammad Al-Qerem[2]

[1] Zarqa University, Zarqa, Jordan
{anabot,ahmad_qerm,hamzeh.aljawawdeh}@zu.edu.jo
[2] Al-Ahliyya Amman University, Amman, Jordan
M.alqerem@ammanu.edu.jo

Abstract. Requirement assessment and prioritization are the crucial requirement engineering processes in the software development domain. However, software developers face challenges with various types and categories of requirements due to generality, time, and cost. However, Various techniques are proposed for requirement prioritization, such as hierarchical analytic process (AHP), MOSCOW, cumulative voting (CV), bubble sort, and binary search tree (BST). These techniques are still unreliable for prioritizing a considerable number of requirements which takes a long time. Therefore, there is a crucial need for a reliable and consistent approach that helps deal with functional and non-functional requirements assessment and prioritization, such as the Evidential Reasoning (ER) approach. The study findings showed how the requirements assessment and prioritization process improved according to their weights, evaluation grades, and belief degrees. In addition to the decision matrices that lead to high-quality decisions.

Keywords: Requirement prioritization · Architecture assessment · Evidential Reasoning (ER) · Multi-criteria decision making (MCDM)

1 Introduction

Software quality prioritization is a necessary process that allows the developers to validate its requirements in later development stages. Different prioritization techniques with varying abilities of objective, subjective, quantitative, and qualitative results are utilized for such purposes. Some of these techniques follow a step-by-step guideline for designing the system's architecture or identifying the weak points. However, the most accurate way to assess software architecture could be by evaluating and prioritizing the quality attributes of that system. This way of assessment allows the system stakeholders to decide if the requirements meet their needs and identify strengths and weaknesses. This study uses a Multi-Criteria Decision Making (MCDM) technique to obtain information on the system's quality attributes to get the required assessment results. MCDM

---

[*] Supported by Zarqa University.

© The Author(s), under exclusive license to Springer Nature Switzerland AG 2023
N. Nedjah et al. (Eds.): ICSPN 2021, LNNS 599, pp. 191–199, 2023.
https://doi.org/10.1007/978-3-031-22018-0_17

techniques are usually used in business studies and can be implemented in software engineering to assess software requirements and prioritization. The Evidential Reasoning (ER) approach as an MCDM technique was used to accomplish the goal of this study. The ER approach is one MCDM technique consisting of a multilevel assessment structure for assessing and prioritizing software quality attributes based on different levels. In addition, it uses an extended decision matrix to describe software attributes and sub-attributes using belief structure. The study organized as follows: Sect. 2 represents the most popular prioritization techniques in addition to the ER approach. In Sect. 3, the study methodology is given and a case study is explained in section four. Eventually, study is concluded in the last section.

2   Background

The recent immense prevalence and use of software applications have increased the complexity of these applications. Most software developing firms are turning to open-source systems to reduce system complexity and incubate with changing requirements [1]. Software architecture assessment and requirement prioritization are vital, ensuring the system functionality accuracy before building it. Also, the development team can manage the systems' functional and non-functional requirements to fulfill users' needs to reduce the development costs and time [2–4]. Software development methods start with the software feasibility study, requirement collection, and analysis. The requirement prioritization process is a challenging activity in the requirement engineering stage which requires using a reliable prioritization technique [5]. A. Requirement Prioritization Techniques Many prioritization techniques are proposed by researchers in the domain of software development depending on the software size, complexity, and other variables. These techniques allow decision-makers to analyze the software requirements to assign unique symbols to indicate their priority and importance. The need for highly skilled developers to implement these techniques to satisfy the users' expectations is considered challenging. Also, using such methods for requirement prioritization requires extensive communication with the stakeholders [4]. According to [6], requirement prioritization consists of three successive phases as follows:

1. The preparation phase is where the developer structures the requirements depending on the prioritization technique. In addition, the selected team or team leader is provided with all the required information.
2. The execution phase, where decision-makers prioritize the requirements depending on the provided information from stage 1. Then, the criteria for evaluation identified and verified by the development team.
3. The presentation phase presents the selected prioritization technique, including its work principle and the final execution results.

2.1   Requirement Assessment and Prioritization Techniques

1) The Analytic Hierarchy Process (AHP)
AHP is among the most MCDM common techniques utilized for software requirement prioritization by comparing each pair of requirements to determine their priority [7]. The comparison process of the requirements depends on the pairwise

comparison by identifying the number of the requirement pairs using n(n − 1)/2 [3, 8, 9]. The prioritization process starts by building up a multilevel hierarchy of the attributes and their sub-attributes to allow the pairwise comparison for this hierarchy. Then, a scale from 1 (equal importance) to 9 (Extreme difference in importance) specifies the critical importance of enabling the decision-makers to make a decision that consumes time due to redundancies using such a technique [10, 11].

2) Bubblesort

The bubble sort technique is one of the basic techniques for sorting and comparing software requirements. It starts by comparing the first pair of requirements to identify the requirement with more properties than the other and continues until the end of the requirements [12, 13]. This technique work principle is nearly similar to the AHP work principle in pairwise comparisons, which is also considered time-consuming [6].

3) Binary Search Tree (BST)

BST is a tree of nodes with two Childs, where each node is considered a candidate requirement [14]. This technique is one of the requirement prioritization techniques that sort requirements in the tree depending on their importance [5]. The requirements on the left are of low significance and have the lowest priority, while the requirements on the right are essential and have the highest priority [14]. This technique work principle starts by constructing the highest node in the tree and assigning a requirement to that node. Then, create the second node with its requirement to be compared with the highest node in the tree. If the priority is lower than the requirement in the node, it is compared with the left one, and so forth. If it has a higher priority than the requirement in the node, it is compared to the right one, and so on. This process carry on until all requirements are in the tree [15]. As mentioned above, there are many prioritization techniques proposed by researchers in the domain of software engineering. The study selected the most popular prioritization techniques to present them.

4) The Evidential Reasoning (ER) approach

ER is a multicriteria decision-making (MCDM) technique that has a multilevel evaluation decision matrix which build-up based on Dempster-Shafer's (D-S) theory of evidence [16, 17]. Additionally, ER utilises belief structure to deal with uncertainties and missing or incomplete information to ensure the quality of the outcomes in the decision matrix [18–21]. The multilevel matrix contains a set of attributes distributed on different levels in the hierarchy. Each attribute in the hierarchy is evaluated depending on its grades and degrees of belief.

3   Methodology

This section proposes the evidential reasoning (ER) approach for software requirement prioritization. ER helps decision-makers to prioritize software requirements in situations of multiple and conflicting requirements. An extended decision matrix to describe the requirements depending on their belief structure. For instance, the results of the evaluation grades for software are assessed using the grades $H1, H2, H3, ..., Hn$, described as Worst, Poor, Average, Good, Excellent. The distributed assessment results for software requirements are assessed by the software engineer as (Worst, 0%), (Poor, 0%),

(Average, 0%), (Good, 30%), (Excellent, 70%), this assessment means that the assessed requirements are Excellent with 70% of belief degree and Good with 30% of belief degree. The evaluation grades capture the different kinds of uncertainties such as obscurity in spontaneous judgments and provide precise degrees of belief structure [20] due to Dempster-Shafer's theory and its basis of decision theory to aggregate valid belief degrees [15]. As mentioned, the ER technique represents software requirements in a hierarchy for evaluation and prioritization. The requirements in the hierarchy are broken down into sub-attributes in the lower level to be measured. Suppose we have L basic Requirements at the lower level in the hierarchy A_i ($i = 1, \ldots, L$) associated with the general Requirement concept y, K alternatives O_j ($j = 1, \ldots, K$) and N for evaluation grades $H_n(n = 1, \ldots, N)$ for each sub-attribute where $S A(O)$ is given as follows:

$$S\left(A_i\left(o_j\right)\right) = \{(H_n, \beta_n, i(O_i))\} \tag{1}$$

where $n = (1, \ldots, N)$, $i = (1, \ldots, L)$, and $j = (1, \ldots, K)$ $(\beta_n, i(O_j))$ represents the degree of belief in the alternative O_j which can be assessed by the nth grade of the ith sub-attribute [12]. Sub-attribute might have its evaluation grades, which could differ from other Requirements in the hierarchy [19]. ER algorithm can be described by transforming the belief degrees into masses where $m_{(n,i)}$ and $m_{(H,i)}$ can be calculated using underneath equations:

$$m_{n,i} = \omega_i \beta_{n,i} \tag{2}$$

$$m_{H,i} = 1 - \sum_{n-1}^{N} m_{n,i} = 1 - \omega_i \sum_{n=1}^{N} \beta_{n,i} \tag{3}$$

Suppose the weight of ith sub-attribute $m_{(n,i)}$ is given by $\omega = (\omega_1, \omega_2, \omega_3, \ldots, \omega_i)$. So, the probability mass represents the nth evaluation grade H_n of the ith sub-attribute. The residual probability mass $m_{(H,i)}$ is unassigned to any grade after assessing the ith sub-attribute [16].

$$m_{H,i} = \bar{m}_{H,i} + \tilde{m}_{H,i} \tag{4}$$

for $i = 1, \ldots, L$ and $\sum_{i=1}^{L} \omega_i = 1$. Assigns the evaluation grades $H = \{H_1, H_2, \ldots, H_N\}$ to the probability masses and L Requirements are aggregated to generate the combined belief degree for each evaluation grade H_n. The unassigned evaluation grades H_n can be calculated as follows:

$$\bar{m}_{H,i} = 1 - \omega_i \quad and \quad \tilde{m}_{H,i} = \omega_i(1 - \sum_{n=1}^{N} \beta_{n,i}) \tag{5}$$

$\bar{m}_{H,i}$ for calculating the relative importance of the ith sub-attribute and $\tilde{m}_{H,i}$ is used for the incomplete information of the ith sub-attribute. Therefore, $m_{H,I(L)} = \bar{m}_{H,I(L)} + \tilde{m}_{H,I(L)}$, $n = (1, \ldots, N)$ are the combined probability assignments by aggregating all original probability masses using the aggregation of the following ER algorithm [7, 16, 22]:

$$\{H_n\} : m_{n,I(i+1)} = K_{I(i+1)}[m_{n,i+1}m_{n,I(i)} + m_{n,i+1}m_{H,I(i)} + m_{H,i+1}m_{n,I(i)}], n = (1, \ldots, N) \tag{6a}$$

$$\{H\} : m_{H,I(i)} = \tilde{m}_{H,I(i)} + \bar{m}_{H,I(i)} \tag{6b}$$

$$\tilde{m}_{H,I(i+1)} = K_{I(i+1)}[\tilde{m}_{H,I(i)}\tilde{m}_{H,(i+1)} + \bar{m}_{H,(i+1)} + \tilde{m}_{H,I(i)}\bar{m}_{H,(i+1)}\tilde{m}_{H,I(i)}] \tag{6c}$$

$$\bar{m}_{(H,I(i+1))} = K_{(I(i+1))}[\bar{m}_H\bar{m}_{H,I(i),i+1}] \tag{6d}$$

$$K_{I(i+1)} = [1 - \sum_{t=1}^{N}\sum_{i=1,j\neq t}^{N} m_{t,I(i)}m_{j,i+1}]^{-1}, \; where \; i = (1,2,...,L-1) \tag{6e}$$

$$H : \beta_H = \frac{\tilde{m}_{n,I(L)}}{1 - \bar{m}_{H,I(L)}} \tag{7a}$$

$$H_n : \beta_n = \frac{m_{n,I(L)}}{1 - \bar{m}_{H,I(L)}}, n = 1,...,N \tag{7b}$$

where β_n is the belief degree to the L Requirements which can be assessed by H_n and β_H is the residual belief degrees unassigned to any H_n, which proves that $\sum_{n=1}^{N}\beta_n + \beta_H = 1$ [16].

Therefore, the final distribution evaluation for O_j can be produced by aggregating L Requirements as follows:

$$S_{(O_j)} = \{(H_n\beta_n(O_j)), n-1,...,N\} \tag{8}$$

To compute the average of $S(O_j)$ for an individual output of H_n, suppose that H_n is denoted by $u(H_n)$ as follows:

$$u(O_j) = \sum_{n=1}^{N}\beta(O_j)u(H_n) \tag{9}$$

β_n indicates the probability lower limit of the evaluated substitutional O_j to H_n and the upper limit can be computed by $(\beta_n + \beta_H)$ [11,20,24,26]. Also, in case there are uncertainties such as missing or incomplete information; they can be described by max, min, and average score of $S(A^*)$ as follows:

$$u_{max}(O_j) = \sum_{N-1}^{n=1}\beta_n(O_j)u(H_n) + (\beta_N(O_j) + \beta_H(O_j)u(H_N)) \tag{10}$$

$$u_{min}(O_j) = (\beta_1(O_j) + \beta_H(O_j))u(H_1) + \sum_{n=2}^{N}\beta_n(O_j)u(H_n) \tag{11}$$

$$u_{avg}(O_j) = \frac{u_{max(O_j)}u_{min(O_j)}}{2}, \quad where \quad u(H_n + 1) \geq u(H_N) \tag{12}$$

When the assessments of $S(A_i(O_j))$ in the decision matrix are complete, the result of $\beta_H(O_j) = 0$ and $u(S(O_j)) = u_{max(O_j)} = u_{min(O_j)} = u_{avg(O_j)}$. Note that these mathematical equations assess characterization rather than Requirements aggregation. The key difference between ER framework and other MCDM techniques is that ER transforms various types of evaluation information in order to assess both functional and non-functional requirement [23]. In addition, information uncertainties of the assessed requirements are treated using Dempster-Shafer's (D-S) theory of evidence [16,27].

4   Case Study and Results

In this section, the ER technique is utilized to evaluate an online computer shop's non-functional requirements or quality attributes where the functional requirements might be considered later. The significant attributes of such shops are usability, efficiency, security, reliability, and maintainability, which are difficult to be assessed. Therefore, these requirements should be aggregated in the lower level of the hierarchy evaluation, as shown in Fig. 1.

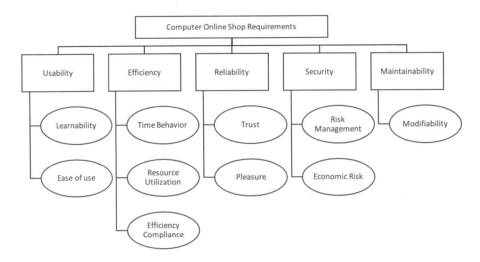

Fig. 1. Computer online store attributes and sub-attributes ER hierarchy.

The degrees of belief for the sub-attributes of the main attributes can be represented as follows:

Time Behavior = (*Average*, 0.4), (*Good*, 0.5)
Resource Utilization = (*Excellent*, 1.0)
Efficiency Compliance = (*Excellent*, 0.4), (*Good*, 0.6).

The assessment of the sub-attributes is completed when the value of the belief degree equals 1 and incomplete if the value less than 1. For instance, the efficiency sub-attributes assessment of time behavior $0.4 + 0.5 = 0.9 < 1$ is incomplete, while resource utilization and efficiency compliance assessment are complete [17]. The number in the brackets denotes the degree of belief to which a sub-attribute evaluated as (Good 0.6) means (Good to a degree of 60%) [7]. The assessment for the efficiency attribute after aggregation can be done by specifying the belief degrees for the sub-attributes in the hierarchy using Eq. 3. For instance, belief degrees for the three sub-attributes are represented as below.

$\beta_{3,1} = 0.4, \quad \beta_{4,1} = 0.5$
$\beta_{4,2} = 1.0$
$\beta_{4,3} = 0.4.$

To calculate the basic probability masses $m(n, i)$ when the three quality attributes are of equal importance $\omega_1 = \omega_2 = \omega_3 = 3$. Then combined probability masses can be combined using Eqs. (6a)–(6e), Suppose $m(n, I(1)) = m(n, 1)$ for $n = 1, \ldots, 5$ to calculate it for time behavior and resource utilization.

First, the combined probability mass for the grade H_n should be generated by aggregating $\left(A_i\left(O_j\right)\right)$; $\tilde{m}_{H,I(2)}$ for the combined probability mass for H due to the possible incompleteness in $S\left(A_i\left(O_j\right)\right), \bar{m}_{H,I(2)}$ for the relative importance of A_i and O_j.

To calculate the combined probability masses, let $m_{n,I(i)}$ ($n = 1, \ldots, 5$), $\tilde{m}_{H,I(i)}$ and $\bar{m}_{H,I(i)}$. Therefore, the values of the first two sub-attributes as follows:

$K_{I(2)} = 1.045, \quad$ Where $m_{H,i} = \tilde{m}_{H,i} + \bar{m}_{H,i}$, and $(i = 1, 2, 3)$, thus
$m_{1,I(2)} = 0, \quad m_{2,I(2)} = 0, \quad m_{3,I(2)} = 0.090, \quad m_{4,I(2)} = 0.4098, \quad m_{5,I(2)} = 0$
$\tilde{m}_{H,I(2)} = 0.0232, \quad \bar{m}_{H,I(2)} = 0.464.$

By combining the results of the first two sub-attributes with efficiency compliance to get it's value:

$K_{I(3)} = 1.126, \quad$ Where $m_{H,I(2)} = 0.4872$
thus, $m_{1,I(3)} = 0, \quad m_{2,I(3)} = 0, \quad m_{3,I(3)} = 0.0675, \quad m_{4,I(3)} = 0.0731, \quad m_{5,I(3)} = 0.1097$
$\bar{m}_{H,I(3)} = 0.3483.$

From (7a) and (7b), the combined degrees of belief can be calculated by:

$\beta_n = \frac{m_{n,I(3)}}{1-\bar{m}_{H,I(3)}} = 0, \quad n = 1, 2$

$\beta_3 = \frac{m_{3,I(3)}}{1-\bar{m}_{H,I(3)}} = \frac{0.0675}{1-0.3483} = 0.1035$

$\beta_4 = \frac{m_{4,I(3)}}{1-\bar{m}_{H,I(3)}} = \frac{0.0731}{1-0.3483} = 0.1121$

$\beta_5 = \frac{m_{5,I(3)}}{1-\bar{m}_{H,I(3)}} = \frac{0.1097}{1-0.3483} = 0.1683$

$\beta_H = \frac{\bar{m}_{H,I(3)}}{1-\bar{m}_{H,I(3)}} = \frac{0.0174}{1-0.3483} = 0.0266$

The assessment for the efficiency by aggregating its sub-attributes is done by (8) as follows:

$S(O_j) = \left\{(H_n, \beta_n(O_j)), \quad n - 1, \ldots, N\right\}$

The final assessment for the efficiency requirement is generated as follows:

$S(efficiency) = S(timebehaviour \oplus resourceutilization \oplus efficiencycompliance)$
$= \{(average, 0.1035), (good, 0.1121)(excellent, 0.1683)\}.$

5 Conclusion

Software requirement prioritization increases the software quality, reduces the time required for the development process, and enhances user experience by selecting requirements that fit their needs. In this study, the ER technique framework has been represented for requirement prioritization. This technique uses non-linear relationships between the aggregated attributes and the main attributes in the hierarchy, increasing the analyzed data quality. ER utilized for solving different problems in different research studies. For instance, Zhang and Deng (2018), and Dong et al. (2019) analyzed fault diagnostic problems in an uncertain environment using the ER technique

[4,5,25]. Akhoundi et al. (2018), used the ER to evaluate potential wastewater reuse options [2,8]. Ng and Law (2020), employed ER to examine user preferences by analyzing affection words in social networking [9]. Tian et al. (2020), integrated the ER approach with probabilistic linguistics to solve multi-criteria decision-making problems by taking into consideration the psychological preferences of the decision-makers. The researchers who used the ER approach for assessing these applications assured that the assessment results are consistent and reliable [13]. Moreover, ER uses improved decision rules depending on the subjective judgments and degrees of belief to confirm the consistency and reliability of the results [1,6]. Therefore, ER utilized for analysing and synthesising the safety of marine system, assessing motorcycle, car assessment, electrical appliance selection, and many more applications [3,14,18,21]. Traditional decision methods might not provide the precise number of assessments that might cause data missing, and the average numbers of the outputs could be provided instead of the accurate number.

Acknowledgements. This research is funded by the Deanship of Research and Graduate Studies at Zarqa University /Jordan.

References

1. Abayomi-Alli, A.A., Misra, S., Akala, M.O., Ikotun, A.M., Ojokoh, B.A., et al.: An ontology-based information extraction system for organic farming. Int. J. Semant. Web Inf. Syst. (IJSWIS) **17**(2), 79–99 (2021)
2. Akhoundi, A., Nazif, S.: Sustainability assessment of wastewater reuse alternatives using the evidential reasoning approach. J. Clean. Prod. **195**, 1350–1376 (2018)
3. Chin, K.S., Yang, J.B., Guo, M., Lam, J.P.K.: An evidential-reasoning-interval-based method for new product design assessment. IEEE Trans. Eng. Manag. **56**(1), 142–156 (2009)
4. Do, P., et al.: Developing a Vietnamese tourism question answering system using knowledge graph and deep learning. Trans. Asian Low-Resource Lang. Inf. Process. **20**(5), 1–18 (2021)
5. Dong, Y., Zhang, J., Li, Z., Hu, Y., Deng, Y.: Combination of evidential sensor reports with distance function and belief entropy in fault diagnosis. Int. J. Comput. Commun. Control **14**(3), 329–343 (2019)
6. Fu, C., Xue, M., Chang, W., Xu, D., Yang, S.: An evidential reasoning approach based on risk attitude and criterion reliability. Knowl.-Based Syst. **199**, 105947 (2020)
7. Huynh, V.N., Nakamori, Y., Ho, T.B., Murai, T.: Multiple-attribute decision making under uncertainty: the evidential reasoning approach revisited. IEEE Trans. Syst. Man Cybern. Part A: Syst. Hum. **36**(4), 804–822 (2006)
8. Lv, L., et al.: An edge-Ai based forecasting approach for improving smart microgrid efficiency. IEEE Trans. Ind. Inform. (2022)
9. Ng, C., Law, K.M.: Investigating consumer preferences on product designs by analyzing opinions from social networks using evidential reasoning. Comput. Ind. Eng. **139**, 106180 (2020)
10. Pashchenko, D.: Fully remote software development due to covid factor: results of industry research (2020). Int. J. Softw. Sci. Comput. Intell. (IJSSCI) **13**(3), 64–70 (2021)
11. Shafer, G.: A Mathematical Theory of Evidence, vol. 42. Princeton University Press (1976)
12. Tian, Z.P., Nie, R.X., Wang, J.Q.: Probabilistic linguistic multi-criteria decision-making based on evidential reasoning and combined ranking methods considering decision-makers' psychological preferences. J. Oper. Res. Soc. **71**(5), 700–717 (2020)

13. Voola, P., Babu, V.: Study of aggregation algorithms for aggregating imprecise software requirements' priorities. Eur. J. Oper. Res. **259**(3), 1191–1199 (2017)
14. Wang, J., Yang, J., Sen, P.: Safety analysis and synthesis using fuzzy sets and evidential reasoning. Reliab. Eng. Syst. Saf. **47**(2), 103–118 (1995)
15. Wang, Y.M., Yang, J.B., Xu, D.L.: Environmental impact assessment using the evidential reasoning approach. Eur. J. Oper. Res. **174**(3), 1885–1913 (2006)
16. Xu, D.L., Yang, J.B.: Intelligent decision system based on the evidential reasoning approach and its applications. J. Telecommun. Inf. Technol. 73–80 (2005)
17. Xu, L., Yang, J.B.: Introduction to Multi-criteria Decision Making and the Evidential Reasoning Approach, vol. 106. Manchester School of Management Manchester (2001)
18. Yang, J., Xu, D.: Knowledge based executive car evaluation using the evidential reasoning approach (1998)
19. Yang, J.B.: Rule and utility based evidential reasoning approach for multiattribute decision analysis under uncertainties. Eur. J. Oper. Res. **131**(1), 31–61 (2001)
20. Yang, J.B., Liu, J., Wang, J., Sii, H.S., Wang, H.W.: Belief rule-base inference methodology using the evidential reasoning approach-RIMER. IEEE Trans. Syst. Man Cybern. Part A: Syst. Hum. **36**(2), 266–285 (2006)
21. Yang, J.B., Singh, M.G.: An evidential reasoning approach for multiple-attribute decision making with uncertainty. IEEE Trans. Syst. Man Cybern. **24**(1), 1–18 (1994)
22. Yang, J.B., Xu, D.L.: Nonlinear information aggregation via evidential reasoning in multi-attribute decision analysis under uncertainty. IEEE Trans. Syst. Man Cybern. Part A: Syst. Hum. **32**(3), 376–393 (2002)
23. Yang, J.B., Xu, D.L.: On the evidential reasoning algorithm for multiple attribute decision analysis under uncertainty. IEEE Trans. Syst. Man Cybern. Part A: Syst. Hum. **32**(3), 289–304 (2002)
24. Yen, J.: Generalizing the Dempster-Schafer theory to fuzzy sets. IEEE Trans. Syst. Man Cybern. **20**(3), 559–570 (1990)
25. Zhang, H., Deng, Y.: Engine fault diagnosis based on sensor data fusion considering information quality and evidence theory. Adv. Mech. Eng. **10**(11), 1687814018809184 (2018)
26. Zhang, X.X., Wang, Y.M., Chen, S.Q., Chen, L.: An evidential reasoning based approach for GDM with uncertain preference ordinals. J. Intell. Fuzzy Syst. **37**(6), 8357–8369 (2019)
27. Zhang, Z.J., Yang, J.B., Xu, D.L.: A hierarchical analysis model for multiobjective decision-making. IFAC Proc. **22**(12), 13–18 (1989)

Analysis of Digital Twin Based Systems for Asset Management on Various Computing Platforms

Yanhong Huang[1], Ali Azadi[2], Akshat Gaurav[3,4(✉)], Alicia García-Holgado[2],
and Jingling Wang[5]

[1] Hubei Key Laboratory of Inland Shipping Technology, School of Navigation,
Wuhan University of Technology, Wuhan, China
yhhuang@whut.edu.cn
[2] University of Salamanca, Salamanca, Spain
{ali.azadi,aliciagh}@usal.es
[3] Ronin Institute, Montclair, USA
akshat.gaurav@ronininstitute.org
[4] Lebanese American University, Beirut 1102, Lebanon
[5] School of Science and Technology, Hong Kong Metropolitan University, Hong Kong, China
s1245831@study.hkmu.edu.hk

Abstract. Any analogous physical object, often a company's resources, may have a digital twin. Its foundation is made up of asset-specific data items, which are then commonly improved using semantic technologies and analysis/simulation environments. The DT paves the way for comprehensive asset management, from basic monitoring to fully autonomous operation. The idea relies heavily on covering the whole asset lifetime. The digital twin should provide extensive networking for its data to enable sharing and exchange across all lifecycle participants. In this paper, we analyze the concept of digital twin in detail. Along with this, we analyze the evaluation of digital twin technology over the years.

Keywords: Digital twin · Cyber physical system · Cloud computing · Fog computing

1 Introduction

The Digital Twin (DT) model represents a new phase in the development of the digital world [7,8]. The term "Digital Twin" was first coined in Apollo program; in which at least two identical spacecraft were developed to replicate the circumstances of a voyage to the moon [1]. Therefore, we can say that digital twins open up a whole new world of possibilities for the analysis of physical objects through their digital forms [1,2]. The digital twine concept is applicable in a wide range of domains; due to this, there are many definitions available for it [1,16]. According to some authors [3,10], digital twin is an asset that stores the digital footprint of a physical object. However, according to some authors, a digital twin is a virtual or physical replica of a real-world object that can be used to track and analyze changes in that object [11]. But some authors, by simulating the behavior of their physical counterparts, Digital Twin (DT) models are used to improve a wide range of manufacturing procedures [14, 16]. In simple terms, digital twins are digital copies of physical assets that are identical in every manner and are constantly up to date with their "real" counterparts. To fully realize the DT paradigm, it is vital to integrate the life of the entire asset, which is made possible by DT data that link the physical and virtual twin [4, 7]. Also, the digital twin may be simulated in order to study the underlying physical asset's behavior.

Nowadays, researchers are proposing the use of digital twins in different domains to replicate the behavior of physical assets. The Digital Twin is an example of a digital model of a particular entity in a Cyber-Physical Production System (CPPS) that can replicate both the item's static and dynamic properties. Some of the models it stores and maps are executable simulation models of the underlying physical object. It's important to note that not every model can be run in the real world, therefore the Digital Twin is more than simply a digital representation of a physical object. In this sense, an asset might be a depiction of something to be created or a pre-existing entity in the physical world [1, 12]. In the Industry 4.0 context, a Digital Twin (DT) may be a virtual depiction of anything from a component in a manufacturing facility, to a product being assembled/manufactured, to the complete production system. DTs may have several degrees of abstraction and can also represent other virtual processes, including other DTs, in a hierarchical fashion. More specifically, a DT maintains a state copy of the entity it represents. In-process analysis of the represented object may be improved by using the previously saved state. More so, the DT is the hub through which all communications with the entity being represented are routed. Aggregating state display and analysis at a single location allows Industry 4.0 agents to monitor and act, system-wide [5, 17]. Digital twin also used with smart transportation systems. A digital twin is a simulation model of a vehicle or system that is built using the available physical models, sensor updates, foot history, etc., to reflect the same conditions and behaviors as the corresponding flying twin [1, 13, 15].

Synchronization with the physical asset, active data collecting from the physical environment, and the capacity for simulation are the three primary features necessary for a Digital Twin. An intelligent digital twin should have all the features of a digital twin, as well as the features of a visual intelligence. A new perspective on technology will emerge from an intelligent and healing real-time physical item. The concept of a digital twin seems to accord with what was just said. The Internet of Things (IoT) and big data have come together to provide a powerful tool that can be used to any problem. A digital twin is a replica of a physical object that may take any shape necessary to address the problems faced by the original [9, 11]. While it is true that digital twins are most useful for tackling problems involving several interconnected systems, entities, and pieces of machinery, this technology may be used to any kind of data management or control challenge. When combined with Edge or Fog computing, Digital Twins may help solve the problem of network latency and poor connection. As a result of the advantages it provides in terms of resource sharing, component reuse, and the development of a data-centric system, most cloud service providers are eagerly anticipating the advent of cloud-based services dedicated to digital twins.

From above discussion, it is clear that with its ability to construct a simulation model of the physical model, the Digital Twin (DT) has emerged as a major tool for real-time monitoring of dynamic circumstances in the physical model. The concept of the "Digital Twin" is only beginning to gain traction at this time. However, to this day, discussions have taken place over how to put this idea into practise and how it should be constructed.

2   Research Methodology

To trace the development of digital twin technologies, a systematic literature study was performed. While drafting this paper, the following steps were taken: The procedure included picking databases, picking Queries, reading research articles, and evaluating those studies. Since there are fewer journals indexed in the Web-of-Science database than in Scopus, a subset of Scopus is used in this study to lessen the likelihood of missing relevant articles during the search process.

2.1   Eligibility Criteria

We include the research papers that present techniques and algorithms for the development of digital twins. To illustrate the global scope of current research, the map also includes English-language publications that will appear between November 2019 and November 2021.

2.2   Restrictions

A limited number of publications were rejected from consideration since they did not fit the research focus. As an example, we do research only in computer science and not in any other discipline.

2.3   Data Source

Using the Scopus bibliographic database, the data was gathered in August 2022. The following two keywords were included in the search strategy to answer the research question.

– Digital Twin

2.4   Search Query Selection

In order to get the information from Scopus database, we used the following query at different screening levels:

– **Stage 1**: included only papers written in English

   "(TITLE-ABS-KEY("digital twin")) AND (LIMIT-TO (LANGUAGE, "English"))"

– **Stage 2**: Include only papers from "computer science" domain,

"(TITLE-ABS-KEY("digital twin")) AND (LIMIT-TO (LANGUAGE ,
"English")) AND (LIMIT-TO (SUBJAREA , "COMP")))"

– **Stage 3**: Include only "Articles" and "conference papers",

(TITLE-ABS-KEY("digital twin")) AND (LIMIT-TO (LANGUAGE , "English"
)) AND (LIMIT-TO (SUBJAREA , "COMP")) AND (LIMIT-TO (DOCTYPE ,
"cp") OR LIMIT-TO (DOCTYPE , "ar"))

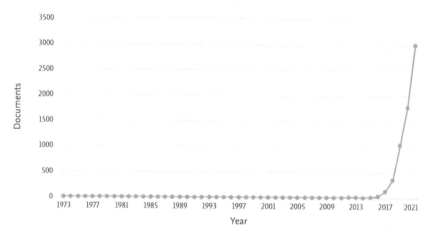

(a) Paper Published in Digital Twin

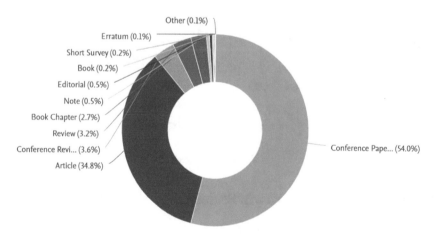

(b) Important types

Fig. 1. Analysis of digital twin sector

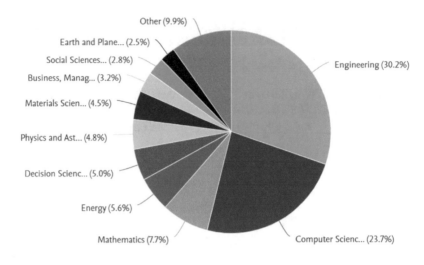

(c) Important Types

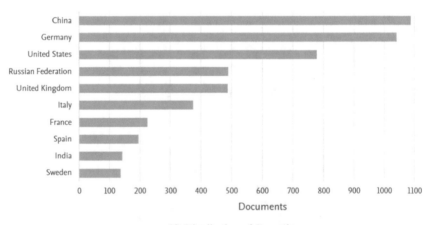

(d) Distribution of Countries

Fig. 1. (*continued*)

3　Evaluation of Digital Twin Sector

In order to get information about the development of digital twin sector, we categorized this section into the following subsections.

3.1　Analysis Document Distribution

In this subsection, we give details about the scientific distribution of the research papers. To obtain the information of the article, we find highly cited articles related to the development of the digital twin sector. Figure 1a presents the number of papers related to digital twin over the years. From Fig. 1a it is clear that after 2017 there is an exponential growth in papers in the field of digital twin. Hence, we can say that this topic is still in its development phase.

3.2 Analysis Topic and Type

Other then number of paper publications, topics, and type of paper are good parameters to understand the development of the digital twin sector. Figure 1b and c presents the important types and topics of paper published under the digital twin theme. From Fig. 1b it is clear that 54% articles published in the digital twin sector are conference articles. In addition, most of the computer science (23.7%) researchers are contributing to the development of new technologies and theories for digital twins.

3.3 Analysis of Country Distribution

The distribution of researchers by nations is also a significant and beneficial component in analyzing the development in the field of digital twin. This metric indicates the effectiveness of researchers in a country. Figure 1d represents the distribution of countries according to the total number of publications. From Fig. 1d it is clear that *China, Germany, United States, Russia, and UK* researchers are actively working in the field of digital twin.

4 Theoretical and Practical Implications

In this subsection, we will explain the implications of our study. From the above section it is clear that digital twins allow digital models to update themselves automatically, partially or completely based on the physical twin. Relegating that functionality to the digital twin will allow one to minimize what is at the plant and enable dynamic updates of software, practices, and policies on scalable cloud native platforms. One of the most exciting examples of where digital twins may be employed is in the MetaVerse, with human digital twins. In addition, we envision that digital twins can have embedded control systems when paired with a cyber physical system, yielding significant performance and configurable advantages. In addition to that, the Digital Twin paradigm is a quintessential artifact of Industry 4.0 and encompasses many of the integration and security challenges present in the Industry 4.0 domain. Hence, we can say that Digital twins are poised to play a vital role in the industry 4.0 era.

4.1 Characteristics of Digital Twin

Users' lack of familiarity with Digital Twin's features and characteristics prevents them from seeing the technology's full potential. Therefore, in this subsection, we will present some of the features and characteristics [1].

- In order to present the digital identity of a physical object, there is a need of unique ID to identify.
- As digital twin stores complete life cycle of physical object; there is a need of a version manager to, which stores the life cycle checkpoints.

– The interface between the digital twin and the physical twine should be smooth and properly synchronized.
– A network is required so that different digital twins can communicate with each other.

5   Conclusion

A digital twin (DT) is a virtual representation of a physical system or industrial process that is accurate on numerous scales of detail. Manufacturing sectors today must overcome the problem of meeting the rising consumer demand for individualized goods while simultaneously reducing lead times. In turn, this requires that they have adaptable manufacturing systems that can be effectively reconfigured on the fly to meet the changing demands of the market. Systems are constantly being upgraded as a result of shorter lifecycles and rising market needs. A production system's digital twin may help with the difficulty of building reconfigurable systems by allowing for the realization and testing of reconfiguration scenarios in a simulated setting. The time required to recommission the production system is reduced, allowing for increased system availability. Therefore, we can say that the digital twin technology helps to improve the digital economy.

References

1. Ashtari Talkhestani, B., Jung, T., Lindemann, B., Sahlab, N., Jazdi, N., Schloegl, W., Weyrich, M.: An architecture of an intelligent digital twin in a cyber-physical production system. at - Automatisierungstechnik **67**(9), 762–782 (2019). https://doi.org/10.1515/auto-2019-0039
2. Biesinger, F., Weyrich, M.: The facets of digital twins in production and the automotive industry. In: 2019 23rd International Conference on Mechatronics Technology (ICMT), pp. 1–6. IEEE, Salerno, Italy (2019). https://doi.org/10.1109/ICMECT.2019.8932101
3. Casillo, M., et al.: A deep learning approach to protecting cultural heritage buildings through IoT-based systems. In: 2022 IEEE International Conference on Smart Computing (SMART-COMP), pp. 252–256. IEEE (2022)
4. Casillo, M., et al.: A situation awareness approach for smart home management. In: 2021 International Seminar on Machine Learning, Optimization, and Data Science (ISMODE), pp. 260–265. IEEE (2022)
5. Dar, A.W., Farooq, S.U.: A survey of different approaches for the class imbalance problem in software defect prediction. Int. J. Softw. Sci. Comput. Intell. (IJSSCI) **14**(1), 1–26 (2022)
6. Deveci, M., et al.: Personal mobility in metaverse with autonomous vehicles using q-rung orthopair fuzzy sets based OPA-RAFSI model. IEEE Trans. Intell. Transp, Syst. (2022)
7. Dietz, M., Putz, B., Pernul, G.: A distributed ledger approach to digital twin secure data sharing. In: Foley, S.N. (ed.) Data and Applications Security and Privacy XXXIII, vol. 11559, pp. 281–300. Springer, Cham (2019). https://doi.org/10.1007/978-3-030-22479-015
8. Feng, H., Chen, D., Lv, Z.: Blockchain in digital twins-based vehicle management in VANETs. IEEE Trans. Intell. Transp. Syst. (2022)
9. Kar, R.: To study the impact of social network analysis on social media marketing using graph theory. Int. J. Softw. Sci. Comput. Intell. (IJSSCI) **14**(1), 1–20 (2022)

10. Kritzinger, W., Karner, M., Traar, G., Henjes, J., Sihn, W.: Digital twin in manufacturing: a categorical literature review and classification. IFAC-PapersOnLine **51**(11), 1016–1022 (2018)
11. Pushpa, J., Kalyani, S.: The fog computing/edge computing to leverage digital twin. In: Advances in Computers, vol. 117, pp. 51–77. Elsevier (2020). https://doi.org/10.1016/bs.adcom.2019.09.003
12. Riyahi, M., et al.: Multiobjective whale optimization algorithm-based feature selection for intelligent systems. Int. J. Intell. Syst. (2022)
13. Shafto, M., Conroy, M., Doyle, R., Glaessgen, E., Kemp, C., LeMoigne, J., Wang, L.: Modeling, simulation, information technology & processing roadmap. Natl. Aeronaut. Space Admin. **32**(2012), 1–38 (2012)
14. Singh, A., et al.: Distributed denial-of-service (DDOS) attacks and defense mechanisms in various web-enabled computing platforms: issues, challenges, and future research directions. Int. J. Semant. Web Inf. Syst. (IJSWIS) **18**(1), 1–43 (2022)
15. Srivastava, A.M., Rotte, P.A., Jain, A., Prakash, S.: Handling data scarcity through data augmentation in training of deep neural networks for 3D data processing. Int. J. Semant. Web Inf. Syst. (IJSWIS) **18**(1), 1–16 (2022)
16. Stacchio, L., Perlino, M., Vagnoni, U., Sasso, F., Scorolli, C., Marfia, G.: Who will trust my digital twin? Maybe a clerk in a brick and mortar fashion shop. In: 2022 IEEE Conference on Virtual Reality and 3D User Interfaces Abstracts and Workshops (VRW), pp. 814–815. IEEE, Christchurch, New Zealand (2022). https://doi.org/10.1109/VRW55335.2022.00258
17. Tarneberg, W., Skarin, P., Gehrmann, C., Kihl, M.: Prototyping intrusion detection in an industrial cloud-native digital twin. In: 2021 22nd IEEE International Conference on Industrial Technology (ICIT), pp. 749–755. IEEE, Valencia, Spain (2021). https://doi.org/10.1109/ICIT46573.2021.9453553

IoT Data Validation Using Blockchain and Dedicated Cloud Platforms

Marco Carratù[1], Francesco Colace[1], Brij B. Gupta[2,3(✉)], Francesco Marongiu[1], Antonio Pietrosanto[1], and Domenico Santaniello[1]

[1] DIIn, University of Salerno, Fisciano (SA), Italy
{mcarratu,fcolace,fmarongiu,apietrosanto,dsantaniello}@unisa.it
[2] International Center for AI and Cyber Security Research and Innovations, & Department of Computer Science and Information Engineering, Asia University, Taichung 413, Taiwan
bbgupta@asia.edu.tw
[3] Lebanese American University, Beirut 1102, Lebanon

Abstract. The Internet of Things world has grown exponentially in recent years, projecting the future toward an increasingly automated reality in which human presence results less and less and decisions about processes are made directly by machines, which exchange information, operating to maximize a goal. Such complex and automated systems pose quite a few challenges from the standpoint of securing cyber-physical spaces, particularly tampering with the data of specific components and/or sensors, which can go a long way in impacting the operation of the entire system and causing damage. This work aims to reduce the possibility of data tampering by using Blockchain systems integrated into cloud platforms for data collection and analysis so that in the case of critical decisions, this information can be verified in a certified way and thus proceed safely. An application case study based on ThingsBoard and HyperLedger Busu technologies was proposed showing how these technologies can integrate and collaborate to achieve effective results.

Keywords: Internet of Things · Industry 4.0 · Blockchain · Data certification · Data tampering · HyperLedger

1 Introduction

The Internet of Things (IoT) was introduced by Kevin Ashton in 1999 and is a neologism referring to the extension of the Internet to the world of objects. The idea behind this concept is the presence of a large number of objects, equipped with sensors, actuators, and connectivity, that can exchange information about their surroundings and communicate with each other through the Internet [1,2]. Many people are often using IoT devices without even knowing it. Year after year it is becoming more and more established as an emerging technology. This also brings with it risks concerning the security of these objects [3,4]. The devices are constantly exchanging information, which, if stolen, can create major problems. To prevent this, special protection techniques are introduced [5–7]. All elements, belonging to the IoT world, connect to the network,

N. Nedjah et al. (Eds.): ICSPN 2021, LNNS 599, pp. 208–216, 2023.
https://doi.org/10.1007/978-3-031-22018-0_19

but they do not have to be electronic devices, they can be any kind of object, work, machine, or plant. The basic characteristic is that it has sensors and is capable of transmitting or receiving data [8, 9]. They become intelligent tools as they take an active role in the system, so they do not just passively listen to what is happening around them but can command some actuators to make changes to the system itself [10, 11]. There are different models of communication between devices in this paradigm [12]:

- **Device-to-Device**: The device-to-device communication model represents two or more devices that are directly connected and communicate with each other, such as smartphones. This model has based on the security of the devices themselves, which means that manufacturers need to make more investments to implement these approaches and ensure trust on both sides.
- **Device-to-Cloud**: The IoT device connects directly to a cloud service. This approach uses traditional communication methods such as Wi-Fi or Ethernet connections to establish a connection between the device and the IP network connected to the cloud. It allows data to be accessed from any location at any time. Very often the service offered, and the devices are from the same manufacturer, this is because incompatibility problems may arise when trying to integrate devices from different manufacturers.
- **Device-to-Gateway**: The IoT device connects through a Gateway to the cloud service, the latter then acts as an intermediary. This communication model is suitable for smart objects that require remote configuration capabilities and temporal interactions.

Application Enablement Platforms (AEP), commonly known as IoT Platforms, are one of the main elements of IoT-based systems, they enable the control, management, and analysis of data produced by devices belonging to different domains and applications [13].

It is estimated that there are more than 300 different IoT platforms globally publicly known, each one with different capabilities and features. They are mainly divided into commercial and open source, depending on its distribution license. A few important properties are considered when choosing which platform to use, including scalability, support for heterogeneity, and stability. The platform must enable interoperability between devices and applications, considering that, year after year, the number of objects connected to the Internet is increasing more and more. Therefore, to include these aspects, the platform must provide a way to connect IoT devices communicating in different network protocols and data structures, but most importantly, it must ensure a low response time. The most common metrics used to measure scalability of a platform are throughput and response time. System stability is measured using multiple factors such as CPU utilization, occupied memory, and tolerance in drop rate [14]. One of the leading platforms on the market is called ThingsBoard, which enables the development and management of IoT projects. It provides the user with the ability to register, manage, and monitor multiple devices. It also provides APIs for server-side applications to send commands to devices and vice versa. Various operations can be performed on the platform, for example, it is possible to create any number of devices, thanks to which the analysis, telemetry, and collection of certain data, through the use of IoT sensors, can be implemented. Each device is uniquely identified, on the platform. The sensors can recognize the objects. Data are fetched from sensors and sent to the ThingsBoard cloud, which receives this information and processes it through Rule Engines, a framework useful for building event-based flows.

Modern networking technologies have contributed deeply to the way IoT devices can produce, send, and collect data from external world [15]. Exchange messages efficiently are one of the major problems in IoT, the advances in networking field have contributed to the implementation of a wide range of new messaging protocols which allowed IoT devices to work with specific environment conditions. Indeed, the performances of IoT devices may vary depending on the application and the use case, choosing the wrong protocol method could seriously impact on overall system performances [16]. Instead, network traffic and latency can be reduced by using an appropriate messaging protocol, and thus increases the reliability of the IoT application associated [17]. However, there is no universal formula that can be used anywhere. Usually, IoT systems are heterogeneous environments which require high degree of interoperability between devices, so multiple protocol must be used. The choice of an appropriate messaging protocol depends on several factors, including the requirements of the IoT application, the capabilities of the software, device, or hardware, and the average data size [18].

Message Queue Telemetry Transport (MQTT) is a protocol, running on the Transport Control Protocol, used for communication within an IoT environment, which was created by IBM as a lightweight machine-to-machine communication method [19]. It is suitable in constrained environments, such as when a device has few memory resources or a limited processor [20]. This protocol is based on a server-client system with a publish-subscribe architectural model in which the server, called a broker, sends updates to MQTT clients. Clients will send messages, containing an argument or value, to the broker, without direct interaction between client and client. The latter can decide to subscribe to a particular topic and publish messages regarding that topic; it will then be the server's job to forward the information to each client that has decided to subscribe to that specific topic. In this way, client devices do not have to impulsively request updates. This reduces node resource consumption [21].

2 Proposed Methodology

The sensors are generally connected to control boards that are responsible for reading the data and transmitting it to a collector generally in the cloud. These boards can have very different computational powers, they can be true computing centers such as, for example, Raspberry boards or more generally Single Board Computers (SBCs), or they can have very limited capabilities such as ESPs. In any case, however, they are all sufficient to perform brief operations on the data before transmitting it. Thus, the proposed approach prevents any manipulation of data on the network, even if it has to be sent in the clear due to the limitations of IoT devices. The first step in the methodology is to assign each device public and private key pair. The public keys are registered within a Smart Contract on Blockchain. In this sense, the SC has the function of being the authority for the exchange of public keys and thus the main actor for data validation and certification in the future [22].

Then a public key signature algorithm compatible with the cryptographic primitives used by the blockchain to sign and verify transactions is programmed within the sensor control board. When a device needs to send data to the cloud platform it affixes, using its private key, a digital signature and sends the data over the network, also in plaintext. At this point, the cloud platform saves the data as received in the most efficient and useful data structures to store information related to the IoT world, but in parallel, it uses the blockchain to validate the data and be sure that it is genuine. The verification process must be done asynchronously to the storage of information because the rate of data coming from the IoT world is too high to be compatible with the verification and certification time of a blockchain. If uncertified information is identified during the verification process, it will then be retrospectively discarded by a synchronization process (Fig. 1).

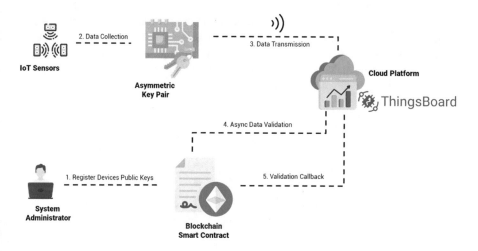

Fig. 1. Main architecture design

2.1   Experimental Case

An experimental case was carried out to test the functionality of the proposed approach. In particular, two main technologies were used: ThingsBoard as a data collection and management platform, and HyperLedger Busu for the creation of a private blockchain used for Smart Contract execution and thus data certification.

Since HyperLedger Busu is based on Ethereum technology, it was chosen to use Keccak-256, the cryptographic primitive underlying the Ethereum blockchain, as the signature function to ensure immediate interoperability not only with the experimental test environment but also for future integration with external blockchain systems.

2.2   Smart Contract Definition

The Smart Contract defined on the blockchain has a dual function: to operate as a Public Key Infrastructure (PKI) for the exchange and management of public keys. A system administrator is responsible for entering and revoking permissions to public keys. A hash table is then used to store the state of each key, while a verification function checks the signature of the data passed in as input and makes sure that the public key from which it came has been properly registered and authorized (Table 1).

2.3   Data Validation

The data are, therefore, collected by a cloud-based platform that takes care of verifying them asynchronously through the blockchain and making changes to the database after the fact if anomalies are found. The ThingsBoard platform, a cloud specifically designed for massive data collection from the Internet of Things world, was used for the case study. One of the particularities of this platform is the use of Rule Chains, which is the ability to manipulate input data to the platform in real-time using a visual graph programming approach. ThingsBoard allows custom nodes to be programmed in a way that extends the operation of Rule Chains (Fig. 2).

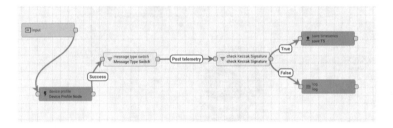

Fig. 2. Data validation rule chain

Therefore, an ad-hoc node for the Rule Chain system, which interacts directly with the blockchain via the Web3 library, was built for data validation. When unverified data is detected an additional node in the chain takes care of marking the data as invalid and then discarded the rest of the computations.

The history of data and validations is still saved in the cloud platform and made accessible in case of further verification or analysis (Fig. 3).

Table 1. Smart contract verification function.

```
function verify(uint256 _data, uint8 _v, bytes32 _r,
bytes32 _s) public returns (address) {
        bytes memory message = bytes(abi.en-
codePacked(Strings.toString(_data)));
        bytes32 _messageHash = toEthSignedMessage-
Hash(message);
        address _recovered = recover(_messageHash, _v,
_r, _s);
        require(_authorizedKeys[_recovered], "Data is not
genuine!");

        return _recovered;
    }

    function toEthSignedMessageHash(bytes memory s) in-
ternal pure returns (bytes32) {
        return keccak256(abi.encodePacked("\x19Ethereum
Signed Message:\n", Strings.toString(s.length), s));
    }

    function recover(bytes32 hash, uint8 v, bytes32 r,
bytes32 s) internal pure returns (address) {
        if (uint256(s) >
0x7FFFFFFFFFFFFFFFFFFFFFFFFFFFFFFF5D576E7357A4501DDFE92F
46681B20A0) {
            return address(0);
        }
        if (v != 27 && v != 28) {
            return address(0);
        }

        address signer = ecrecover(hash, v, r, s);
        if (signer == address(0)) {
            return address(0);
        }

        return signer;
    }
```

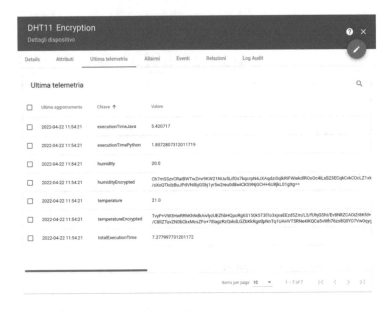

Fig. 3. ThingsBoard telemetry prototype

3 Conclusions

The issue of data certification is certainly one of the most significant in the Internet of Things. The possibility of an external actor intruding into the communication channels and thus altering in some way the data sent by IoT devices opens dangerous scenarios, which in the perspective of automation of industries and daily life are not acceptable [23]. Several efforts have been made to solve these problems, in particular, the use of Blockchain technology offers a valuable tool for data certification operations [24].

As shown by experimental results, the use of this technology introduces a high degree of reliability, however, reaction times are incompatible with what is properly the IoT world. The proposed approach aims to solve this issue by going directly within the data stream by integrating low-level interactions with the blockchain itself. The extension of the cloud platform for validation is equally crucial by going to divide data collection and validation into two distinct phases.

The impact that this approach can have at the industrial level can be substantial when looking specifically at the entire Industry 4.0, in which machines communicate autonomously to achieve a common purpose. In this scenario, blockchain handles the immutable certification of data by enabling the infrastructure to make decisions, even critical ones, about certain actions in a completely autonomous and secure way.

References

1. Atzori, L., Iera, A., Morabito, G.: The Internet of Things: a survey. Comput. Netw. **54**(15), 2787–2805 (2010). https://doi.org/10.1016/j.comnet.2010.05.010
2. Lu, J., Shen, J., et al.: Blockchain-based secure data storage protocol for sensors in the industrial Internet of Things. IEEE Trans. Ind. Inform. **18**(8), 5422–5431 (2022). https://doi.org/10.1109/TII.2021.3112601
3. Wang, W., Xu, H., Alazab, M., Gadekallu, T.R., Han, Z., Su, C.: Blockchain-based reliable and efficient certificateless signature for IIoT DEVICES. IEEE Trans. Ind. Inform. **18**(10), 7059–7067 (2022). https://doi.org/10.1109/TII.2021.3084753
4. Mamta, et al.: Blockchain-assisted secure fine-grained searchable encryption for a cloud-based healthcare cyber-physical system. IEEE/CAA J. Autom. Sinica **8**(12), 1877–1890 (2021). https://doi.org/10.1109/JAS.2021.1004003
5. Al-Fuqaha, A., Guizani, M., Mohammadi, M., Aledhari, M., Ayyash, M.: Internet of Things: a survey on enabling technologies, protocols, and applications. IEEE Commun. Surv. Tutor. **17**(4), 2347–2376 (2015). https://doi.org/10.1109/COMST.2015.2444095
6. Lin, J., Yu, W., Zhang, N., Yang, X., Zhang, H., Zhao, W.: A survey on Internet of Things: architecture, enabling technologies, security and privacy, and applications. IEEE Internet Things J. **4**(5) (2017). https://doi.org/10.1109/JIOT.2017.2683200
7. Colace, F., de Santo, M., Marongiu, F., Santaniello, D., Troiano, A.: Secure medical data sharing through blockchain and decentralized models. In: Studies in Computational Intelligence, vol. 1030 (2022). https://doi.org/10.1007/978-3-030-96737-6_13
8. Casillo, M., Colace, F., Lorusso, A., Marongiu, F., Santaniello, D.: An IoT-Based System for Expert User Supporting to Monitor, Manage and Protect Cultural Heritage Buildings, vol. 1030 (2022). https://doi.org/10.1007/978-3-030-96737-6_8
9. Tewari, A., et al.: Secure timestamp-based mutual authentication protocol for IoT devices using RFID tags. Int. J. Semant. Web Inf. Syst. (IJSWIS) **16**(3), 20–34 (2020)
10. bin Zikria, Y., Yu, H., Afzal, M.K., Rehmani, M.H., Hahm, O.: Internet of Things (IoT): operating system, applications and protocols design, and validation techniques. Future Gener. Comput. Syst. **88** (2018). https://doi.org/10.1016/j.future.2018.07.058
11. Liang, Y., et al.: PPRP: preserving-privacy route planning scheme in VANETs. ACM Trans. Internet Technol (2020, accepted). https://doi.org/10.1145/3430507
12. Tschofenig, H., Arkko, J., Thaler, D., McPherson, D.: Architectural Considerations in Smart Object Networking (2015)
13. Fahmideh, M., Beydoun, G.: Big data analytics architecture design–an application in manufacturing systems. Comput. Ind. Eng. **128** (2019). https://doi.org/10.1016/j.cie.2018.08.004
14. Yokotani, T., Sasaki, Y.: Comparison with HTTP and MQTT on required network resources for IoT (2017). https://doi.org/10.1109/ICCEREC.2016.7814989
15. Gubbi, J., Buyya, R., Marusic, S., Palaniswami, M.: Internet of Things (IoT): a vision, architectural elements, and future directions. Future Gener. Comput. Syst. **29**(7), 1645–1660 (2013). https://doi.org/10.1016/j.future.2013.01.010
16. Naik, N.: Choice of effective messaging protocols for IoT systems: MQTT, CoAP, AMQP and HTTP (2017). https://doi.org/10.1109/SysEng.2017.8088251
17. Sisinni, E., Saifullah, A., Han, S., Jennehag, U., Gidlund, M.: Industrial internet of things: challenges, opportunities, and directions. IEEE Trans. Ind. Inform. **14**(11), 4724–4734 (2018). https://doi.org/10.1109/TII.2018.2852491
18. Bandyopadhyay, S., Bhattacharyya, A.: Lightweight Internet protocols for web enablement of sensors using constrained gateway devices (2013). https://doi.org/10.1109/ICCNC.2013.6504105

19. Hunkeler, U., Truong, H.L., Stanford-Clark, A.: MQTT-S - a publish/subscribe protocol for wireless sensor networks (2008). https://doi.org/10.1109/COMSWA.2008.4554519
20. Thangavel, D., Ma, X., Valera, A., Tan, H.X., Tan, C.K.Y.: Performance evaluation of MQTT and CoAP via a common middleware (2014). https://doi.org/10.1109/ISSNIP.2014.6827678
21. Dinculeană, D., Cheng, X.: Vulnerabilities and limitations of MQTT protocol used between IoT devices. Appl. Sci. (Switzerland) **9**(5) (2019). https://doi.org/10.3390/app9050848
22. Casillo, M., Castiglione, A., Colace, F., de Santo, M., Marongiu, F., Santaniello, D.: COVID-19 data sharing and organization through blockchain and decentralized models. In: CEUR Workshop Proceedings, vol. 2991, pp. 128–140 (2021)
23. Anthi, E., Williams, L., Slowinska, M., Theodorakopoulos, G., Burnap, P.: A supervised intrusion detection system for smart home IoT devices. IEEE Internet Things J. **6**(5), 9042–9053 (2019). https://doi.org/10.1109/JIOT.2019.2926365
24. M. Casillo, Colace, F., et al.: Decentralized approach for data security of medical IoT devices. In: CEUR Workshop Proceedings, vol. 3080 (2021)

Security on Social Media Platform Using Private Blockchain

Geerija Lavania(✉) and Gajanand Sharma

JECRC University, Jaipur, India
girija47lavania@gmail.com

Abstract. Blockchain could be a decentralized, shared and open progressed record that's utilized to record trades over various PCs so that any expound record can't be balanced retroactively without the alter of each following square. It gives us a modern framework to record a open history of exchanges without depending on any centralized trusted party. The property of decentralization, openness, permission less and tamper-resistant has pulled in significant intrigued in both scholastic and mechanical communities. In this paper we work on social media security using block chain. It the growth of the social media platforms, raise the need of the securing the data communicated. Now, the data which is communicated can be secured in number of ways, but the end-points of social media applications are not secure. In the paper we propose some technology using Blockchain for security of social media platform framework, which can ensure both information security and privacy while simultaneously increasing the trust of the public sectors. We tried to work on securing the end app, by making the innovative concept of use authentication using the Zig-Zag arrangement for security of social media platform.

Keywords: Blockchain · Security · Privacy and Zig-Zag · Social media platform

1 Introduction

Blockchain innovation can possibly change collaborations between states, organizations and residents in a way that was inconceivable only 10 years ago [1,2]. Although often listed with technologies such as artificial intelligence (AI) or IoT (Internet of Things), this technology is unique in its fundamental nature. Not at all like other innovations that have the potential to provide totally unused administrations to citizens and other partners alike, Blockchain has the potential to make strides as of now existing forms to open unused sources of productivity and esteem [3–5]. Social media are a source of communication between the information proprietor (information generator) and viewers (end clients) for online communications that make virtual communities utilizing online social systems. A social arrange could be a social chart that speaks to a relationship among clients, organizations, and their social exercises [7–9]. These clients, organizations, bunches, etc., are the hubs, and the relationships between the clients, organizations, bunches are the edges of the chart [11–13]. An online social system is

N. Nedjah et al. (Eds.): ICSPN 2021, LNNS 599, pp. 217–226, 2023.
https://doi.org/10.1007/978-3-031-22018-0_20

a web platform used by conclusion clients to form social systems or connections with other individuals that have similar views, interface, exercises, and/or real-life associations [14–16,37].

At display Blockchain is, perhaps, very possibly the most empowering innovation, and it has an amazingly high potential to alter the display plans of activity. It is fundamentally a spread data base of records, or a open record, everything being rise to (or computerized events) that have been executed and separated between taking an interest parties. Each trade within the open record is affirmed by the understanding of a greater portion of the members' handling control within the entirety system [17–20]. Blockchain has the potential to altogether alter the way monetary educate, for illustration, work nowadays. As an early prove note that Bitcoin is one of the foremost commonly utilized applications based on Blockchain innovation. Too, due to the alluring characteristic of decentralization, persistency, timelessness, and audit ability, it has assist far-reaching suggestions in a wide run of businesses, divisions, and application zones, such as Internet of Things (IoT) [6,21–23], restorative health, and smart contracts, separately. As a result, blockchain innovation has gotten significant consideration in both the academic and the industrial communities [4,24–26].

As a promising innovation to attain decentralized agreement, blockchain isn't as it were restricted in crypto currency but too basic to undertakings and makes uncommon openings for Computerized Wellbeing, Supply Chain, Vitality, IoT, Fintech, etc. Although blockchain can keep up a secure open record by all members through the dispersed organize, how to apply in numerous mechanical divisions adaptable is still a huge challenge [5,27,29,30].

The blockchain is an appropriated data base of records of all trades that have been executed and partitioned between taking portion parties. Each trade is checked by most of individuals of the system. It contains each and each record of each trade [31,32, 34,40]. Bitcoin is the foremost well known computerized cash and is an outline of the blockchain. The most objective of online social media is to share substance with greatest clients [28,35]. Clients utilize platform, such as Facebook, Twitter, and LinkedIn, to distribute their schedule exercises. In some cases, online social platform clients share information around themselves and their lives with companions and colleagues. Be that as it may, in these published data, a few of the uncovered substance through the online social platform are private and thus ought to not be published at all. Regularly, clients share a few parts of their everyday life routine through status overhauls or the sharing of photos and recordings. Right now, different platform clients utilize Smartphone to require pictures and make recordings for sharing through online social platform. This information can have area data and some metadata embedded in it. Platform benefit suppliers collect a run of information around their clients to offer personalized services, but it can be utilized for commercial purposes. In expansion, users' information may too be provided to third parties, which lead to protection spillages [33,37].

2 Backgrounds

The context of this document for social networking sites, blockchain technology provides the technology to implement a highly privacy protection decentralized network.

2.1 Blockchain Technology

A square in a blockchain is a progressed ensured to store the data and bolt it for forever. The data included on the square is changeless, i.e. Data can't be changed. Areas are used to store factions of significant trades that are clearly laid and encrypted into a Tree structure. So every area makes a connection the block chain technology besides remembering the cryptographic hash square. A chain is made up by affiliated area. This incremental communication verifies this same accuracy of the previous work, right back to the fundamental area, which is known as the starting area. The nonce is chosen by the diggers to address a cryptographic enigma for making the taking after square within the chain. It is known as Verification of work. To ensure the genuineness of a square and the data contained in it, the square is for the most part carefully checked [7].

Presently and once more isolated squares can be conveyed at the same time, making an impermanent fork. Regardless a safe and sound hash-based history any blockchain, for example, includes a foreordained analysis for having scored various types of a collection of interactions so that the one with best score can be selected over others. Roaming rectangles are those that have been not selected for having joined the chain. Every now and again, shipmates who encourage these same data sets have different varieties of a set of interactions. They keep hands down the foremost vital scoring interpretation of the information set known to them. When a companion obtains a greater variability (generally the original modification with a separate unutilized rectangle would include), those who increase or completely replace their own statement dataset as well as re-transmit the improved performance to their own associates. There'll never be a continual assurance that a context of the present study is always in the best support of the collection of interactions.

When a companion obtains a higher-scoring variation (typically the original adjustment with a single unused square included), people expand or overwrite about their claim database and re-transmit the improvement to their companions. There will never be a constant guarantee that a specific segment will always be in the best support of the collection of interactions [39]. Each piece in a blockchain is distinguished by a hash esteem created by ordinarily utilizing the secure hash cryptographic algorithm-256 bits and it provide the sender and receiver side using this algorithm provide security of social media platform [38]. The blockchain technology allows use of highly safe and confidentiality distributed structures in which transactions really aren't governed by 3rd parties. Using the block chain technology, historical data and current data stored in a fixed storage area of frames (i.e., record) that's also sent over the system in an unquestionable and immutable manner. The blockchain innovation improves data protection and security by fumbling and dispersing information across the entire network. This research paper proposes a system of a decentralized framework utilizing the blockchain innovation, which can guarantee both data security and security whereas at the same time expanding the believe of the open segments.

Squares can be recognized by their square height (Block No.) and piece header hash. The information within the square is identified through a computerized calculation known as a Hash work. It not as it were locks the information to be seen by the members within the Blockchain but moreover makes the information permanent. Each square has its hash function. In Blockchain, when the data has been recorded, it won't be changed.

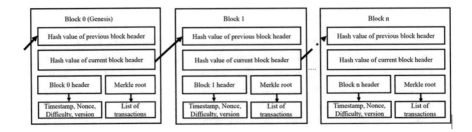

Fig. 1. These squares can be recognized by their square height (Block No.) and their hash value

Blockchain works like a computerized legitimate official with timestamps to undertake not to change information (Fig. 1).

As a result of this feature, Bitcoin method does not involve a middleman or a competent authority since it is decentralized and distributed. Private keys have been delegated to smart contracts attendees so how they can join and authorize transactions with making people caring [36–39].

2.2   The Upsides of Distributed Ledger Technology

We have gained a great deal of knowledge about Blockchain technology. Let's take a look at what it has to offer

Security and privacy
In a traditional relational database, you have had to believe that perhaps the administrator isn't going to be changing the data. Even so, as for Block chain technology, shifting or specific developments seems to be impossible; this same relevant information inside the Block - chain is unchanging; this can be erased or rectified.

Accountability
Centrally controlled structures are complicated, so although Block chain technology (a decentralized framework) is fully transparent. Institutions and infrastructure improvements could use crypto currency technologies to design a fully decentralized system that eradicates a need for a centralized specialist, growing the platform's transparency.

High level of accessibility
Unlike highly centralized structures, Blockchain is a decentralized P2P agreement that is easily accessible due to its decentralized essence. As everyone in the Blockchain network is on a mentoring system and also has a desktop trying to run, even if one peer starts going down, other peers continue to operate.

Increased security
Some other notable gain of Blockchain is its data integrity. All Crypto currency transactions are highly secured as well as provide honesty, so invention is did think to do is provide high security. As a consequence, instead of reliance on a 3rd party, you tend to believe data encryption computation.

2.3 Security Issue in Social Media Platform

In the decentralized identity and access management space, social media networks are now at a crossing point between something being monolithic patented technology implementations as well as run the application. But what were the specifications for a system to be categorised as just an identifier? The following are the characteristics of identity management, all of which have been now managed to meet by Social Media networks:

Personal data storage

Identity management refers to the management of data which describes a user's individuality. Social Networking sites, more than any other IT system in the world, satisfies this necessity. Fb, not Photos, is indeed the biggest treasure trove of individual pictures on the internet. The biggest number of specific profile pictures on the world is stored in data storage facilities of Social Media providers, not in such a govt identity registration system (at least literally not a single we're cognizant of...) or even one of the much-heralded Federalized Web Services.

Techniques for maintaining private information and how this is considered

Authentication systems cannot just store personal data; those who also manage it, allowing users to search, transmit, and perspective it. One of primary role of social networks is to improve communication. Those who focus on providing user-friendly tools that allow users to define how their personal characteristics are showcased in considerable detail, in both related to visual layout and metadata displayed [40]. They as well offer sophisticated tools for consumers to browse for it and advertising companies to mine profile information.

Credential-based authentication and authorization to personal data

This is probably the most important criterion. Consumers should have authority about who has access to which parts of their private information in just about any authentication scheme. The majority of time, this is based on identifying whether the individual trying to access the data needs to meet definite criteria (and has credentials to prove this). A dataset of medical records, for instance, might only enable user access that can prove those who are cardiologists in the same way. This function has become more commonly accessible on social media sites [40].

Tools for determining whom have authenticated personal details

Often this identity management involve data tracking tool that allows the user seeing who has accessed personal details. Even though users wish to stay anonymised when perusing other online profiles, the above features is commonly not implemented fully in Social Networks. Even so, a few Interpersonal Networks enable profile tracking to also be installed, and many Social Networking sites offer additional indeed very thorough unnamed statistical data on profile page connectivity [40].

2.4 Security over Block Chain

The framework rewards relevant comment creative's and content creators to crypto currency on the a daily basis. Regardless of the fact that such a reward system must have been created to inspire people to contribute to large concentration, our assessment of the underpinning virtual currency communication and coordination on the crypto currency illustrates that the more than 16% of virtual currency transactions in Steemit are being sent to custodians arrested on suspicion of becoming bots, and also the existence of an underlying supply chain network for the bots, both of which indicate a significant misuse of the current Steemit reward system. Our research aims to shed light on the current condition of this new blockchain-based social media platform, including the effectiveness of its architecture as well as the operation of the general agreement protocols and reward system. The platform is restarted on a routine basis [36]. The blockchain is indisputably an innovative field in recent years, particularly in the field of information systems, due to the decentralized platform and blockchain technology.

A peer-to-peer network is a network that allows users to connect with each other. There is an impressive and historic event held. There is possibility for blockchain to be used by a wide range of organizations. Allow the development of such a reliable and consistent, secure, as well as efficient system In the feature, there seems to be an unchanging system. However there are still issues which need to be discussed, a few of people have been addressed, as well as the new technological notion of the public blockchain is becoming steadier. Regardless of the fact that it has a number of benefits, it has some security issues, which we've discussed in this article. The regulator must discuss the corresponding regulatory hurdles for the new edge technology, and institutions must be ready to adopt blockchain technology, that may have a lower effect on the current system [39,40].

3   Authentication Using the Zig-Zag Arrangement for Security of Social Media Platform

With the growth of social media platforms, there seems to be a greater must to protect the information transferred. Even though the data that would be communicated can now be protected in a number of ways, the end-points of social media platforms really aren't. We tried to secure the end app in the proposed concept by developing a creative concept of use verification using a Zig-Zag agreement or sliding of images to form a specific pattern in combination with the mobile number hash pattern and the IMEI no of the mobile number. In this concept suppose we are the selected images and the app will fetch the SHA-512 code of these images which will be unique for these images. And the number will be unique for a particular image. This concept is use for making more secure social media and secure user authentication.

In this case we are taking the concept of the grid of the images, which can extend to the level of 5×4, which can be designed as per the flexibility of the mobile screen (Fig. 2).

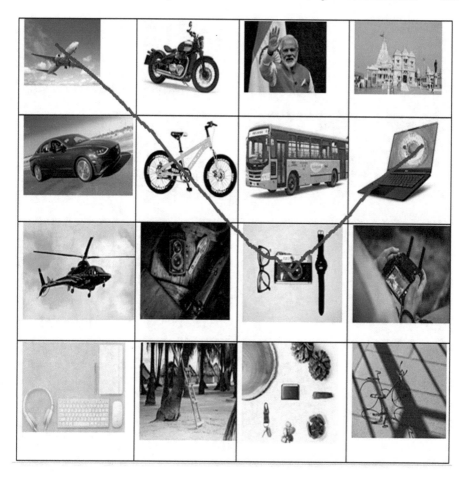

Fig. 2. Zig-Zag arrangement of the grid of the images for security of mobile

Now, you move your figure or mouse in such a way that the following images are selected (Fig. 3).

Now, what the logic behind is that is, the images which are covered with the red line indicates that, these are the selected images and the app will fetch the SHA-512 code of these images which will be unique for these images and then we will fetch the IMEI number and extract the last two digits from it, e.g. if IMEI number is 86888099168067 then last two digits are 67, then we will extract the 67 characters from each of the image SHA-512 code and combine them with the IMEI number. After that the resultant pattern will be sorted in the database and used for the process of the validation of the user.

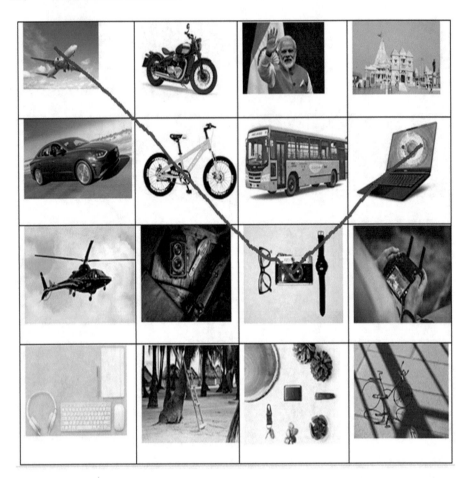

Fig. 3. Zig-Zag arrangement and these are the selected images from red line

4   Conclusion

This paper proposed a security authentication in social media platform using zig-zag arrangement by block chain technology. Using blockchain technology; this article proposes a zig-zag framework that can implement for only make ensure safety and privacy in the social media platform. Encryption technology, data integrity, and the distributed coordination and monitoring provided by blockchain technology could provide required privacy and security in social media platform using block chain algorithmic, according with new framework it provide User side security using Zig-zag pattern. We can increase the security as well as make it more difficult to break the security by using this pattern.

5   Future Work

Implementing such a framework and then further exploring its full potential in a real-world environment is the active future work. Blockchain technology to complete and enhance the security and privacy of personal data on zig-zag patterns in social media platforms. This is another some other technique of social media platform security identity verification is the zig-zag pattern, which is created each time when the user registers in and arithmetical key generate in every time of log of users. In the future work we will analysis the concept and implementation of this work.

References

1. Mamta, et al.: Blockchain-assisted secure fine-grained searchable encryption for a cloud-based healthcare cyber-physical system. IEEE/CAA J. Autom. Sinica **8**(12), 1877–1890 (2021). https://doi.org/10.1109/JAS.2021.1004003
2. Lu, J., Shen, J., et al.: Blockchain-based secure data storage protocol for sensors in the industrial Internet of Things. IEEE Trans. Ind. Inform. **18**(8), 5422–5431 (2022). https://doi.org/10.1109/TII.2021.3112601
3. IC-BCT 2019. Springer
4. Kabashkin, I.: Chapter 5 Risk Modelling of Blockchain Ecosystem. Springer (2017)
5. Blömeke, S., Mennenga, M., Herrmann, C., Kintscher, L., et al.: Recycling 4.0. In: Proceedings of the 7th International Conference on ICT for Sustainability (2020)
6. Raj, M.G., Pani, S.K.: Chaotic whale crow optimization algorithm for secure routing in the IoT environment. Int. J. Semant. Web Inf. Syst. (IJSWIS) **18**(1), 1–25 (2022)
7. Magdy, N., Barakat, M., Mokhtar, B.: IoT systems internal mapping using RTT with the integration of blockchain technology. In: 2019 Novel Intelligent and Leading Emerging Sciences Conference (NILES) (2019)
8. Farooque, M., Jain, V., Zhang, A., Li, Z.: Fuzzy DEMATEL analysis of 6 barriers to blockchain-based life cycle assessment in China. Comput. Ind. Eng. (2020)
9. Blockchain and Trustworthy Systems. Springer (2020)
10. Murugajothi, T., Rajakumari, R.: Blockchain based data trading ecosystem. Proc. Mater. Today (2020)
11. Magal-Royo, T., de Siqueira Rocha, J.M., Mascarell, C.S., Somavilla, et al.: Chapter 16 Cybersecurity and Electronic Services Oriented to E-Government in Europe. IGI Global (2021)
12. Implementing a blockchain from scratch: why, how, and what we learned. EURASIP J. Inf. Secur. (2019)
13. Zhai, S., Yang, Y., Li, J., Qiu, C., Zhao, J.: Research on the application of cryptography on the blockchain. J. Phys.: Conf. Ser. (2019)
14. Costa, F.Z.D.N., de Queiroz, R.J.: A blockchain using proof-of-download. In: 2020 IEEE International Conference on Blockchain (Blockchain)
15. Visconti, R.M.: Chapter 16 Blockchain Valuation: Internet of Value and Smart Transactions. Springer (2020)
16. Gupta, R., Pal, S.K.: Introduction to Algorithmic Government. Springer (2021)
17. Cryptofinance and Mechanisms of Exchange. Springer (2019)
18. Bistarelli, S., Mazzante, G., Micheletti, M., Mostarda, L., Sestili, D., Tiezzi, F.: Ethereum smart contracts: analysis and statistics of their source code and opcodes. Internet Things (2020)

19. Niranjanamurthy, M., Nithya, B.N., Jagannatha, S.: Analysis of blockchain technology: pros, cons and SWOT. Cluster Comput. (2018)
20. Morabito, V.: Business Innovation Through Blockchain. Springer (2017)
21. Blockchain: The India Strategy by NitiAyog (2020). Available at: https://www.niti.gov.in/sites/default/files/2020-01/Blockchain_The_India_Strategy_Part_I.pdf
22. Blockchain in global health - an appraisal of current and future applications (2020)
23. Blockchain in global health - an appraisal of current and future applications (2020)
24. A survey on security and privacy issues of blockchain technology (2018)
25. An Overview of Blockchain Technology: Architecture, Consensus, and Future Trends (2017). Available at: https://www.researchgate.net/publication/318131748_An_Overview_of_Blockchain_Technology_Architecture_Consensus_and_Future_Trends
26. State of blockchain q1 2016: blockchain funding overtakes bitcoin (2016). Available at: https://www.coindesk.com/state-of-blockchain-q1-2016/
27. Security implications of blockchain cloud with analysis of block withholding attack. In: 2017 17th IEEE/ACM International Symposium on Cluster, Cloud and Grid Computing (CCGRID), p. 458 (2017)
28. Sahoo, S.R., et al.: Security issues and challenges in online social networks (OSNs) based on user perspective. Computer and Cyber Security, pp. 591–606 (2018)
29. Ai, C., Han, M., Wang, J., Yan, M.: An efficient social event invitation framework based on historical data of smart devices. In: 2016 IEEE International Conferences on Social Computing and Networking (SocialCom), pp. 229–236. IEEE (2016)
30. Back, A., et al.: Hashcash – a denial of service counter-measure
31. Barnas, N.: Blockchains in national defense: trustworthy systems in a trustless world. Blue Horizons Fellowship, Air University, Maxwell Air Force Base, Alabama
32. Cai, Z., He, Z., Guan, X., Li, Y.: Collective data-sanitization for preventing sensitive information inference attacks in social networks. IEEE Trans. Depend. Secure Comput. 15, 577–590 (2016)
33. Usha Rani, M., Saravana Selvam, N., Jegatha Deborah, L.: An improvement of yield production rate for crops by predicting disease rate using intelligent decision systems. Int. J. Softw. Sci. Comput. Intell. (IJSSCI) 14(1), 1–22 (2022)
34. Capurso, N., Song, T., Cheng, W., Yu, J., Cheng, X.: An android-based mechanism for energy efficient localization depending on indoor/outdoor context. IEEE Internet Things J. 4, 299–307 (2017)
35. Chen, F., Deng, P., Wan, J., Zhang, D., Vasilakos, A.V., Rong, X.: Data mining for the internet of things: literature review and challenges. Int. J. Distrib. Sens. Netw. 11, 431047 (2015)
36. Li, C., Palanisamy, B.: Incentivized blockchain-based social media platforms: a case study of steemit. In: WebSci '19, June 30–July 3, Boston. MA, USA (2019)
37. Ali, S., Islam, N., Rauf, A., Din, I.U., Guizani, M., Rodrigues, J.J.: Privacy and Security Issues in Online Social Networks in Future Internet (2018)
38. Elisa, N., Yang, L., Chao, F., Cao, Y.: A framework of blockchain-based secure and privacy-preserving E-government system. Wirel. Netw. 1–11 (2018). https://doi.org/10.1007/s11276-018-1883-0
39. Nakamoto, S.: Bitcoin: a peer-to-peer electronic cash system (2008). https://www.bitcoin.org/bitcoin.pdf
40. Ramalingam, V., Mariappan, D., Premkumar, S., Ramesh Kumar, C.: Chapter 4 Distributed Computing in Blockchain Technology. Springer (2022)

Plant Disease Detection using Image Processing

Anupama Mishra[1], Priyanka Chaurasia[2], Varsha Arya[3,4(✉)],
and Francisco José García Peñalvo[5]

[1] Himalayan School of Science & Technology, Swami Rama Himalayan University,
Dehradun, India
[2] Ulster University, Northern Ireland, UK
p.chaurasia@ulster.ac.uk
[3] Insights2Techinfo, Bangalore, India
varshaarya21@gmail.com
[4] Lebanese American University, Beirut 1102, Lebanon
[5] Department of Informatics and Automation, University of Salamanca,
Salamanca, Spain
fgarcia@usal.es

Abstract. Plant diseases have an impact on the development of their particular species, hence early detection is crucial. Numerous Machine Learning (ML) models have been used for the identification and classification of plant diseases, this field of study now appears to have significant potential for improved accuracy. In order to identify and categorise the signs of plant diseases, numerous developed/modified ML architectures are used in conjunction with a number of visualisation techniques. Additionally, a number of performance indicators are employed to assess these structures and methodologies.

Keywords: Plant disease · Machine learning

1 Introduction

India may be a cultivated country and concerning seventieth of the Population depends on agriculture. Farmers have large range of diversity for selecting numerous appropriate crops and finding the suitable pesticides for plant [4,6,8]. Hence, harm to the crops would end in large loss in productivity and would ultimately have an effect on the economy. Leaves being the foremost sensitive a locality of plants show illness symptoms at the earliest. The crops should be monitored against diseases from the terribly 1st stage of their life-cycle to the time they are ready to be harvested [10,12,14]. Initially, the manoeuvre accustomed monitor the plants from diseases was the quality eye observation that is a long technique that desires specialists to manually monitor the crop fields. inside the recent years, variety of techniques are applied to develop automatic and semi-automatic illness detection systems and automatic detection of the

N. Nedjah et al. (Eds.): ICSPN 2021, LNNS 599, pp. 227–235, 2023.
https://doi.org/10.1007/978-3-031-22018-0_21

diseases by simply seeing the symptoms on the plant leaves makes it easier yet as cheaper. These systems have to this point resulted to be quick, cheap and a lot of correct than the quality technique of manual observation by farmers In most of the cases illness symptoms ar seen on the leaves, stem and fruit. The plant leaf for the detection of illness is taken under consideration that shows the illness symptoms. There are several cases wherever farmers do not have a completely compact data concerning the crops and conjointly the illness which will get affected to the crops. This paper could also be effectively utilised by farmers thereby increasing the yield instead of visiting the skilled and obtaining their recommendation. the most objective is not solely to discover the illness mistreatment image process technologies [17,18,20]. It conjointly directs the user on to Associate in Nursing e-commerce web site wherever the user should buy the medicine for the detected illness by comparison the rates and use befittingly in step r with the directions given. Greenhouse conjointly referred to as a building, or, if with enough heating, a Hodr house, could also be a structure with walls and roof created mainly of clear material, like glass, inside that plants requiring regulated atmospheric condition ar fully grown. As greenhouse farming is gaining a lot of importance currently a day's, this paper helps the greenhouse farmers in Associate in Nursing economical means. numerous techniques may be accustomed review the illness detection and discuss in terms of varied parameters. As per Figs. 1, 2 and 4, fungi, bacteria, and viruses causes most plant illnesses. Disease symptoms are obvious signs of infection [22–24]. Plant diseases cause visible spores, mildew, or mould and leaf spot and yellowing. Fungi cause plant diseases. Fungi infect plants by stealing nutrients and breaking down tissue. Plant diseases are prevalent. Plants show disease symptoms or effects. Fungi infections cause leaf patches, yellowing, and birdseye spots on berries. Some plant illnesses appear as a growth and mould on the leaves.The paper is organized into the following sections. 1st section provides a quick introduction to the importance of illness detection. Second section discusses the current work disbursed recently throughout this space and conjointly reviews the techniques used. Section 3 includes methodologies utilized in our paper. Section 4 shows the experimantal details, Sect. 5 discussed the results and performance analysis and lastly, Sect. 6 concludes this paper along with future directions.

2   Related Work

Alternia leaf spot, brown spot, mosaic disease, grey spot, and rust impact apple productivity. Current analysis lacks proper and timely detection of apple diseases to ensure apple trade health. SSD, DSSD, and R-SSD are object detection algorithms with two parts: the pre-network model, which extracts basic options. The opposite is a multi-scale feature map-using auxillary structure [1,2]. There are many machine learning techniques that are used to solve many real world problems [3]. Using square geometrician distances, Kmeans segmentation divides the leaf picture into four groups. Color Co-occurrence technique is used to extract colour and texture features [7]. Abuse classification uses neural network detection and rule-based backpropagation. System disease detection and

Fig. 1. Leaf infected by bacteria

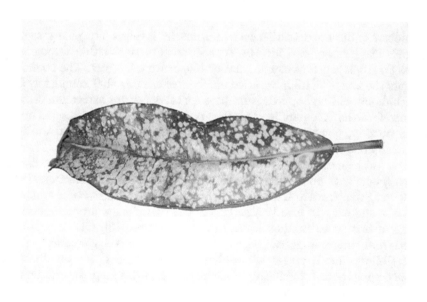

Fig. 2. Leaf infected by virus

categorization accuracy was 93%. Observe leaf fungus on fruit, vegetable, cereal, and industrial crops. Every crop is grown differently [8]. For fruit crops, k-means agglomeration is the segmentation method employed, with texture options focusing on ANN [9] and closest neighbour algorithms to achieve an overall average accuracy of 90.6%. For vegetable crops, chan-vase segmentation, native binary patterns for texture feature extraction, SVM, and closest neighbour classification achieved an overall average accuracy of 87.9%. Mistreated grab-cut formula divides industrial crops. Ripple-based feature extraction has been utilised as a classifier with an 84.8% average accuracy. Mistreatment kmeans cluster and smart edge detector separate cereal crops. Extract colour, shape, texture, colour texture, and random rework. SVM Associate in Nursing closest neighbour classifiers achieved 83.6% accuracy. A processed image of a cold plant leaf shows its health. Their strategy is to limit Chemicals to the morbid cold plant. MATLAB extracts features and recognises images. This paper is preprocessed. Filtering, edge detection, morphology. Laptop vision expands the image classification paradigm. Here, a camera captures images and LABVIEW creates the GUI [12,19]. The FPGA and DSP-based system monitors and manages plant diseases. The FPGA generates plant picture or video for viewing and labelling. The DSP TMS320DM642 processes and encrypts video/image data. Single-chip nRF24L01 pair. Knowledge transfer uses 4 GHz sender. It uses multi-channel wireless connection to reduce system cost and has two data compression and transmission methods.

3 Proposed Methodology

The process of disease detection system primarily involves four phases as shown in Fig. 3. The primary part involves acquisition of pictures either through smart devices [11,16,21] such as camera and mobile or from internet. The second part segments the image [5,15] into varied numbers of clusters that completely different techniques will be applied. Next part contains feature extraction strategies and therefore the last part is regarding the classification of diseases. Imaging In this portion, plant leaf photographs are gathered using digital media like cameras, mobile phones, etc. with required resolution and size. Internet-sourced photos are also acceptable. The applying system developer loves image data formation. Image data boosts the classifier's effectiveness in the detection system's final stage. Segmentation This component simplifies an illustration so it's more significant and easy to investigate. This is the basic image processing strategy because of feature extraction. k-means agglomeration [13], Otsu's algorithmic method, and thresholding can segment images. k-means agglomeration organises possibilities into K categories. Minimizing the distances between objects and clusters completes the classification. Highlighting In this step, alternatives from the interest space must be extracted. These choices authenticate an image's meaning. Color, shape, and texture are supported. Most researchers want to use texture to detect plant diseases. Gray-level co-occurrence matrix (GLCM),

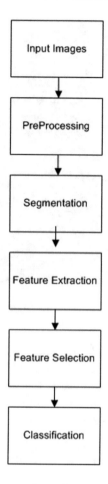

Fig. 3. Approach to classify the disease

colour cooccurrence approach, spatial greylevel idependence matrix, and bar graph-based feature extraction may be used to construct the system. GLCM classifies textures. Classification The classification section checks if the image is healthy. Some works classify unhealthy images into various disorders. For classification, MATLAB needs a classifier package routine. Researchers have used KNN, SVM, ANN, BPNN, Naive Bayes, and call tree classifiers in recent years. SVM is a popular classifier. SVM is an easy-to-use and reliable classifier.

4   Experimental Details

4.1   Experiment Setup and Data Sets

To implement the Machine Learning methods, the experiment was performed in the environment, which includes Python 3.10.4 64 bit on Jupyter Notebook

Fig. 4. Leaf infected by fungal infection

in Visual Studio Code with CPU: 11th Gen Intel(R) Core(TM) i5-1135G7 with Clock Speed @ 2.40 GHz–3.32 GHz , GPU RAM: 16GB DDR4, Storage: 1TB HDD with 256GB SSD. The scikit-learn, Matplot, Numpy, and Panda libraries were used through out the experiment and performance evaluation. The data set is collected from the email spam folder and normal mails.

5   Result and Performance Evaluation

Coaching and testing are distinct. One is in a research lab, where the model is tested with a constant dataset for training and testing. Field condition is the contrary, meaning our model was tested with $64000 world photos (land). Since the lighting circumstances and backdrop features of the $64000 field samples are different, our model may have a poor accuracy compared to the accuracy values in the science lab. To counteract this, we included a variety of photos in the training part (heterogeneity). For the evaluation of the performance in classification machine learning, we have the following metrics:

RECALL: how many spam emails recalled from all spam emails.

PRECISION: what is the ratio of email correctly classified as spam.

ACCURACY: it measures how many observations, both positive and negative, were correctly classified.

F1-Score: it combines precision and recall into one metric. The higher the score the better our model is.

ROC Curve: It is a chart that visualizes the tradeoff between true positive rate (TPR) and false positive rate (FPR). Basically, for every threshold, we calculate TPR and FPR and plot it on one chart. Of course, the higher TPR and the lower FPR is for each threshold the better and so classifiers that have curves that are more top-left side are better (Tables 1 and 2).

Table 1. Confusion matrix

Random forest		Naive Bayes	
842	1	834	9
25	271	2	294

Table 2. Classification report

Classifiers	Precision	Recall	F1 score	Accuracy
Random Forest	0.97	1.00	0.98	97.71
Naive Bayes	1.00	0.99	0.99	99.03

The accuracy of period detection of apple plant disease victimisation deep learning approach supported improved convolution neural networks is a smaller amount compared to the planned system as a result of it detects multiple diseases in an exceedingly single system.

6 Conclusion

This planned work is concentrates on the accuracy values throughout the $64000 field circumstances, and this work is reinforced by having many disease photographs. Therefore, an application that was developed for the detection of disease-affected plants and healthy plants has been completed. In general, this process is carried out from the ground up, and the results are very accurate. The work that has to be done in the long term is to increase the number of photographs that are present within the preset information and to update the design so that it is more accurate in accordance with the dataset.

Acknowledgement. This research was partially funded by the Spanish Government Ministry of Science and Innovation through the AVisSA project grant number (PID2020-118345RB-I00).

References

1. Jiang, P., Chen, Y., Liu, B., He, D., Liang', C.: Period detection of apple leaf diseases mistreatment deep learning approach supported improved convolutional neural networks **7**, 06 (2019)
2. Behera, T.K., et al.: The NITRDrone dataset to address the challenges for road extraction from aerial images. J. Sig. Process. Syst. 1–13 (2022)
3. Jain, A.K., et al.: A survey of phishing attack techniques, defence mechanisms and open research challenges. Enter. Inf. Syst. **16**(4), 527–565 (2022)
4. Zhou, R., Kaneko, S., Tanaka, F., Kayamori, M., Shimizu, M.: Disease detection of Cercospora Leaf Spot in sugar beet by strong templet matching. Comput. Phys. Sci. Agri. **108**, 58–70 (2014)
5. Chopra, M., Singh, S.K., Sharma, A., Gill, S.S.: A comparative study of generative adversarial networks for text-to-image synthesis. Int. J. Softw. Sci. Comput. Intell. (IJSSCI) Bouarara, H.A. **14**(1), 1–12 (2022)
6. Barbedo, J.G.A., Godoy, C.V.: Automatic classification of Soybean diseases supported digital pictures of leaf symptoms', SBI AGRO (2015)
7. Barbedo, J.G.A.: A review on the most challenges in automatic disease identification supported visible vary images. Biosyst. Eng. **144**, 52–60
8. Bashish, D.A., Braik, M., Ahmad, S.B.: A framework for detection and classification of plant leaf and stem diseases. Int. Conf. Signal Image Process 113–118 (2010)
9. Ahmad, I., et al.: Ensemble of 2D Residual Neural Networks Integrated with Atrous Spatial Pyramid Pooling Module for Myocardium Segmentation of Left Ventricle Cardiac (2022)
10. Punajari, J.D., Yakkundimath, R., Byadgi, A.S.: Image process based mostly detection of fungous diseases in plants. Int. Conf. Data Commun. Technol. **46**, 1802–1808 (2015)
11. Cvitić, I., Peraković, D., Periša, M. et al: Ensemble machine learning approach for classification of IoT devices in smart home. Int. J. Mach. Learn. Cyber. **12**, 3179–3202 (2021). https://doi.org/10.1007/s13042-020-01241-0
12. Husin, Z.B., Aziz, A.H.B.A.: Ali Yeon Bin Md Shakaff Rohani Binti S Mohamed Farook, Feasibility Study on Plant Chili unwellness Detection mistreatment Image process Techniques. In: 2012 Third International Conference on Intelligent Systems Modelling and Simulation
13. N-Gram-Codon and Recurrent Neural Network (RNN) to Update Pfizer-BioNTech mRNA Vaccine. Int. J. Softw. Sci. Comput. Intell. (IJSSCI) **14**(1), 1–24
14. Zhang, C., Wang, X., Li, X.: Design of observation and management disease system supported DSP&FPGA. In: 2010 Second International Conference on Networks Security, Wireless Communications and Trusted Computing
15. Akilandeswari, J., Jothi, G., Dhanasekaran, K., Kousalya, K., Sathiyamoorthi, V.: Hybrid firefly-ontology-based clustering algorithm for analyzing tweets to extract causal factors. Int. J. Semant. Web Inf. Syst. (IJSWIS) **18**(1), 1–27 (2022)
16. Khoudja, M.A., Fareh, M., Bouarfa, H.: Deep embedding learning with autoencoder for large-scale ontology matching. Int. J. Semant. Web Inf. Syst. (IJSWIS) **18**(1), 1–18 (2022)
17. Omrani, E., Khoshnevisan, B., Shamshirband, S., Saboohi, H., Anuar, N.B., Nasir, M.H.N.: Potential of radial basis function based support vector regression for apple unwellness detection. J. Measuring 233–252 (2014)

18. Gharge, S., Singh, P.: Image process for soybean unwellness classification and severity estimation, rising analysis in computing, data, communication and Applications 493–500 (2016)

19. Zou, L., Sun, J., Gao, M. et al.: A novel coverless information hiding method based on the average pixel value of the sub-images. Multimed. Tools Appl. **78**, 7965–7980 (2019). https://doi.org/10.1007/s11042-018-6444-0

20. García-Peñalvo, F.J., et al.: Application of artificial intelligence algorithms within the medical context for non-specialized users: the CARTIER-IA patform. Int. J. Interact. Multimedia Artif. Intell. **6**(6), 46–53 (2021). https://doi.org/10.9781/ijimai.2021.05.005

21. Alsmirat, M.A. et al.: Accelerating compute intensive medical imaging segmentation algorithms using hybrid CPU-GPU implementations. Multimedia Tools Appl. **76**,(3), 3537–3555 (Feb 2017). https://doi.org/10.1007/s11042-016-3884-2

22. García-Peñalvo, F.J. et al.: KoopaML: a graphical platform for building machine learning pipelines adapted to health professionals. Int. J. Interact. Multimedia Artif. Intell. In Press

23. Pari Tito, F., García-Peñalvo, F.J., Pérez Postigo, G.: Bibliometric analysis of media disinformation and fake news in social networks. Revista Universidad y Sociedad **14**(S2), 37–45 (2022)

24. García-Peñalvo, F.J.: Developing robust state-of-the-art reports: systematic literature reviews. Educ. Knowl. Soc. **23**(e28600) (2022). doi: https://doi.org/10.14201/eks.28600

A Deep Learning Based Approach to Perform Fingerprint Matching

Vivek Singh Baghel[1(✉)], Smit Patel[2], Surya Prakash[1], and Akhilesh Mohan Srivastava[1]

[1] Department of Computer Science and Engineering, Indian Institute of Technology Indore,
Indore 453552, India
{phd1801201005,surya,phd1701101001}@iiti.ac.in
[2] Department of Electrical Engineering, Indian Institute of Technology Indore,
Indore 453552, India
ee180002057@iiti.ac.in

Abstract. As we move towards a technological-driven era, the traditional methods of authenticating an individual are becoming redundant and easier to crack. Biometric authentication has emerged as a very promising authentication method as it uses physiological and behavioral characteristics of the human body, which are unique to every individual. To adopt recent developments in deep learning and apply them to biometric authentication, we propose a novel Vision Transformer (ViT) based Siamese network framework for fingerprint matching in a fingerprint authentication system. Our primary focus is holistic, and an end-to-end pipeline has been constructed and implemented using an ensemble of task-specific algorithms to procure the best possible result from the model. We also endeavor to identify specific problems in the application of ViT to fingerprint matching and used two existing approaches, which are Shifted Patch Tokenization (SPT) and Localized Self Attention (LSA), to tackle those shortcomings effectively. We have considered two variations for the model, namely Intermediate-Merge (I-M) Siamese network and Late-Merge (L-M) Siamese network, and tested the performances on the IIT Kanpur fingerprint database. The obtained results in terms of accuracy, True Positive Rate (TPR), and True Negative Rate (TNR) clearly show the significant performance of the proposed technique for fingerprint matching by means of an approach based on deep learning.

Keywords: Fingerprint matching · Vision Transformer · Siamese networks · Deep learning.

1 Introduction

In today's technological-driven era, protecting identity and confidential data has become extremely potent. The first means to do so were a simple lock and key. As technology advanced, the keys were replaced by passwords, and locks were replaced by sophisticated authentication algorithms. Computerization of verification systems also indicates that spurious authentication can lead to damage on multiple fronts, including data, finance, privacy, and so on. Credential replication, malware, and retracing

© The Author(s), under exclusive license to Springer Nature Switzerland AG 2023
N. Nedjah et al. (Eds.): ICSPN 2021, LNNS 599, pp. 236–247, 2023.
https://doi.org/10.1007/978-3-031-22018-0_22

techniques have made it possible to retrieve passwords and abate the novelty of such methods [21, 27].

In order to overcome the limitations of the traditional password protection regime, biometric authentication systems [20, 25] have been widely used. In such systems, a unique biological characteristic of an individual is used to verify one's identity. Biometrics comprise a broad category of traits, which are physiological and behavioral biometric traits, including fingerprint, face, iris, ear, voice, and so on. Amongst all of these, the most popular and widely-used biometric characteristic is the fingerprint. A fingerprint contains a unique identification feature pattern for humans, consisting of ridges and valleys. The unique features, i.e., minutiae points, extracted from these ridge patterns are used to authenticate an individual. Fingerprint-based authentication systems are popular because of their scalability, convenience, and ease of usage compared to other biometric traits.

In the conventional fingerprint matching process, tedious preprocessing is involved in matching two fingerprint images. Accurate extraction of minutiae points is dependent upon the image processing algorithms or physical techniques, and based on the similarity of the pre-selected features; the algorithm determines a match. These methods are computationally inefficient and not very reliable [8, 29]. On the contrary, when machine learning, particularly deep learning [10, 23, 24], is employed for fingerprint matching, a majority of preprocessing is eliminated, and the results procured are also significant. This is because the biometric features that are extracted and learned using deep learning models have superior discriminative ability for inter-class samples and high similarity for intra-class samples. Moreover, deep learning-based approaches are driven by the data they are trained on; hence, given a quality dataset, deep learning algorithms tend to effectively fit and replicate the results meticulously. In this paper, the presented work has several attributes which make it novel in the domain of biometrics, particularly fingerprint matching using a deep learning-based approach. To the best of our knowledge, the following are the salient unique attributes of this paper.

- This is the first work to employ and implement ViTs [9] for fingerprint matching in a Siamese setting for learning similarity between the two inputs and also propose variations in its configuration.
- We have incorporated crucial modifications (SPT and LSA) that are inspired from [17], which mainly tackle the problems faced by the model due to limited dataset size and inefficient working of the self-attention mechanism. This further increases the overall performance.
- We also propose an end-to-end pipeline for fingerprint matching and create custom and problem statement-focused dataset augmentation flow by drawing insights from similar works.

The rest of the paper is organized as follows. Section 2 discusses some of the previous work related to fingerprint authentication and ViTs. Various steps of the proposed model are discussed in Sect. 3. The experimental results are shown in Sect. 4. At last, the paper is concluded in Sect. 5.

2   Related Work

Ever since biometric authentication and verification have taken the place of traditional authentication methods, researchers worldwide have deeply invested in making the algorithms more secure, reliable, and fail-proof. Fingerprint authentication systems, being at the heart of biometric systems, are evolving with time and rapidly advancing technology. Typically, there are two primary components of fingerprint authentication, which are fingerprint feature extraction, and fingerprint similarity detection.

Initially, common fingerprint features such as minutiae points, ridge patterns, and singular points were extracted as potential features, and matching was performed by directly comparing these features. In [22], a fingerprint matching algorithm has been proposed, where the minutiae points are utilized as the key discriminatory features. To enhance the performance, Jain et al. [15] have combined level-3 features such as ridges and pores with fingerprint patterns and minutiae to perform a holistic comparison using more fingerprint details. In [14], the authors have proposed a hybrid approach where minutiae points and features from the texture of the fingerprint are used for determining the similarities between the gallery and probe fingerprint images.

With the advent of artificial intelligence, researcher shifted their focus from traditional feature mapping to more sophisticated machine learning and deep learning algorithms. A fingerprint-based genetic algorithm has been proposed by [26] which finds the optimal vector transformation between two different fingerprints. In [13], a Hidden Markov Model-based approach has been devised for detecting the similarity between the gallery and probe fingerprint images. Liu et al. [18] have designed a deep learning model, namely Finger ConvNet, as a novel approach to improve the speed and accuracy of fingerprint matching.

Siamese networks basically comprise two exactly the same arms, which are trained in parallel. A pair of images are fed to this network, which returns the similarity between the input images as an output. In this regard, Chowdhury et al. [5] have proposed a multi-scale dilated Siamese neural network for fingerprint images, where the work mainly focused on showing the importance of minutiae containing patches over the without minutiae patches. Alrashidi et al. [2] have applied adversarial learning and used multi-sensor data in a Siamese setting for the fingerprint matching. Vision Transformer (ViT) was introduced by the Google Brain team in [9] as a powerful architecture for image classification. Since then, it has been adopted and modified for various tasks such as image encoding [30] and moving object detection [3].

This paper intends to harness the superior image feature learning capability of the ViTs and apply it to the task of fingerprint matching in a Siamese network setting. To the best of our knowledge, this is the first work that employs ViT for similarity detection between the gallery and probe fingerprint images.

3   Proposed Technique

This paper proposes a novel method for fingerprint matching by utilizing the Siamese network architecture. In the Siamese network, two identical parallel networks work for feature extraction. In the proposed technique, ViT is used as the arm of the Siamese network. In addition, SPT and LSA have been utilized to solve the problem of the

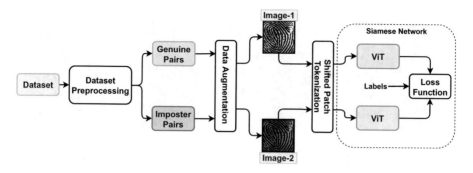

Fig. 1. Block diagram representing various steps involved in the training process of the proposed model

limited number of images available in the fingerprint database and the inefficiency of the self-attention mechanism, respectively. The complete architecture of the proposed model has been depicted in Fig. 1. Further, the various steps involved in the proposed technique are discussed as follows.

3.1 Data Preprocessing

The fingerprint alignment is one of the crucial processes in fingerprint-based authentication systems. Therefore, we have utilized the singular point in order to roughly align the fingerprint images before moving towards the further processing of the data. A singular point is defined as a point where ridge curvature in a fingerprint image is maximum. It is important to note that commonly, there is only one singular point (i.e., core point) for every fingerprint; hence, they are popularly used as a feature for registration and identification. In the proposed alignment algorithm, the singular point coordinates of all samples for each subject are extracted and stored. Now, the arithmetic mean of all the extracted singular points' coordinates is calculated and stored separately for every subject. Essentially, for each subject, the new designated singular point is the average calculated, which is mathematically shown for every i^{th} subject as follows.

$$\left(x_{i_{new}}, y_{i_{new}}\right) = \left(\sum_{j=1}^{n} \frac{x_{ij}}{n}, \sum_{j=1}^{n} \frac{y_{ij}}{n}\right) \tag{1}$$

where $j \in [1, n]$ and $n = 4$ for the database utilized in the experimentation. After obtaining the updated singular points, the images are translated such that the new singular point of the image lies at the coordinates procured. As seen in Fig. 2(a), the original image is on the left, and the translated image, according to the aforementioned algorithm, is on the right. The undefined portion of the image after alignment (the dark portion in Fig. 2(a)) is cropped out to avoid learning from pixels by the model which does not exist.

Since the translation of fingerprint images is not uniform for samples of a subject, the size of images in the dataset becomes erratic and cannot be used directly for training

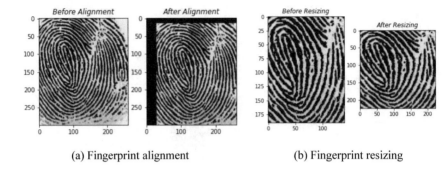

(a) Fingerprint alignment   (b) Fingerprint resizing

Fig. 2. An example to represent (a) An original image and image after singular point alignment, and (b) A fingerprint image before and after preprocessing

the model. To resolve this issue, we first determine the dimensions of the minimum portion in an image, which is common to all images in the entire dataset. Further, that intersect portion is extracted from every image regardless of its index. Following this, the image is resized to the dimensions 224×224 as prescribed in [9]. In Fig. 2(b), the input and output fingerprint images are shown after the data preprocessing step in the pipeline.

3.2   Vision Transformer Model

Vision Transformer (ViT) was first introduced in [9]. The ViT outperformed the state-of-the-art models in image classification on popular databases. In ViT, the input image is divided into square patches to execute a form of lightweight windowed attention. In order to get here from Natural Language Processing (NLP), each patch in the image problem is analogous to a word in the language problem. These patches are flattened and passed via a single feed-forward layer to produce a linear patch projection for feeding into the Transformer. The ViT only uses the Encoder setup from the original Transformer architecture [28], and the internal organization is more or less the same to retain the performance. The embedding matrix E, as indicated in [9], is randomly generated and contained in this feed-forward layer.

One issue with Transformers is that the order of a sequence is not automatically guaranteed because data is sent in at once rather than timestamp-wise as in Recurrent Neural Networks (RNNs) and Long Short-Term Memory (LSTM). To overcome this, the original Transformer work [28] recommends using positional encoding/embedding, which place the inputs in a specific order. Hence for ViT, the positional embedding matrix E_{pos} is randomly generated and added to the concatenated matrix containing the patch and embedding projections. Moreover, to enable the output of a single probability rather than a sequence of vectors, the ViT model [9] adopted the approach from Bidirectional Encoder Representations from Transformers (BERT) [6] and concatenated a learnable $[class]$ parameter with the patch projections. If the $[class]$ token is not used during the process, only positional embedding is added to the output of visual tokens. The Transformer Encoder's outputs are then used to predict the image class probability using a Multilayer Perceptron (MLP) layer. The input features effectively capture the core of the image, making the MLP head's classification task easier. The Transformer provides multiple outputs, but the MLP head only receives the output relating to the particular [*class*] embedding, and the other outputs are disregarded. The output gives the probability distribution of the classes of the image. Furthermore, as ViT doesn't use Convolutional Neural Networks (CNNs), it structurally lacks locality inductive bias (weights or assumptions of the model to learn target and generalize beyond train data) as compared to CNNs, and they require a very large amount of training data to obtain satisfactory visual representation. Since our dataset is limited to a few hundred test subjects, using the traditional ViT approach yields low accuracy. Hence, Shifted Patch Tokenization (SPT) [17] has been utilized to solve the problem due to the small size of training data. In addition, Locality Self Attention (LSA) [17] has also been utilized to handle the problem of inefficient functioning of the self-attention mechanism for images in Transformer. Further, the above-discussed ViT is used in a Siamese setting in the proposed technique as shown in Fig. 1, which is further discussed as follows.

3.3   Siamese Network with ViT

A Siamese neural network is a modification of Artificial Neural Networks (ANNs) that are widely used for machine learning. It is an artificial neural network that uses the same weights while working in tandem on two different input vectors to compute comparable output vector. Essentially, the two arms have the same configuration with the same parameters and weights, and parameter updating is mirrored across both sub-networks as depicted in Fig. 3. There are mainly three different types of Siamese networks that are classified on the basis of their merging position in the architecture [11]: (1) Late-Merge Siamese Networks, (2) Intermediate-Merge Siamese Networks, and (3) Early-Merge Siamese Networks. We have used two of these three variations of Siamese networks for our experiment, as shown in Fig. 3. In these variants of the Siamese network, we have used a ViT architecture in place of a traditional fully connected CNN or RNN so that the power of ViT can be harnessed for similarity detection between fingerprint images as shown in the Siamese network module of Fig. 1. In the Siamese network, we pass two input image vectors in the model and procure an output probability at the tail, which is used to determine the label of the image pair, i.e., genuine pair (label as 1) or an imposter pair (label as 0/−1) as shown in Fig. 4. Further, the aforementioned two variations of the Siamese network used in the proposed model are discussed.

Late-Merge Siamese Network: In this Siamese network setting, the network's left and right arms remain distinct throughout the model. Hence, for the two input vectors, let's say v_1 and v_2, two outputs are received, say f_1 and f_2. Further, the loss function, i.e., the Contrastive loss function in our case, takes two embedding vectors as input and equates the Euclidean distance between them if the label is 1 or behaves like Hinge loss function [12] if the label is 0. The expression for Contrastive loss function is given in Eq. 2, where y is the label, f_1 and f_2 are the output vectors, m is the margin, and $d = ||f_1 - f_2||_2$ is the Euclidean distance between f_1 and f_2.

$$L(f_1, f_2, y) = yd^2 + (1 - y)\max(m - d, 0)^2 \qquad (2)$$

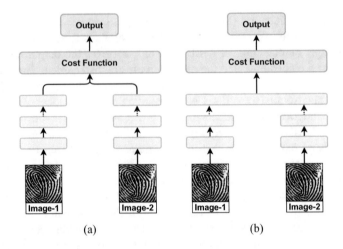

Fig. 3. Representation of two different types of Siamese networks, (a) Late-Merge Network, and (b) Intermediate-Merge Siamese Network

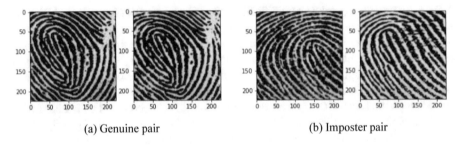

(a) Genuine pair   (b) Imposter pair

Fig. 4. Representation of genuine and imposter pairs formed for the training purpose after data preprocessing

It is important to note that the output layer of the late-merge network has two neurons, i.e., it outputs two different probabilities with respect to the genuine and imposter class. Further, we also train our model using the Cosine Embedding loss function for empirical purposes. The formula for the same is given as follows.

$$L(f_1, f_2, y) = \begin{cases} 1 - \cos(f_1, f_2), & y = 1 \\ \max(0, \cos(f_1, f_2) - m), & y = -1 \end{cases} \tag{3}$$

Intermediate-Merge Siamese Network: In this architecture, two input vectors/images v_1 and v_2 are given separately to the model. They follow the pipeline separately while sharing the weights up until the penultimate layer in the dense network of the Siamese architecture. The absolute difference between the two internal vectors is calculated before the output layer of the Siamese network and then fed to the final layer. We have utilized the Binary Cross-Entropy (BCE) loss function in this architecture for training the model. Mathematically, the loss function can be seen in Eq. 4. In the case of an intermediate-merge Siamese network, the output layer has a single neuron, i.e., it outputs a single probability.

$$L(p, y) = -(y \log(p) + (1 - y) \log(1 - p)) \tag{4}$$

4   Experimental Analysis

This section discusses the experimental analysis of the proposed Siamese network-based architecture for fingerprint matching. To perform the experimentation, we have utilized the IIT Kanpur fingerprint database [1]. In this database, the number of subjects is 1378, where each subject contains 4 samples. The genuine and imposter pairs are formed by combining the samples from the same subject and different subjects, respectively, after performing the alignment as discussed in Sect. 3.1. As the fingerprint database size is insufficient to train the deep learning-based architecture like ViT, data augmentation has been utilized, which is discussed as follows.

4.1   Data Augmentation

Since deep learning models are usually data hungry, different methods are applied to increase the size of the dataset while maintaining the quality of the original set. Data augmentation refers to approaches that are used to increase the amount of data by adding slightly changed copies of current data or creating new synthetic data from existing ones. When training a machine learning or deep learning model, data augmentation functions as regularization and also helps to reduce overfitting. In our proposed technique, two of the following three image augmentation approaches have been selected randomly to perform the data augmentation for training purposes.

- Addition of random Gaussian noise
- Random change in image brightness
- Random change in image contrast.

Each data augmentation approach has the probability of 1.0 for execution, and the generated new pairs are assigned the same label as that of the parent pairs. The augmented new pairs are then appended to the original image pair dataset. These augmented genuine and imposter pairs are then used to train the proposed model.

4.2   Training of the Proposed Model

There are a lot of hyper-parameters and functions involved when training a complex deep learning model. Since experimenting with every permutation and combination of every parameter is not feasible, some parameters are fixed based on empirical data and an existing work [4]. In the proposed model, we have split the dataset into train and test data in the ratio 85-15, where the 15% validation data has been taken from the train set during the training process. The optimizer used for training is $AdamW$ [19] as it is an improvement over $Adam$ by incorporating variable weight decay. In order to break the

Table 1. Comparative study between Late-Merge (L-M) and Intermediate-Merge (I-M) Siamese network

Siamese network	Loss function	Accuracy	TPR	TNR
L-M	Cosine Embedding	0.572	0.387	0.723
L-M	**Contrastive**	**0.697**	**0.817**	**0.628**
I-M	BCE	0.5	0.25	0.691
I-M	BCEWithLogits	0.52	0.32	0.677

Bold entries represent the best results achieved in respective combination of parameter

Table 2. Performance of the proposed model on various batch sizes

Batch size	Accuracy	TPR	TNR
32	0.470	0.617	0.408
64	0.575	0.696	0.448
128	**0.697**	**0.817**	**0.628**
256	0.607	0.704	0.562

Bold entries represent the best results achieved in respective combination of parameter

plateau for validation accuracy, a scheduler is utilized to reduce the learning rate. The model has been trained for 10 epochs on Nvidia Tesla V100 32GB GPU with a learning rate of 0.001 and multiple batch size configurations.

As it has been discussed in Sect. 3.3, the proposed model has been trained on two different variations of Siamese settings, i.e., Intermediate-Merge (I-M) and Late-Merge (L-M) Siamese network. Both of these networks are trained with two loss functions each, i.e., Contrastive loss and Cosine Embedding loss for I-M and BCE loss and BCE-WithLogits loss for L-M. Essentially, BCEWithLogits loss combines the BCE loss with a Sigmoid layer in one single class.

4.3   Results Analysis

It is aforementioned that the proposed model has been trained by means of two different settings of Siamese network and various loss functions as depicted in Table 1. In addition, the proposed model is trained by considering the different batch sizes and embedding sizes as shown in Tables 2 and 3, respectively. It can be inferred from Table 1 that the late-merge Siamese model using the Contrastive loss function outperforms all models having other loss functions. Here, all other parameters are kept the same during this experiment. In particular, it is evident that the late-merge Siamese network performs significantly better than intermediate-merge Siamese network, regardless of the loss function used. This is because merging at a later stage helps the model exploit the input images separately, thereby allowing the model to be more discriminative while learning features [11]. In contrast, merging at an intermediate position allows the model

Table 3. Performance of the proposed model on various embedding sizes

Embedding size	Batch size	Accuracy
256	128	0.618
512	**128**	**0.697**
768	128	0.652
1024	128	0.607

Bold entries represent the best results achieved
in respective combination of parameter

to exploit the input image features initially, and then these fused features are further used. This reduces the discriminative ability of the ViT in focus.

Furthermore, as we know that hyper-parameters play an important role in determining the final performance of the model; among these, it is highly crucial to select a proper batch size. It can be seen from Table 2 that when the model is trained with a batch size of 128, it produces the best results. It is also well-established that using larger batch sizes leads to significant degradation in the model quality as these methods tend to converge sharply, leading to poor generalization [16]. This phenomenon accounts for the lesser accuracy of the model with batch size 256. On the other hand, using smaller batch sizes causes the gradient estimation to become noisy and have a large variance. Moreover, using smaller batch sizes causes overfitting [7] on the minibatch distribution instead of the actual dataset distribution, which results in decreased accuracy. This explains the poor performance of the model with 32 (very poor) and 64 batch sizes. Furthermore, keeping all parameters constant, we have also tested for the most suitable embedding sizes for the model, and the results are given in Table 3. Even though 768 is the most widely used embedding for transformers, in our approach, it is not the most suitable embedding. The results indicate that empirically, 512 is the most optimal embedding size for our model. Finally, it can be observed by all these results analysis that the proposed model has performed pretty well in performing the fingerprint matching.

5   Conclusions

Deep learning is one of the most promising areas of research, which has shown a lot of potential, especially in areas such as computer vision, robotics, biometrics, and computer graphics. However, it has been relatively less explored in the field of fingerprint biometrics. With the advent of powerful models like Vision Transformers, there is a lot of scope for research and innovation in fingerprint biometric research. In this paper, we have proposed a completely novel pipeline employing Vision Transformers in multiple Siamese settings for the task of fingerprint matching. We have also incorporated Shifted Patch Tokenization and Locality Self Attention to eliminate our problem statement's limited dataset availability issue and inefficient functioning of the self-attention mechanism, respectively. The proposed model has been tested on the IIT Kanpur fingerprint database. The obtained results in terms of accuracy, TPR, and TNR are significantly good for performing the fingerprint matching. To the best of our knowledge, this is the first work involving Vision Transformers for extracting and matching fingerprint features. In the future, this work can be further extended to enhance the performance.

References

1. Ali, S.S., Ganapathi, I.I., Mahyo, S., Prakash, S.: Polynomial vault: a secure and robust fingerprint based authentication. IEEE Trans. Emerg. Topics Comput. **9**(2), 612–625 (2021)
2. Alrashidi, A., Alotaibi, A., Hussain, M., AlShehri, H., AboAlSamh, H.A., Bebis, G.: Crosssensor fingerprint matching using siamese network and adversarial learning. Sensors **21**(11) (2021)
3. Arnab, A., Dehghani, M., Heigold, G., Sun, C., Lucic, M., Schmid, C.: Vivit: a video vision transformer. In: Proceedings of IEEE/CVF International Conference on Computer Vision, pp. 6816–6826 (2021)
4. Bergstra, J., Bardenet, R., Bengio, Y., K'egl, B.: Algorithms for hyper-parameter optimization. In: Proceedings of Advances in Neural Information Processing Systems, vol. 24, pp. 1–9 (2011)
5. Chowdhury, A., Kirchgasser, S., Uhl, A., Ross, A.: Can a CNN automatically learn the significance of minutiae points for fingerprint matching? In: Proceedings of IEEE Winter Conference on Applications of Computer Vision (WACV), pp. 340–348 (2020)
6. Devlin, J., Chang, M.W., Lee, K., Toutanova, K.: BERT: Pre-training of deep bidirectional transformers for language understanding. In: Proceedings of the 2019 Conference of the North American Chapter of the Association for Computational Linguistics: Human Language Technologies, vol. 1 (Long and Short Papers), pp. 4171–4186. Association for Computational Linguistics (2019)
7. Dietterich, T.: Overfitting and undercomputing in machine learning. ACM Comput. Surv. **27**(3), 326–327 (1995)
8. Do, P., et al.: Developing a vietnamese tourism question answering system using knowledge graph and deep learning. Trans. Asian Low-Resour. Lang. Inf. Process. **20**(5), 1–18 (2021)
9. Dosovitskiy, A., Beyer, L., Kolesnikov, A., Weissenborn, D., Zhai, X., Unterthiner, T., Dehghani, M., Minderer, M., Heigold, G., Gelly, S., Uszkoreit, J., Houlsby, N.: An image is worth 16x16 words: Transformers for image recognition at scale. In: Proceedings of International Conference on Learning Representations (2021)
10. Elmisery, A.M., et al.: Cognitive privacy middleware for deep learning mashup in environmental IOT. IEEE Access **6**, 8029–8041 (2017)
11. Fiaz, M., Mahmood, A., Jung, S.K.: Deep siamese networks toward robust visual tracking. In: Visual Object Tracking with Deep Neural Networks. IntechOpen (2019)
12. Gentile, C., Warmuth, M.K.K.: Linear hinge loss and average margin. In: Kearns, M., Solla, S., Cohn, D. (eds.) Proceedings of Advances in Neural Information Processing Systems, vol. 11, pp. 1–7 (1998)
13. Guo, H.: A hidden markov model fingerprint matching approach. In: Proceedings of International Conference on Machine Learning and Cybernetics, vol. 8, pp. 5055–5059 (2005)
14. Jain, A., Ross, A., Prabhakar, S.: Fingerprint matching using minutiae and texture features. In: Proceedings of International Conference on Image Processing, vol. 3, pp. 282–285 (2001)
15. Jain, A.K., Chen, Y., Demirkus, M.: Pores and ridges: high-resolution fingerprint matching using level 3 features. IEEE Trans. Pattern Anal. Mach. Intell. **29**(1), 15–27 (2007)
16. Keskar, N.S., Mudigere, D., Nocedal, J., Smelyanskiy, M., Tang, P.T.P.: On large-batch training for deep learning: generalization gap and sharp minima. In: Proceedings of International Conference on Learning Representations, pp. 1-16 (2017)
17. Lee, S.H., Lee, S., Song, B.C.: Vision transformer for small-size datasets (2021)

18. Liu, Y., Zhou, B., Han, C., Guo, T., Qin, J.: A novel method based on deep learning for aligned fingerprints matching. Appl. Intell. **50**(2), 397–416 (2020)
19. Loshchilov, I., Hutter, F.: Decoupled weight decay regularization. In: Proceedings of International Conference on Learning Representations, pp. 1–18 (2019)
20. Maltoni, D., Maio, D., Jain, A.K., Prabhakar, S.: Handbook of fingerprint recognition, 2nd edn. Springer Publishing Company, Incorporated (2009)
21. Marechal, S.: Advances in password cracking. J. Comput. Virol. **4**(1), 73–81 (2008)
22. Prabhakar, S., Jain, A., Wang, J., Pankanti, S., Bolle, R.: Minutia verification and classification for fingerprint matching. In: Proceedings of International Conference on Pattern Recognition, vol. 1, pp. 25–29 (2000)
23. Sahoo, S.R., et al.: Hybrid approach for detection of malicious profiles in twitter. Comput. Electr. Eng. **76**, 65–81 (2019)
24. Sedik, A., et al.: Efficient deep learning approach for augmented detection of coronavirus disease. Neural Comput. Appl. **34**(14), 11423–11440 (2022)
25. Srivastava, A.M., Rotte, P.A., Jain, A., Prakash, S.: Handling data scarcity through data augmentation in training of deep neural networks for 3d data processing. Int. J. Semant. Web Inf. Syst. (IJSWIS) **18**(1), 1–16 (2022)
26. Tan, X., Bhanu, B.: Fingerprint matching by genetic algorithms. Pattern Recogn. **39**(3), 465–477 (2006)
27. Tembhurne, J.V., Almin, M.M., Diwan, T.: Mc-dnn: fake news detection using multi-channel deep neural networks. Int. J. Semant. Web Inf. Syst. (IJSWIS) **18**(1), 1–20 (2022)
28. Vaswani, A., Shazeer, N., Parmar, N., Uszkoreit, J., Jones, L., Gomez, A.N., Kaiser, L., Polosukhin, I.: Attention is all you need. In: Proceedings of Advances in Neural Information Processing Systems, vol. 30, pp. 1–11 (2017)
29. Yang, J., Xie, S., Yoon, S., Park, D., Fang, Z., Yang, S.: Fingerprint matching based on extreme learning machine. Neural Comput. Appl. **22**(3), 435–445 (2013)
30. Zhang, P., Dai, X., Yang, J., Xiao, B., Yuan, L., Zhang, L., Gao, J.: Multi-scale vision longformer: a new vision transformer for high-resolution image encoding. In: Proceedings of the IEEE/CVF International Conference on Computer Vision, pp. 2998–3008 (2021)

Convolutional Neural Network and Deep One-Class Support Vector Machine with Imbalanced Dataset for Anomaly Network Traffic Detection

Kwok Tai Chui[1](✉), Brij B. Gupta[2,3](✉), Hao Ran Chi[4], and Mingbo Zhao[5]

[1] Hong Kong Metropolitan University (HKMU), Hong Kong, China
jktchui@hkmu.edu.hk
[2] International Center for AI and Cyber Security Research and Innovations, & Department of Computer Science and Information Engineering, Asia University, Taichung 413, Taiwan
bbgupta@asia.edu.tw
[3] Lebanese American University, Beirut 1102, Lebanon
[4] Instituto de Telecomunicações, 3810-193 Aveiro, Portugal
ytchr@av.it.pt
[5] School of Information Science & Technology, Donghua University, Shanghai 200051, China
mzhao4@dhu.edu.cn

Abstract. Anomaly detection of network traffic is important for real-time network monitoring and management. The research challenges for anomaly network traffic detection (ANTD) are attributable to the nature of highly imbalanced dataset of abnormal samples and poor generalization. Generating additional training data or undersampling does not perform well with highly imbalanced dataset. The drives to the common formulation of ANTD as an one-class classification problem. Convolution neural network is utilized for feature extraction. This is followed by a deep one-class support vector machine classifier with customized kernel via multiple kernel learning to address the issue of generalization and overfitting. The deep architecture leverages the performance of traditional support vector machine to a large extent. Performance evaluation reveals that the proposed algorithm achieves accuracy of 97.5%. Ablation studies show that our algorithm enhances the accuracy by 2.52–15.2%. Compared with existing work, our algorithm enhances the accuracy by 3.07%. Several future research directions are discussed for further exploration and analysis.

Keywords: Anomaly detection · Convolutional neural network · Deep support vector machine · Imbalanced dataset · Network traffic · One-class classification

© The Author(s), under exclusive license to Springer Nature Switzerland AG 2023
N. Nedjah et al. (Eds.): ICSPN 2021, LNNS 599, pp. 248–256, 2023.
https://doi.org/10.1007/978-3-031-22018-0_23

1   Introduction

The abnormalities of network traffic contain crucial information for network management on potential malicious and sudden traffic events. Recently, many machine learning algorithms were proposed for anomaly network traffic detection (ANTD), in which the overview can be referred to review articles [1,2]. A major portion of network traffic events is normal event that forms the nature of imbalanced network traffic dataset. By contrast, the abnormalities of network traffic are minority samples in the dataset. Building a classification model with imbalanced dataset usually yields biased classification towards the majority class [3–5]. There are two common approaches to respond to the issue of imbalanced dataset (i) generating additional training data for the minority classes using data augmentation algorithms [6–8] or data generation algorithms such as generative adversarial network [9,10,13]; and (ii) undersampling of the majority class [11,12,15]. However, former approach cannot significantly reduce the imbalanced ratio whereas the latter approach sacrifices the availability of the samples in the majority class.

Formulating the ANTD as one-class classification problem is considered to take the advantage from highly imbalanced dataset given the availability of a vast amount of normal samples. The rest of the paper is organized as follows. Section 1 is organized with three subsections to cover a literature review, a summary of research limitations, and a highlight of our research contributions. Section 2 presents the methodology. It is followed by the performance evaluation of proposed algorithm and comparison with existing works in Sect. 3. At last, a conclusion is drawn along with future research directions.

1.1   Literature Review

To provide a fair comparison between various algorithms, only related works employed MAWILab dataset in performance evaluation and analysis are included in the discussion. In [14], a hybrid approach was proposed to fuse a multi-channel generalized likelihood ratio test (MCGLRT) and multi-scale decomposition (MSD) for ANTD. It achieved area under receiver operating characteristic curve (AUC) of 92.1% that improved the performance by 13.8–91.6% compared with other algorithms using principal component analysis, wavelet packet transform, switched subspace-projected basis, and randomized basis approaches. Another work [16] proposed a hybrid long short-term memory (LSTM) and fully connected network (FCN) for ANTD. It achieved AUC of 98.2%. A preliminary study was conducted in [17] using convolutional neural network for ANTD. The performance evaluation extracted 84000 samples from the population of the dataset (11,649,810). The average accuracy was 67% in a five-round repetitions. LSTM was also adopted with the incorporation of an OCSVM [18]. The achieved AUC of 94.6%. Author in [20] proposed a model for Handling Data Scarcity Through Data Augmentation in Training of Deep Neural Networks for 3D Data Processing.

1.2   Research Limitations

The research limitations of the existing works are summarized as follows (i) Existing works did not employ cross-validation to examine the issue of potential model overfitting and to better fine-tune the ANTD model; (ii) Some existing works [14–17] formulated the ANTD problem as multi-class classification problem where biased classification is observed towards the majority class; and (iii) Lack of investigation on the impact of imbalanced ratio on the model performance.

1.3   Research Contributions

In this research work, we have proposed a convolutional network network-based deep one-class support vector machine (CNN-DOCSVM) algorithm for ANTD. The research contributions are highlighted as follows: (i) The OCSVM is extended to deep architecture to further enhance the classification performance; (ii) customized kernel is designed using multiple kernel learning (MKL) in order to enhance the generalization and overfitting; (iii) performance analysis on the impact of imbalanced ratio on the model performance; (iv) Enhancement of accuracy by 3.07% compared with existing work; (v) Ablation studies show that our algorithm enhances the accuracy by 2.52–15.2%.

2   Methodology

This section is divided into three parts (i) overview of the architecture of CNN-DOCSVM; (ii) components of CNN; and (iii) DOCSVM with MKL.

2.1   Overview of CNN-DOCSVM

Figure 1 shows the architecture of CNN-DOCSVM. It consists of two modules (i) Feature extraction using CNN; and (ii) One-class classification model using DOCSVM with customized kernel via MKL. In regard to the CNN algorithm, the design is based on 1x1 convolution layers to reduce the model complexity [19,21,22]. It can be a much simpler design when one-class classification problem is formulated with sufficient samples. Inspired by [23], a global average pooling is adopted replace a traditional fully connected layer because global average pooling takes the advantage in native convolution structure by linking the categories with the feature maps. The OCSVM classification model is a deep architecture that extends the model performance. It is worth noting that typical kernel functions such as linear, polynomial, radial basis, and sigmoid kernels are for general purpose instead of a specific application. Therefore, we also customize the kernel function via MKL [24,25] for optimal performance.

2.2   Design and Components of CNN for Feature Extraction

The major components of the CNN algorithm are convolutional layers, maximum pooling layers, and a global average pooling layer. Their roles are explained as follows.

Convolutional layers: The layers contain many filters where the filter parameters are to be designed during model training. Convolution is applied between each filter and the inputs to form a feature map.

Maximum pooling layers: The layers manage the pooling operations to measure the maximum value in each patch of the feature map. This aims at finding the most representative feature in each patch.

Global average pooling layer: The layer is intended to replace a fully connected layer in traditional architecture. It takes the advantage to generate one feature map for each category.

To obtain better optimal solutions for the design of CNN, grid search, as a common approach (lightweight compared with the exhaustive search) is applied to obtain the set of hyperparameters [26, 27].

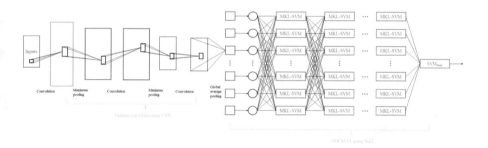

Fig. 1. Architecture of CNN-DOCSVM.

2.3   ANTD Classification using DOCSVM with Customized Kernel via MKL

The DOCSVM customizes the kernel via MKL on four typical kernel functions namely linear K_{linear}, sigmoid $K_{sigmoid}$, third order polynomial K_{3poly}, and radial basis function K_{rbf}, along with the possibilities for extension using two kernel properties including linearity and product.

The formulations of the kernel functions are defined as follows.

$$K_{linear} = \langle x_i, x_j \rangle \tag{1}$$

$$K_{sigmoid} = \tanh(x_i, x_j + c) \tag{2}$$

$$K_{3poly} = (x_i, x_j + d)^3 \tag{3}$$

$$K_{rbf} = e^{\frac{\|x_i - x_j\|^2}{2\sigma}} \tag{4}$$

for some input vectors x_i and x_j, and hyperparameters c, d, and σ.

Applying the kernel properties to some Mercer's kernels K_1 and K_2, new kernel can be designed:

$$K_{linearity} = K_1 + K_2) \tag{5}$$
$$K_{product} = K_1 \bullet K_2) \tag{6}$$

The abovementioned four kernels in Eqs. (1)–(4) are Mercer's kernels that can derive various new kernels based on kernel properties in Equations (5)–(6). Heuristic approach is then applied to solve the MKL design in the building blocks of Fig. 1 [28].

3   Results

The result is comprised of the performance evaluation of the CNN-DOCSVM, and the performance comparison between CNN-DOCSVM and existing works.

3.1   Performance Evaluation of the CNN-DOCSVM

MAWILab [29] was selected as benchmark dataset for the ANTD classification problem. K-fold cross-validation is adopted to better analyze the issue of model overfitting and fine tune the hyperparameters. Typically, K=5 [30,31] or K=10 [32,33] is chosen for general applications. With the availability of sufficient training samples and the relatively lightweight model (based on 1x1 convolutional layers and support vector machine), K=10 is thus selected.

Table 1 summarizes the average accuracy (5-fold cross-validation) of the ANTD models using CNN-DOCSVM (MKL), CNN-DOCSVM (K_{linear}), CNN-DOCSVM ($K_{sigmoid}$), CNN-DOCSVM (K_{3poly}), CNN-DOCSVM (K_{rbf}), CNN-OCSVM (K_{linear}), CNN-OCSVM ($K_{sigmoid}$), CNN-OCSVM (K_{3poly}), and CNN-OCSVM (K_{rbf}). Some observations are drawn as follows:

- CNN-DOCSVM (MKL) yields the best accuracy of 97.5%. It outperforms other algorithms by 2.52–15.2%.
- CNN-DOCSVM outperforms CNN-OCSVM that implies the enhancement of model performance with deep architecture. The average improvement of accuracy is 8.43%.
- Among various kernel functions, the ranking of accuracies (descending order) is MKL>K_{rbf}>K_{3poly}>$K_{sigmoid}$>K_{linear}.

3.2   Performance Comparison between CNN-DOCSVM and Existing Works

Table 2 compares the performance between our work and existing works with the evaluation metrics algorithm, dataset, multi-class ANTD, one-class ANTD, cross-validation, and result. The discussion is made in each metric:

Table 1. Performance evaluation of CNN-DOCSVM and CNN-OCSVM with various kernels

Algorithm	Kernel function	Accuracy (%)
CNN-DOCSVM (MKL)	MKL	97.5
CNN-DOCSVM (K_{linear})	K_{linear}	91.2
CNN-DOCSVM ($K_{sigmoid}$)	$K_{sigmoid}$	93.9
CNN-DOCSVM (K_{3poly})	K_{3poly}	94.6
CNN-DOCSVM (K_{rbf})	K_{rbf}	95.1
CNN-OCSVM (MKL)	MKL	90.8
CNN-OCSVM (K_{linear})	K_{linear}	84.6
CNN-OCSVM ($K_{sigmoid}$)	$K_{sigmoid}$	85.9
CNN-OCSVM (K_{3poly})	K_{3poly}	86.6
CNN-OCSVM (K_{rbf})	K_{rbf}	87.7

Table 2. Performance comparison between our work and existing works.

Work	Algorithm	Dataset	Multi-class ANTD	One-class ANTD	Cross-validation	Result
[14]	MCGLRT and MSD	MAWILab 2019	Yes	No	No	AUC of 92.1%
[16]	LSTM and FCN	MAWILab 2018, 2019, and 2020	Yes	No	No	AUC of 98.2%
[17]	CNN	MAWILab 2021 (0.72%)	Yes	No	No	Average accuracy of 67%
[18]	LSTM and OCSVM	MAWILab 2020	No	Yes	No	Accuracy of 94.6%
Our work	CNN-DOCSVM	MAWILab 2020	No	Yes	Yes	Accuracy of 97.5%

Algorithm: Traditional deep learning algorithms were used in existing works [14–17]. SVM and deep SVM were used in [18] and our work, respectively.

Dataset: MAWILab is a regularly updated dataset since 2001. Although different patterns of abnormalities may observe, the issue of imbalanced dataset remains across the years with dominated normal class.

Multi-class ANTD: Existing works [14–17] focused on multi-class classification problem.

One-class ANTD: Existing work [18] and our work formulated the ANTD as one-class classification problem.

Cross-validation: Existing works [14–18] did not employ cross-validation that cannot well examine the issue of potential overfitting and better fine-tune the trained model.

Result: Considered identical dataset, our work enhances the accuracy by 3.07% compared with [18].

4   Conclusion and Future Research Directions

In this paper, a convolutional network network-based deep one-class support vector machine algorithm is proposed for anomaly network traffic detection with highly imbalanced dataset. Performance evaluation and analysis show that the algorithm achieves accuracy of 97.5% with 5-fold cross-validation. Ablation studies confirm the contributions of the deep support vector machine and kernel selection that further enhance the model accuracy by 2.52–15.2%. Compared with existing work, the accuracy improvement is 3.07%.

Future research directions are recommended (i) generating extensive training data in the abnormality classes; (ii) evaluating the algorithm with more benchmark dataset; and (iii) formulating the problem with multi-class classification problem.

References

1. Abbasi, M., Shahraki, A., Taherkordi, A.: Deep learning for network traffic monitoring and analysis (NTMA): a survey. Comput. Commun. **170**, 19–41 (2021)
2. Ahmad, S., Jha, S., Alam, A., Alharbi, M., Nazeer, J.: Analysis of intrusion detection approaches for network traffic anomalies with comparative analysis on botnets (2008–2020). Secur. Commun. Netw. **2022**, 1–11 (2022)
3. Tarekegn, A.N., Giacobini, M., Michalak, K.: A review of methods for imbalanced multi-label classification. Pattern Recognit. **118**, 1–12 (2021)
4. Zhang, T., Chen, J., Li, F., Zhang, K., Lv, H., He, S., Xu, E.: Intelligent fault diagnosis of machines with small & imbalanced data: a state-of-the-art review and possible extensions. ISA Trans. **119**, 152–171 (2022)
5. Al-Ayyoub, M., et al.: Accelerating 3D medical volume segmentation using GPUs. Multimedia Tools Appl. **77**(4), 4939–4958 (2018)
6. Chlap, P., Min, H., Vandenberg, N., Dowling, J., Holloway, L., Haworth, A.: A review of medical image data augmentation techniques for deep learning applications. J. Med. Imaging Radiat. Oncol. **65**(5), 545–563 (2021)
7. Lewy, D., Mańdziuk, J.: An overview of mixing augmentation methods and augmentation strategies. Artif. Intell. Rev. **2022**, 1–59 (2022)
8. Gupta, B.B., Badve, O.P.: GARCH and ANN-based DDoS detection and filtering in cloud computing environment. Int. J. Embed. Syst. **9**(5), 391–400 (2017)
9. Chui, K.T., et al.: An MRI scans-based Alzheimer's disease detection via convolutional neural network and transfer learning. Diagnostics **12**(7), 1–14 (2022)
10. Chui, K.T., Lytras, M.D., Vasant, P.: Combined generative adversarial network and fuzzy C-means clustering for multi-class voice disorder detection with an imbalanced dataset. Appl. Sci. **10**(13), 1–19 (2020)
11. Lee, W., Seo, K.: Downsampling for binary classification with a highly imbalanced dataset using active learning. Big Data Res. **28**, 1–19 (2022)
12. Ren, J., Zhang, Q., Zhou, Y., Hu, Y., Lyu, X., Fang, H., Li, Q.: A downsampling method enables robust clustering and integration of single-cell transcriptome data. J. Biomed. Inf. **130**, 1–10 (2022)
13. Elmisery, A.M., et al.: Cognitive privacy middleware for deep learning mashup in environmental IoT. IEEE Access **6**, 8029–8041 (2017)

14. Huang, L., Ran, J., Wang, W., Yang, T., Xiang, Y.: A multi-channel anomaly detection method with feature selection and multi-scale analysis. Comput. Netw. **185**, 1–10 (2021)
15. Tembhurne, J.V., Almin, M.M., Diwan, T.: Mc-DNN: fake news detection using multi-channel deep neural networks. Int. J. Semant. Web Inf. Syst. (IJSWIS) **18**(1), 1–20 (2022)
16. Sahu, S.K., Mohapatra, D.P., Rout, J.K., Sahoo, K.S., Pham, Q.V., Dao, N.N.: A LSTM-FCNN based multi-class intrusion detection using scalable framework. Comput. Electric. Eng. **99**, 1–19 (2022)
17. Liu, X.: An abnormal network traffic detection method on MAWILab dataset based on convolutional neural network. In: 2022 IEEE 2nd International Conference on Electronic Technology. Communication and Information (ICETCI), pp. 1233–1235. IEEE, Changchun, China (2022)
18. Li, Y., Xu, Y., Cao, Y., Hou, J., Wang, C., Guo, W., Cui, L.: One-class LSTM network for anomalous network traffic detection. Appl. Sci. **12**(10), 1–16 (2022)
19. Yu, C., Xiao, B., Gao, C., Yuan, L., Zhang, L., Sang, N., Wang, J.: Lite-hrnet: A lightweight high-resolution network. In: Proceedings of the IEEE/CVF Conference on Computer Vision and Pattern Recognition, pp. 10440–10450. IEEE, Nashville, Tennessee (2021)
20. Srivastava, A.M., Rotte, P.A., Jain, A., Prakash, S.: Handling data scarcity through data augmentation in training of deep neural networks for 3D data processing. Int. J. Semant. Web Inf. Syst. (IJSWIS) **18**(1), 1–16 (2022)
21. Shin, Y., Park, J., Hong, J., Sung, H.: Runtime support for accelerating CNN models on digital DRAM processing-in-memory hardware. IEEE Comput. Architect. Lett. **21**(2), 33–36 (2022)
22. Sahoo, S.R., et al.: Hybrid approach for detection of malicious profiles in twitter. Comput. Electric. Eng. **76**, 65–81 (2019) ISSN 0045-7906. https://doi.org/10.1016/j.compeleceng.2019.03.003
23. Gong, W., Chen, H., Zhang, Z., Zhang, M., Wang, R., Guan, C., Wang, Q.: A novel deep learning method for intelligent fault diagnosis of rotating machinery based on improved CNN-SVM and multichannel data fusion. Sensors **19**(7), 1–37 (2019)
24. Chui, K.T., Lytras, M.D., Liu, R.W.: A generic design of driver drowsiness and stress recognition using MOGA optimized deep MKL-SVM. Sensors **20**(5), 1–20 (2020)
25. Chui, K.T., Liu, R.W., Zhao, M., De Pablos, P.O.: Predicting students' performance with school and family tutoring using generative adversarial network-based deep support vector machine. IEEE Access **8**, 86745–86752 (2020)
26. Masud, M., et al.: Pre-trained convolutional neural networks for breast cancer detection using ultrasound images. ACM Trans. Internet Technol. (TOIT) **21**(4), 1–17 (2021)
27. Liu, J., Lin, M., Zhao, M., Zhan, C., Li, B., Chui, J.K.T.: Person re-identification via semi-supervised adaptive graph embedding. Appl. Intell. 1–17 (2022)
28. Gönen, M., Alpaydın, E.: Multiple kernel learning algorithms. J. Mach. Learn. Res. **12**, 2211–2268 (2011)
29. Fontugne, R., Borgnat, P., Abry, P., Fukuda, K.: Mawilab: combining diverse anomaly detectors for automated anomaly labeling and performance benchmarking. In: Proceedings of the 6th International Conference, pp. 1–12. ACM, Philadelphia, USA (2010)
30. Cvitić, I., Peraković, D., Periša, M., Gupta, B.: Ensemble machine learning approach for classification of IoT devices in smart home. Int. J. Mach. Learn. Cybern. **12**(11), 3179–3202 (2021). https://doi.org/10.1007/s13042-020-01241-0

31. Chui, K.T., et al.: Long short-term memory networks for driver drowsiness and stress prediction. In: International Conference on Intelligent Computing and Optimization, pp. 670–680. Springer, Cham, (2020)
32. Chui, K.T., Tsang, K.F., Chung, S.H., Yeung, L.F.: Appliance signature identification solution using K-means clustering. In: IECON 2013–39th Annual Conference of the IEEE Industrial Electronics Society, pp. 8420–8425. IEEE, (2013)
33. Chui, K.T., et al.: A genetic algorithm optimized RNN-LSTM model for remaining useful life prediction of turbofan engine. Electronics **10**(3), 1–15 (2021)

A Comprehensive Comparative Study of Machine Learning Classifiers for Spam Filtering

Saksham Gupta[1(✉)], Amit Chhabra[1], Satvik Agrawal[1], and Sunil K. Singh[2]

[1] Chandigarh College of Engineering and Technology, Sector 26, Chandigarh, India
dev.sakshamgupta@gmail.com
[2] Kalinga Institute of Industrial Technology, Patia, Bhubaneswar, Odisha, India

Abstract. In July 2021, the daily spam count globally reached 283 billion and constitutes 84.12% of the total email volume. The increasing surge in the spam or unsolicited emails that can hamper communication has led to an intrinsic requirement for robust and reliable antispam filters. In recent years, spam filtration and monitoring have become significant concerns for mail and other internet services. Machine learning strategies are being employed to act as safeguards against internet spam. This study provides a systematic survey of spam filtering methods using machine learning techniques. Logistic Regression, Random Forest, Naive Bayes, and Decision Tree methods used for spam filtering have been compared based on precision, recall, and accuracy on a dataset composed of Twitter tweets, Facebook posts, and YouTube comments. The preliminary discussion involves a background study of the related work on spam filtering and the research gaps in the current literature. Further, a detailed discussion on each method has been provided in this study. The results of our experiments indicate that Decision Trees provide the best accuracy at 97.02% and precision at 98.83%, and Logistic Regression has the highest recall at 99.89%.

Keywords: Machine learning · Naive Bayes · Spam · Spam filtering · Decision tree · Random Forest · K-Fold · Logistic Regression · Network Filtering · Spam Prevention

1 Introduction

Information distribution is straightforward and quick in the age of digitalization. Users from all over the world can share information on various platforms. Email is the most efficient and cost-effective method of sharing information. Because of their simplicity, emails are vulnerable to various attacks [1,2]. The host system's privacy may be jeopardized by harmful disguised information in these emails. The attacker considers emails and other information sharing irrelevant and sends unwanted messages to a broad audience. A significant amount of data is necessitated by the email systems. The emails could include malware, rats, and other hazards.

© The Author(s), under exclusive license to Springer Nature Switzerland AG 2023
N. Nedjah et al. (Eds.): ICSPN 2021, LNNS 599, pp. 257–268, 2023.
https://doi.org/10.1007/978-3-031-22018-0_24

Customers can create filters with several email service providers. This strategy is inefficient since it is tough to execute, and people seem to dislike having their email details available to external sources. Our culture has become embedded with the internet of things [3,5]. IoT (Internet of Things) has become increasingly crucial in intelligent cities. With the advent of IoT, spammers are on the rise. The spam rate remains high because the website aids in the disguise of the attacker's identity. Filling the server's storage or capacity may cause the response time to slow down. Every company carefully evaluates the available tools to identify and prevent spam emails in their environment effectively. The allow/blocklist system is one of the most well-known techniques for recognizing and analyzing incoming emails.

According to social networking specialists, 40 percent of social network accounts are exploited for spam. Investigating these highlights can help enhance the detection of these types of emails [3,6]. We can categorize them based on their characteristics. For mail detection, learning-based classifiers are utilized. The detection procedure presumes that the email contains a set of characteristics that distinguish it from legitimate emails. Numerous aspects contribute to the identification process's complexity.

Multiple machine learning techniques can be utilized to create efficient and low-cost spam filters [4,8]; however, these techniques perform with varying accuracies under different circumstances. Current research on spam filtering generally consists of finding model accuracy on a single dataset of multiple independent datasets. This can lead to skewed results as spam does not conform to a single type. In this study, we have collected data not just from Twitter, but also YouTube and Facebook to create a more generalized dataset that better mimics real world spam. This study compares the approaches and principles of machine learning techniques for spam filtering. It also classifies several detection approaches using machine learning algorithms. A broad survey of research gaps in spam filtering has been presented in the paper as well.

Further, this paper has been organized as follows: Section 2 describes the related work and Sect. 3 provides details on our proposed methodology. Section 4 consists of descriptions of machine learning models for spam filtering, followed by results and discussion in Sect. 5. Section 6 concludes the paper and provides future scope.

2   Related Work

Any user or automated system can send bogus or undesired bulk mail. Receiving spam email has become increasingly widespread during the previous decade. Spam email detection relies heavily on machine learning. Researchers employ a variety of models and methodologies. Researchers presented a survey that used a supervised technique to detect emails [4,10].

There is high adoption of varying machine learning and non-machine learning strategies for detecting, blocking, and preventing unwanted email. A collection of attempts to learn email filters is available. The effects of mail on various domains

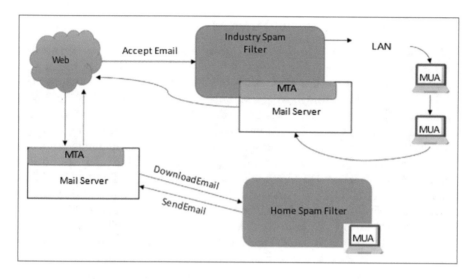

Fig. 1. Client based spam filtering [7]

have been deliberated, and various economic and legal issues surrounding email have been examined [4,16]. Because of its speed and simplicity, ML (Machine Learning) models produce exact results.

Bhuiyan et al. give a review of the current email spam filters. Figure 1 provides the basic model of client-side spam filtering used in web systems to prevent spam. The study presented at the European Conference on Machine Learning and Artificial Intelligence examined a variety of deep learning-based detection algorithms and provided details on the multiple strategies used for spam detection [7].

Table 1. Comparison of different techniques in spam filtering

Research	Algorithm	Datasets	Accuracy
DeBarr & Wechler [9]	Random Forest	Custom	95.2%
Rusland et al. [11]	Naive Bayes Modified	Spam base and spam data	88% on spam base dataset 83% on spam dataset
Xu et al. [12]	SVM, Bayes Net, Naive Bayes	Twitter & Facebook dataset	90% using SVM
Hijawi et al. [14]	Random Forest, Naive Bayes, Decision Tree	Spam Assassin	99.3% using Random Forest
Banday & Jan [15]	K-nearest neighbours, L-SVM	Custom	96.69% on SVM
Olatunji et al. [17]	SVM and ELM classifier	Enron Dataset	94.06%
Zheng et al. [18]	SVM	Social Networks dataset	99.5%

2.1 Research Gaps in Spam Filtering

The domain's research limitations and open research topics are discussed in this section. Tests and algorithms should be trained on actual data rather than artificial data in the future since artificial data-trained models underachieve on real-life data. It is also possible to improve accuracy by adopting hybrid algorithms [13, 19]. Clustering can be improved using dynamic upgrading and clustering algorithms. Email detection can be done using machine learning systems. Experts in linguists and phonetics can collaborate to manually annotate datasets, creating efficient and acceptable spam datasets [20, 27]. GPUs and Programmable Logic Arrays (FPGAs), which have high energy efficiency, flexibility, and real-time capabilities, are utilized for training and testing detection and filtering models. Future studies should focus on the distribution of standard labeled data for scientists to train and use efficient classifiers, as well as the inclusion of other characteristics to the dataset. The message's topic and the communication's text were used as categorization features. For reliable results, manual data selections and capabilities should also be used. Some identifiable research gaps in the current literature are listed as follows:

1. Since most researchers presented their findings in terms of precision, specificity, recall, and other metrics, the spatial challenge of machine learning should be regarded as an evaluation metric.
2. Fast response time can be achieved by applying handcrafted features and a distinct preprocessing phase.
3. For better classification, dynamic modification of value space is required. Current filters are incapable of updating their feature space.
4. If the system's security is maintained, more accurate and reliable outcomes can be achieved, and since many models have a greater false positive rate than required, it must be lowered to be helpful in practical applications.
5. Simple classifiers can achieve image detection, and since spammers utilize images, such classifiers should be used to detect spam. Deep learning methodologies may also be used to study multilingual spam detection.
6. Many researchers rely on traits that they believe are reliable. Such traits are unreliable as different types of spam need to be handled in separate ways, and no general algorithm has been devised yet (Table 1).

3 Methodology

The phrase "spam" was coined in 1981 to refer to unsolicited email, but it became popular in the middle of 1990 when it was seen outside of intellectual and research groups. The development expenditure trick is a well-known example, in which a consumer receives a communication with a proposal that should result in a prize. In today's technological age, the dodger/spammer fabricates a scenario in which the unlucky victim requires financial support so that the scammer can amass a much more significant sum of money, which they would subsequently split. As a result, spam emails are becoming increasingly common. Various companies are

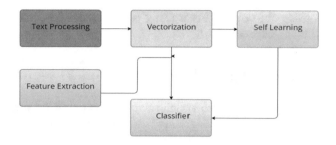

Fig. 2. Standard spam filtering [22]

working on new ways to detect and filter junk mail [21]. This section discusses some methods for comprehending the process.

The regular Misuse Detection Method is a conventional spam filtering system and is implemented using a system of regulations and protocols [22,32]. The method depicted in Fig. 2 is a typical representation of this model. The first step employs content sensors and artificial intelligence algorithms. Backlist filters are then performed on the emails to ensure they were not in the blocklist file. Further, rule-based filters are used to identify the entity which has sent the email. Finally, allocation and task filters are used to allow the account to send mail. Many rules and techniques can ensure safe contact between people and institutions [23]. On a client's machine, various frameworks should be installed. Incoming emails can be received and handled via systems that communicate with client mail agents.

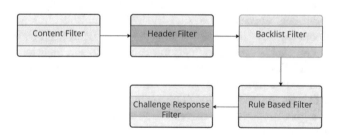

Fig. 3. Case based spam filtering [23]

There is also an anti-spam level known as the enterprise level. With an organizational filtering technique, a network client already does spam filtering effectively and consistently on a network. Existing solutions employ the rule of email ranking [23]. A rating function is defined, and each post is assigned a score. The most common and prominent machine learning methods are instance or sample-based spam filters. Figure 3 depicts the essential characteristics of the case base. During the first step, data is collected by communication and security networks using the collecting method. The critical change follows with preprocessing stages via the client windows environment, outline abstraction, email data classification selection, validating the entire system using vector expression, and categorizing the data into binary classes.

4   Machine Learning Methods for Spam Filtering

Since it enables computers to learn and develop their capabilities without explicit programming, machine learning is among the most crucial and significant areas of artificial intelligence. Making automated tools for data processing and retrieval is machine learning's core objective. The training dataset serves as the starting point for learning [24]. To find patterns from data enough so better judgments may be made in the future based on the user's input, it may be a direct achievement, assessment, example, or feedback. Models for machine learning are created to learn on their own.

Machine learning could be used to discover what is happening online. Machine learning can handle large amounts of information. Although it often gives quicker and more effective results in detecting harmful content, developing its models for high-performance levels can require more prolonged effort and resources. Automation and cognitive technologies can be even more effective at handling large amounts of data.

Labelled data is required for supervised machine learning models. After receiving labelled training data, training models help predict events [25]. The models start by analyzing an available training dataset and developing a mechanism for predicting success ratings. The system can forecast every updated data related to the user just during the training process. The learning program analyzes the outcome of the predicted output and finds flaws so the model can be improved. Many challenges can be solved using this method of learning.

Since the last decade, machine learning has also been used to classify data. This strategy can tackle any classification problem [26]—a decision tree consists of a series of questions. We ask further questions when we obtain a response until, we reach a decision on the record. Based on the data presented, models are built iteratively or repeatedly. The purpose is to forecast the value of a target variable given a sequence of values. Finally, regression and classification problems are solved using a tree structure.

Santos et al. discuss enhancing email classification using randomized forest and active learning. Each email was divided into two pieces using the data from the messages. They discovered inverse document frequencies for all email features. The data inside the training dataset is labelled using clustering [28]. After evaluating the cluster prototype email, the supervised machine learning experiment was carried out. As a result, the confusion matrix is more accurate, according to the research findings.

A fuzzy rule-based tree and the Naive Bayes algorithm can be designed and used to determine what individuals were doing [29]. The degree to which characters are explained or described is unbiased and rational. For email classification, Mamdani rules are utilized. Naive Bayes classifier is employed in such a scenario. Votes are divided into smaller groups using the baking process. The answer gives

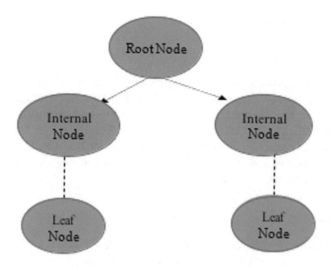

Fig. 4. Decision tree structure [21]

them an ideal weight that may be applied to percentages to improve accuracy. Three hundred fifty of something like the 1,000 emails were spam, and 650 were phishing scams [30].

Decision Trees created using supervised machine learning utilized the Markov model and were used to calculate the chances of events that could happen in a combination to categorize emails as either junk mail or ham [31]. All emails were initially classified as valid by calculating the overall chance of each email using later classed email phrases. Following that, email decision trees were created. There were 5172 total emails on which the analysis was done [33]. From the total of 5172 emails, 2086 were genuine. The Enron dataset provided a feature set that could be used to categorize the emails. Figure 4 represents a simple decision tree that can be calibrated to filter spam.

Based on machine learning techniques, the Multiview method email classification technique was developed [34]. This method concentrates on obtaining more detailed information [35]. Domestic and foreign feature sets make up a double-view dataset. The proposed approach was evaluated using two datasets in a real-world network setting [36]. According to the study's findings, the Multiview model is more accurate than simple email classification.

A Naive classifier is used in the classification algorithm. The Bayes theorem is being used by the Naive Classification model. The worth of one characteristic affects the worth of other features. Because there is no iterative process involved, Nave classifiers are easy to build and deliver good accuracy performance on massive datasets. The Naive Bayes strategy performs better than other approaches [37].

5 Results and Discussion

5.1 Dataset

The dataset we have used is a custom dataset created using data collected from Twitter for 500 tweets using the Twitter API. Data was also scraped from Facebook posts and YouTube comments using Python and BeautifulSoup library. The training data consists of 41,523 words divided into 800 sentences. The collected data were manually labelled as either spam or not spam. Once the data were categorized, we divided the dataset into 80:20 training (640 sentences) and testing (160 sentences).

5.2 Results

By adjusting the predictions supplied by the group's model with the available data, accuracy is utilized to gain a better grasp of such a model used to predict challenges [33]. Therefore, various data combinations must be used to train the model. As shown in Table 2, We have tested four distinct algorithms with different accuracies and out of these four, the decision tree performed with the best accuracy of 96.41%. The accuracies of these models have been shown in Fig. 5.

Table 2. Performance results of ML algorithms

Model	Precision	Recall	Accuracy
Logistic Regression	95.9%	99.89%	96.23%
Random Forest	86.03%	89.92%	86.09%
Naive Bayes	98.2%	90.31%	90.31%
Decision Tree	97.82%	98.54%	96.41%

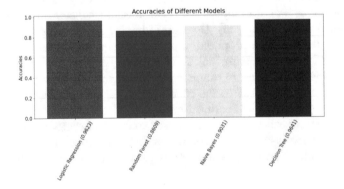

Fig. 5. Model accuracies for ML algorithms

Since the decision tree has the highest accuracy, we have applied K-Fold cross-validation to improve the model accuracy and implement a more generalized model. Table 3 shows the results of the decision tree after performing 5-Fold validation; among these, native 5-Fold validation results in the best accuracy of 97.02 %. Repeated 5-fold has 95.51% accuracy and stratified 5-fold has 96.23% accuracy. Figure 6 shows the accuracy of the decision tree with 5-Fold techniques.

Table 3. Performance results on 5-Fold techniques on decision tree

Model	Precision	Recall	Accuracy
5-Fold	97.82%	98.54%	97.02%
Stratified 5-Fold	97.83%	98.45%	95.51%
Repeated 5-Fold	98.83%	98.54%	96.23%

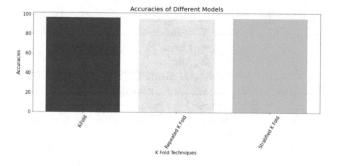

Fig. 6. Decision tree accuracies for 5-Fold strategies

6   Conclusion

As we can see from the results, ML models have very high accuracy when differentiating between spam and valid data blocks. Out of the different strategies, for our mixed dataset, decision tree had the highest accuracy (97.02%) and precision (98.83%), while logistic regression had the highest recall (99.89%). We have seen the different machine learning algorithms and how they perform on a mixed dataset that consists of sentence groups extracted from various sources. We have also provided research gaps in the current spam filtering literature and pointed out significant difficulties in creating a real-time spam filtering system. Our techniques do not have a hundred percent accuracy and, therefore, we are liable to make incorrect classifications. Another limitation comes with the enormous amount of information that can be classified as spam. We can have spam in different formats like text, images, and video, and we could have spam with different intents as some could be advertisements for certain institutions while others could be scams. We conclude by stating that ML algorithms are an efficient, low-cost software-based solution to detect and filter out spam, and while their application may not be perfect, they can still function with high accuracy for specific use cases, and they can be applied in practical scenarios to prevent the propagation of spam. In the future, we plan on implementing deep learning techniques either independently or in conjunction with the existing machine learning methods to achieve higher accuracy. We also intend to investigate hardware-based spam filtering solutions, as this paper primarily focuses on software-based solutions.

References

1. Faris, H., Al-Zoubi, A.M., Heidari, A.A., Aljarah, I., Mafarja, M., Hassonah, M.A., Fujita, H.: An intelligent system for spam detection and identification of the most relevant features based on evolutionary random weight networks. Inf. Fusion **48**, 67–83 (2019)
2. Yu, H.Q., Reiff-Marganiec, S.: Learning disease causality knowledge from the web of health data. Int. J. Semant. Web Inf. Syst. (IJSWIS) **18**(1), 1–19 (2022)
3. Blanzieri, E., Bryl, A.: A survey of learning-based techniques of email spam filtering. Artif. Intell. Rev. **29**, 63–92 (2008)
4. Alghoul, A., Ajrami, S., Jarousha, G., Harb, G., Abu-Naser, S.: Email classification using artificial neural network. Int. J. Acad. Eng. Res. (2018)
5. Sahoo, S.R., et al.: Spammer detection approaches in online social network (OSNs): a survey. In: Sustainable Management of Manufacturing Systems in Industry 4.0, pp. 159–180. Springer, Cham (2022)
6. Gupta, B.B., Badve, O.P.: GARCH and ANN-based DDoS detection and filtering in cloud computing environment. Int. J. Embed. Syst. **9**(5), 391–400 (2017)
7. Udayakumar, N., Anandaselvi, S., Subbulakshmi, T.: Dynamic malware analysis using machine learning algorithm. In: 2017 International Conference on Intelligent Sustainable Systems (ICISS) (2017)
8. Chui KT, et al.: Handling data heterogeneity in electricity load disaggregation via optimized complete ensemble empirical mode decomposition and wavelet packet transform. Sensors **21**(9):3133 (2021). https://doi.org/10.3390/s21093133
9. DeBarr, D., Wechsler, H.: Using social network analysis for Spam Detection. Adv. Soc. Comput. 62–69 (2010)
10. Lu, J., Shen, J., et al.: Blockchain-based secure data storage protocol for sensors in the industrial internet of things. IEEE Trans. Indus. Inf. **18**(8), 5422–5431 (2022). https://doi.org/10.1109/TII.2021.3112601
11. Rusland, N.F., Wahid, N., Kasim, S., Hafit, H.: Analysis of Naive Bayes algorithm for email spam filtering across multiple datasets. In: IOP Conference Series: Materials Science and Engineering, vol. 226, p. 012091 (2017)
12. Xu, H., Sun, W., Javaid, A.: Efficient spam detection across online social networks. In: 2016 IEEE International Conference on Big Data Analysis (ICBDA) (2016)
13. Gupta, B.B.: A lightweight mutual authentication approach for RFID tags in IoT devices. Int. J. Netw. Virtual Organ. (2016)

14. Hijawi, W., Faris, H., Alqatawna, J., Al-Zoubi, A.M., Aljarah, I.: Improving email spam detection using content based feature engineering approach. In: 2017 IEEE Jordan Conference on Applied Electrical Engineering and Computing Technologies (AEECT) (2017)

15. Banaday, M., Jan, T.: Effectiveness and limitations of statistical spam filters. In: arXiv. (2009)

16. Cvitić, I., Peraković, D., Periša, M. et al.: Ensemble machine learning approach for classification of IoT devices in smart home. Int. J. Mach. Learn. Cyber. **12**, 3179–3202 (2021). https://doi.org/10.1007/s13042-020-01241-0

17. Olatunji, S.O.: Extreme learning machines and support vector machines models for email spam detection. In: 2017 IEEE 30th Canadian Conference on Electrical and Computer Engineering (CCECE) (2017)

18. Zheng, X., Zhang, X., Yu, Y., Kechadi, T., Rong, C.: Elm-based spammer detection in social networks. J. Supercomput. **72**, 2991–3005 (2015)

19. Olatunji, S.O.: Extreme learning machines and support vector machines models for email spam detection. In: 2017 IEEE 30th Canadian Conference on Electrical and Computer Engineering (CCECE) (2017)

20. Dean, J.: Large-scale deep learning for building intelligent computer systems. In: Proceedings of the Ninth ACM International Conference on Web Search and Data Mining (2016)

21. Adewole, K.S., Anuar, N.B., Kamsin, A., Varathan, K.D., Razak, S.A.: Malicious accounts: dark of the social networks. J. Netw. Comput. Appl. **79**, 41–67 (2017)

22. Barushka, A., Hájek, P.: Spam filtering using regularized neural networks with rectified linear units. In: AI*IA 2016 Advances in Artificial Intelligence, pp. 65–75 (2016)

23. Gupta, S., Sharma, P., Sharma, D., Gupta, V., Sambyal, N.: Detection and localization of potholes in thermal images using deep neural networks. Multimedia Tools Appl. **79**, 26265–26284 (2020)

24. Zheng, X., Zhang, X., Yu, Y., Kechadi, T., Rong, C.: Elm-based spammer detection in social networks. J. Supercomput. **72**, 2991–3005 (2015)

25. Ferrag, M.A., Maglaras, L., Moschoyiannis, S., Janicke, H.: Deep learning for cyber security intrusion detection: approaches, datasets, and comparative study. J. Inf. Secur. Appl. **50**, 102419 (2020)

26. Kumar, N., Sonowal, S., Nishant: Email spam detection using machine learning algorithms. In: 2020 Second International Conference on Inventive Research in Computing Applications (ICIRCA) (2020)

27. Sharma, R., Sharma, T.P., Sharma, A.K.: Detecting and preventing misbehaving intruders in the internet of vehicles. Int. J. Cloud Appl. Comput. (IJCAC) **12**(1), 1–21 (2022)

28. Santos, I., Penya, Y.K., Devesa, J., Bringas, P.G.: N-grams-based file signatures for malware detection. In: Proceedings of the 11th International Conference on Enterprise Information (2009)

29. Bhuiyan, H., Ashiquzzaman, A., Juthi, T., Biswas, S., Ara, J.: A survey of existing E-Mail spam filtering methods considering machine learning techniques. Global J. Comput. Sci. Technol. (2018)

30. Kumar, S., Singh, S.K., Aggarwal, N., Aggarwal, K.: Evaluation of automatic parallelization algorithms to minimize speculative parallelism overheads: an experiment. J. Discrete Math. Sci. Crypt. **24**, 1517–1528 (2021)

31. Singh, I., Singh, S.K., Kumar, S., Aggarwal, K.: Dropout-VGG based convolutional neural network for traffic sign categorization. Lecture Notes on Data Engineering and Communications Technologies, pp. 247–261 (2022)

32. Ling, Z., Hao, Z.J.: An intrusion detection system based on normalized mutual information antibodies feature selection and adaptive quantum artificial immune system. Int. J. Semant. Web Inf. Syst. (IJSWIS) **18**(1), 1–25 (2022)
33. Singh, I., Singh, S.K., Singh, R., Kumar, S.: Efficient loop unrolling factor prediction algorithm using machine learning models. In: 2022 3rd International Conference for Emerging Technology (INCET) (2022)
34. Singh, S.K.: Linux yourself (2021)
35. Gansterer, W.N., Janecek, A.G., Neumayer, R.: Spam filtering based on latent semantic indexing. In: Survey of Text Mining II, pp. 165–183 (2008)
36. Lee, D., Lee, M.J., Kim, B.J.: Deviation-based spam-filtering method via stochastic approach. EPL (Europhys. Lett.) **121**, 68004 (2018)
37. Wang, J., Katagishi, K.: Image content-based email spam image filtering. J. Adv. Comput. Netw. **2**, 110–114 (2014)

A Novel Approach for Social Media Content Filtering Using Machine Learning Technique

Akshat Gaurav[1,2], Varsha Arya[2,3(✉)], and Kwok Tai Chui[4]

[1] Ronin Institute, Montclair, New Jersey, USA
akshat.gaurav@ronininstitute.org
[2] Lebanese American University, Beirut 1102, Lebanon
[3] Insights2Techinfo, Bangalore, India
varsha.arya@insights2techinfo.com
[4] Hong Kong Metropolitan University (HKMU), Hong Kong, China
jktchui@ouhk.edu.hk

Abstract. A completely interconnected world is possible because of social networks, which allow individuals to interact, share ideas, and organize themselves into digital environments. Both online behavior and online content processing are critical for security applications to be successful. Extremists now have an easy way to spread their ideas and beliefs via social media. Cyberbullying and the spread of false news and phoney reviews on social media are two of the numerous security risks that arise as a result of this. This necessitates the need for the creation of a method to identify and minimize dangerous information on social networks. Risks and the extent of their repercussions are subjectively evaluated in the context of security occurrences. Societal consequences of negative impressions can be dire. Citizens' polls are routinely used to measure these sentiments, but they take a long time and don't adjust well to shifting security dynamics. In light of this, we developed a machine learning-based social media content filtering strategy. In order to train our model to identify fraudulent tweets on the Twitter network, we used the four different machine learning approach to find the malicious comments.

Keywords: Social networks · Twitter · Machine learning

1 Introduction

In addition to the ability to create a personal profile and submit material, online social networks (OSNs) enable users to form a network of relationships with others. People may use these types of online profiles (Fig. 1) and postings to introduce themselves and share details about their personality, opinions, and emotions with the world.

However, with the rise in popularity of social media, spamming has become a major problem. Spam accounts are used for a variety of nefarious purposes,

N. Nedjah et al. (Eds.): ICSPN 2021, LNNS 599, pp. 269–275, 2023.
https://doi.org/10.1007/978-3-031-22018-0_25

including stealing critical information, propagating bogus news, and distributing malware. Users' safety is put at risk by spam accounts. Spam accounts on social media grew by a whopping 355% in the first quarter of 2013, according to a research by Nexgate. Social media spam accounts for around 15% of Twitter's user base, according to a separate research by Nexgate.

Fig. 1. Social media platforms

Exploration of social media for security applications poses problems that no one area or expertise can solve. Social media platforms and networks are often used to disseminate false or misleading information, whether as part of propaganda or disinformation efforts. The absence of social responsibility in many digital platforms encourages a wide range of new types of abuse. This widely available technology is full of bad uses, such as disinformation, propaganda, and false news.

In this context, we create a security model for social media platforms based on machine learning. We use the Twitter network to collect tweets as a way to evaluate the effectiveness of the suggested methodology. The Twitter microblogging platform is a prominent social networking service and the Internet's biggest microblog. Tweets, or 140-character messages, are the primary means through which Twitter users communicate thoughts and information with each other in real time. Research demonstrates that Twitter has become a venue for on-line radicalization and the publishing of hate and extremism promoting material owing to the low publication barrier, lack of strict supervision, anonymity, and extensive penetration. Both Twitter moderators and intelligence and security informatics agents may benefit from automatic detection of hate and extremism-promoting tweets. The sheer amount of tweets produced each day makes manual detection of these tweets and subsequent filtering of information from the raw data almost unfeasible. It is difficult to automatically classify tweets since they are made up of a little amount of text and a lot of noise. Research presented here aims to provide answers to the challenges faced by intelligence and security informatics agents and Twitter moderators in preventing online radicalisation on the Internet's biggest microblogging platform.

This study is separated into the following sections: Sect. 2 presents the literature review, Sect. 3 and Sect. 4 present the methodology and results, respectively. Finally, Sect. 5 concludes the paper.

2 Related Work

More than a billion people across the globe are now linked together via social media platforms like Facebook and Twitter, allowing them to converse and exchange information with one another and within groups [3,5]. Spreading information about contagious illnesses and discussing solutions to human issues such as reducing child trafficking and violence against women may greatly benefit mankind via the use of these platforms. Fake news and privacy violations may also be caused via social media networks, which are widely used to disseminate misinformation. Human behaviour on social media is changing due to the expansion of artificial intelligence (AI) systems, strong machine learning methods, and cyber-attacks [2] on information networks. Artificial intelligence (AI) [8] and cyber security play important roles in social media networks, and this article examines both technologies in detail [7].

In recent years, the proliferation of social media platforms has made it easier for people to engage and openly express their thoughts, which has led to an explosion in the amount of data being generated. As a result, hackers now have easy access to the personal information of social media users, raising new questions about overall security. This [1] article gives an overview of several sentiment analysis methodologies and methods for social media security and analytics. Sentiment analysis may be utilised in a variety of security applications, including deception detection, anomaly detection, risk management, and disaster assistance. Data provenance, mistrust, e-commerce security, breaches of consumer security, market monitoring, trustworthiness, and risk assessment in social media were all the focus of an in-depth research presented recently. There has been a discussion of numerous approaches, methodologies, datasets, and application domains where sentiment analysis is employed. Different machine learning approaches have been employed in the security domains to accomplish sentiment analysis and the outcomes of these techniques are discussed in this paper. Various difficulties, gaps, and recent developments in the subject are discussed, as well as a path ahead for future research.

Social networking is a great way to keep in touch with individuals who share your interests. There are several advantages to joining a virtual community, including the development of social skills, the formation of alliances, the advancement of one's job, and many more. Social media has been a popular means of communication for internet users for the better part of the previous decade. They give a way for people to stay in touch with their loved ones. Users benefit from social media since it allows them to connect with people from all over the world. Social media's ubiquity and usefulness raises concerns about user data security. In this context, the authors in [6] examine a variety of social network assault methods and the consequences they might have. We suggest a machine learning-based solution for improved security arrangements and data security based on the results of our investigation.

Data gathered and inferred from publicly accessible and overt sources is known as Open-Source Intelligence (OSINT). Social media intelligence gleaned from publicly accessible data on Web 2.0 platforms such as Twitter, YouTube, and Facebook is referred to as "Open-Source Social Media Intelligence" (social-networking website). An overview of Intelligence and Security Informatics (ISI) in open-source social media intelligence is presented by the author in [1]. Two case studies in open-source social media intelligence provide an introduction to the technical issues and fundamental machine learning framework, tools, and methodologies. Free-form textual material on social networking platforms is the primary topic of this article. We focus on radicalization and civil disturbance as two of the most crucial uses of this technology. On top of that, we address open research issues, relevant publications and publishing sites; research findings and future directions; and current research.

In today's society, social media has a significant impact on the way people think and act. Some individuals are reluctant to submit their talents on these platforms for a number of reasons, including security, lack of fundamental skills, and lack of information about the different skill sets.. On these platforms, women confront a wide range of threats to their safety. Professional women's social networking site WoKnack. This gives women the chance to show off their skills and educate other women. Registration for female users only, talent classification, post-verification and privacy protection are the focus of this research. Gender identification prior to registration is accomplished by the use of machine learning algorithms to facial image, identity document, and voice-related gender verification. During the procedure, a 91% increase in accuracy was achieved. Natural language processing was used to classify skills, and then post-verification was carried out using these categories. Using the most accurate algorithm results in a 94% success rate. Users' privacy has been protected by adding data anonymization, skill and location grouping [4].

3   Research Methodology

In this section, we go into further depth regarding our methodology. Figure 2 depicts the proposed way of solving the problem. Experimental dataset collecting, training dataset construction, data cleaning, machine learning model application, and performance assessment all fall under the umbrella of the suggested technique, which is composed of many steps. kaggle's twitter dataset is downloaded in phase one. The data collection is made up of tweets from a variety of different people. There are two parts to the dataset, one for training and the other for testing. Both the training and testing datasets include 31962 rows each. In the next step, we filtered out all of the irrelevant words and phrases from the training and testing datasets, respectively. Tweets are classified as malicious or legal in the dataset used for training. Figure 2 shows the phrases used in both malicious and lawful tweets.

Fig. 2. Proposed framework

Algorithm 1 represents the algorithm used in the proposed approach for the identification of malignant comments. The first step of the algorithm is to segment the data set into training and testing data sets. After segmentation, the items in the training data set are sampled. Finally, the sampled items are converted into machine-understandable values and then machine learning algorithms are applied on them.

Algorithm 1: Malicious Content Detection

Input: Social media Post
Output: Post is malicious or not
Begin:
Data base selection:
$\mathbb{D} \rightarrow \{\mathbb{D}_{train}, \mathbb{D}_{testing}\}$
for *each* $I \in \mathbb{D}_{train}$ **do**
  $\quad I \leftarrow I_{lowercase}$
  $\quad I \leftarrow I \neq Numeric$
  $\quad I \leftarrow I_{sample}$
end
$\mathbb{D}_{train} \leftarrow Machine\ learning_{Xgboost}$
$\mathbb{D}_{test} \leftarrow Machine\ learning_{Xgboost}$
Performance evaluation
End

---

4   Results and Discussion

In this subsection, we give the background about the proposed approach. We utilized the Kaggle dataset for the assessment of our suggested technique; this dataset comprises tweets from various individuals. The training data set contains 31962 rows. The training data set is classified as harmful and valid tweets. Distribution of terms according to hate remarks and nonhate comments. The distribution of text in the training data set indicates that the data set is balanced. For the processing of the dataset, we first transform the comments into lowercase and then delete the special character. Then, we sample the data set again for most common terms. Then, lastly, we used the count vectorizer to convert the text into numeric codes. The result of count vectorizer is used to the machine learning model.

The dataset was analysed and the machine learning model was implemented using Keras and Tensorflow on a 64-bit Intel i5 processor with 8 GB of RAM in a Windows 11 environment.

We used traditional statistical metrics such as accuracy, precision, and recall to assess the performance of the machine learning models. The values of statistical parameters are represented in Table 1.

– Accuracy: Assesses the suitability of the machine learning approach.

$$Accuracy = \frac{\delta t_p + \delta t_n}{\delta t} \tag{1}$$

– Recall: It measures the percentage of genuine comments that are not rejected using machine learning algorithms.

$$R = \frac{\delta t_p}{\delta t_p + \delta f_n} \tag{2}$$

– Precision: It calculates the fraction of authentic comments among the total number of unfiltered comments.

$$P = \frac{\delta t_p}{\delta t_p + \delta f_p} \tag{3}$$

Table 1. Variation of statistical parameters of ML techniques

Parameter	Xgboost	Logical regression	Random forest	SVM
Accuracy	0.920	0.9285	0.91	0.9256
Precision	0.93	0.932	0.932	0.921
Recall	0.97	0.98	0.981	0.981

From Table 1 it is clear that logical regression (LR) has the highest accuracy, precision, recall, and f1 score. Hence, we can say that from all the six machine learning techniques logical regression detects malicious comments from the database.

5   Conclusion

Through social networks, people can communicate and express themselves. One of the most popular social media networks is Twitter. People who misappropriate the service, want to jeopardize the security of others, publish malicious URLs, or otherwise attempt to harm others' computers also exist. Unfortunately, Twitter has not found a comprehensive solution to the problem of spam accounts. The problem of spam detection has become crucial. Even though a lot of effort was put into the project, the intended outcomes could not be achieved. Twitter spam detection is the subject of this article. Data from the Kaggle data set is used in the suggested method. For the purpose of identifying spam twits in the database, we used four different machine learning approaches. From the results, it is presumed that logical regression detects the malicious users from the database most efficiently.

References

1. Agarwal, S., Sureka, A., Goyal, V.: Open source social media analytics for intelligence and security informatics applications. Lect. Notes Comput. Sci. (including subseries Lecture Notes in Artificial Intell. Lect. Notes Bioinf.) **9498**, 21–37 (2015)
2. Gaurav, A., et al.: A novel approach for ddos attacks detection in covid-19 scenario for small entrepreneurs. Technol. Forecast. Soc. Change **177**, 121554 (2022)
3. Sahoo, S.R., et al.: Spammer detection approaches in online social network (osns): a survey. In: Sustainable Management of Manufacturing Systems in Industry 4.0, pp. 159–180. Springer (2022)
4. Shanmugarajah, S., Praisoody, A., Uddin, M., Paskkaran, V., De Silva, H., Tharmaseelan, J., Thilakarathna, T.: Woknack-a professional social media platform for women using machine learning approach, pp. 91–96 (2021)
5. Sharma, S., Jain, A.: Role of sentiment analysis in social media security and analytics. Wiley Interdiscip. Rev. Data Mini. Knowl. Discovery **10**(5) (2020)
6. Sharma, V., Yadav, S., Yadav, S., Singh, K.: Social media ecosystem: review on social media profile's security and introduce a new approach. Lect. Notes Netw. Syst. **119**, 229–235 (2020)
7. Thuraisingham, B.: The role of artificial intelligence and cyber security for social media. pp. 1116–1118 (2020)
8. Yu, H.Q., Reiff-Marganiec, S.: Learning disease causality knowledge from the web of health data. Int. J. Semant. Web Inf. Syst. (IJSWIS) **18**(1), 1–19 (2022)

A Comprehensive Review on Automatic Detection and Early Prediction of Tomato Diseases and Pests Control Based on Leaf/Fruit Images

Saurabh Sharma[1(✉)], Gajanand Sharma[1], Ekta Menghani[1], and Anupama Sharma[2]

[1] JECRC University, Jaipur, India
{saurabh.20mhen014,gajanand.sharma,ekta.menghani}@jecrcu.edu.in
[2] Aryabhatta College of Engineering & Research Center, Ajmer, India

Abstract. Recently, deep learning has proven to be extremely effective in solving challenges connected to the identification of plant diseases. Nevertheless, when a model trained on a specific dataset is assessed in new greenhouse settings, poor performance is seen. Because of this, we provide a way to increase model accuracy by utilizing strategies that can enable the model's generalization capabilities be refined to deal with complicated changes in new greenhouse conditions in this paper. In order to build and test a deep learning-based detector, we utilize photos from greenhouses to train and test the detector. To test the system's inference on new greenhouse data, we utilize the characteristics developed in the previous step to identify target classes. So, our model can differentiate data changes that strengthen the system when applied to new situations by having precise control over inter- and intra-class variations. Using the different inference dataset, we review the different target classes with different type of methodology. The researchers in our field of plant disease recognition feel that our study provides useful suggestions for their future work.

Keywords: Deep learning · Convolutional neural network · Tomato leaf disease · Artificial intelligence

1 Introduction

In addition to providing us with food, plants also protect us from harmful radiation. Agricultural output has long been plagued by plant diseases, which are a major obstacle to the long-term expansion of agriculture. China's tomato industry covers about 700 million square meters and is one of the country's most significant cash crops. Tomato infections are a common occurrence due to a variety of environmental variables. Currently, there are at least 20 different forms of tomato infections that have had a negative impact on tomato productivity and

© The Author(s), under exclusive license to Springer Nature Switzerland AG 2023
N. Nedjah et al. (Eds.): ICSPN 2021, LNNS 599, pp. 276–296, 2023.
https://doi.org/10.1007/978-3-031-22018-0_26

quality, resulting in significant economic losses. Preventing and treating illness in tomatoes are therefore crucial for the industry. Disease diagnosis in the past relied mostly on artificial recognition methods, such as

- subjectively assess disease kinds based on years of planting experience of farmers or
- get disease specimen photographs and search the Internet for judgment.

All terrestrial life would cease to exist if not for plants, which also protect the ozone layer from harmful UV rays. Tomatoes are a food-rich plant and a commonly consumed vegetable [1]. About 160 million tons of tomatoes are consumed each year across the world. For agricultural households, the tomato is a substantial factor to decreasing poverty [5]. Since tomatoes are the most nutritious food on the globe, their cultivation and production affect agricultural economies across the world. One of the tomato's many health benefits is its ability to prevent ailments including hypertension, hepatitis, and gingivitis [3]. Because of its extensive usage, there is an increase in tomato demand as well. Small farmers account for more than 80% of agricultural output, yet nearly half of their crops are lost to disease and pests. When it comes to tomato growth, illness and parasitic insects are the primary issues that must be studied.

Identifying pests and diseases manually is time-consuming and expensive. Because of this, it is imperative that farmers have access to automated AI image-based solutions. Due to the availability of relevant software packages or tools, images are now acknowledged as a reliable technique of recognizing illness in image-based computer vision applications. In order to process photos, they use an intelligent image identification technique known as image processing. This technology boosts efficiency, decreases costs, and improves recognition accuracy. Plants are essential to life, yet they face several challenges. Ecological damage can be avoided if a precise and early diagnosis is made. Product quality and quantity deteriorate when diseases aren't identified in a systematic way. As a result, the economy of a country suffers even more [1,2]. According to the United Nations Food and Agriculture Organization (FAO), agricultural production must increase by 70% by 2050 to fulfill world food demand [3,4]. Fungicides and bactericides, on the other hand, have a detrimental influence on the agricultural ecology. As a result, we require disease categorization and detection methods that are both rapid and effective for use in the agricultural environment. It will be possible to create early disease detection systems for tomato plants thanks to advances in disease detection technologies such as image processing and neural networks. When plants are under stress, their output can be decreased by up to half [1,7]. Once the plant has been inspected, it's time to choose the best course of action based on previous experience [5,8]. As a result of the varying educational and professional backgrounds of the farmers, this system lacks scientific consistency. Farmers may misidentify a disease, resulting in a harmful treatment for the plant. Similarly, field trips by domain experts are expensive. Automated illness diagnosis and classification approaches based on photographs are required, as the domain expert position can no longer be performed by humans. The problem of leaf disease must be addressed with a suitable treatment [6,9]. An important part of the season's production costs must be taken into consideration in order to manage tomato disease [10–12]. Tomato mosaics and yellow curvature are two of the most common illnesses found in tomatoes. As a result of the decreased quality and quantity of the final output, they adversely impair plant growth. 80–90% of plant infections are found on the leaves, according to previous studies [15]. It takes a long time to track the farm and identify different kinds of the illness on afflicted plants. Farmers' assessments of the type of plant disease might be incorrect. Because of this, the plant's protection mechanisms may be inadequate and harmful. Reduced processing costs, reduced environmental effect, and reduced loss risk can all be achieved with early detection [16–18]. As technology has advanced, there have been a variety of solutions presented. For the purpose of identifying leaf diseases, the same approaches are employed in this work. The primary goal here is to draw attention to the lesion more so than in other parts of the image. Some issues have been observed, including (1) variations in light and spectral reflectance, (2) low input picture contrast and (3) image size and shape range Contrast, grayscale conversion, picture resizing/cropping/filtering are all pre-processing procedures that may be done before the final image is taken. In the following stage, a picture is broken down into its constituent parts. Areas of interest as contaminated areas in the picture can be identified using these items [22]. Segmentation has a number of flaws, however:

In cases where the lighting conditions aren't the same as in the original pictures, color segmentation isn't successful.

– It is because of the first seed selection that a region is divided into distinct areas.
– It takes an inordinate amount of time to deal with the many types of texture.

Once the sample has been classified, the final step is to decide to which class the sample belongs. One or more of the procedure's input variables are then polled in this manner. When a certain sort of input is sought, the procedure is used. The most difficult classification task is to improve classification accuracy. This is followed by the creation and validation of datasets that are distinct from the training set.

Consult with medical professionals to examine the signs and symptoms of the condition. As a result, a person with extensive professional training may correctly identify plant illnesses. Farmers, on the whole, are uneducated and lacking in professional expertise. It's difficult for them to keep up with contemporary agriculture's output demands since they tend to make poor diagnoses of plant illnesses.

It has been possible in recent years to identify tomato illnesses using computer vision. One of the most significant topics in computer vision is object detection. It is the primary goal of object detection to pinpoint an image's key area of interest and then classify everything within it. Combining object identification models with multiple deep convolutional neural networks [23] (DCNNs), this study is able to not only identify types of tomato illnesses but also pinpoint unhealthy patches so that appropriate therapy may be used.

Four alternative deep convolutional neural networks are used to identify two object identification architectures, namely Faster R-CNN [24] and Mask R-CNN [25]. Automated visual feature extraction using deep convolutional neural networks is used to identify, categorize, and pinpoint sick areas in feature maps. Detecting objects may be accomplished in two ways: with one architecture and with the other. Faster R-CNN and Mask R-CNN both aim to discover and identify infected tomatoes, whereas Mask R-CNN aims to segment diseased tomato lesion regions.

2   Literature Review

Here, we provide a list of the most current publications that are relevant to our study. For visual categorization and object recognition, we lay out several fundamental techniques in deep learning. After that, we'll go through several methods for identifying plant diseases.

2.1   Artificial Intelligence

The systems that utilize techniques that implement artificial intelligence are capable of processing large amounts of information, predicting trends in the identification of employers in a efficient and rapid manner. In fact, there are programs that interpret our natural language and that act as assistants in various activities of the diarrhea, one of the tools that can be used to observe artificial intelligence and the application of Facebook in the process of recognizing employers suggest the label.

2.2   HSL Color Model

The HSL model (Tono, Saturation, Luminance), also known as HSI. The HSL model doubles as a double cone or hexagon. Its vertices correspond to white color and black, its angle refers to the tone, the displacement corresponds to saturation and the white color distance - black corresponds to the spectacle as observed in Cylindrical color model created by stadium engineer Alvy Ray Smith (one of the founders of the PIXAR studio) for digital color management in the decade of 1970. The colors used are the most common than the traditional RGB model used in the process. compensation. Digital image. Number: The numbers of different cylindrical color models are not standardized. The numbers HSL, HSI, HSB and including HSV can be easily exchanged for references to different color models. To increase confusion, get introduced Translated numbers: TCS (Tone, Saturation), TSB (Tone, Saturation, Brightness) and other numbers. In each case, the HSL and HSV numbers are always referred to as the most common and distinct cylindrical models. The remaining sections can refer to any of these models or including other models. In particular, the order of the channel numbers is invented and it is easy to find the HLS number in the HSL's common denominator. Various diagnostic procedures for images, these names are also confusing. For example, in Photoshop, the HSL model that is analyzed here is called HSB and only included in the color selector, but not in the model for color processing.

2.3   Conversion from RGB Color Model to HSL Color Model

The RGB value must be expressed as a number between 0 and 1, MAX is equal to the maximum value of the RGB value and MIN is equal to the minimum value of these values.

2.4   Image Processing

The processing and analysis of images has been developed for the three major problems related to them.

– Digitize and codify the image to facilitate the transmission, presentation and promotion of the image.
– Improve the restoration of images to explain their content with the facility.
– Describe and segment images.

The set of image processing methods is divided into three categories that require continuation:

– Algorithm in the space domain. Refers to the method that we also consider a group of additional pixels.
– Algorithm in the domain of frequency. In general, these methods are applied to the coefficients produced by the Fourier transform.
– Characterization extraction algorithm. focuses on the analysis of images to extract items, borders, shapes and regions of interest.

2.5   Digital Processing

The basic principles of digital image processing have been established for many years, but have not been realized due to the lack of computers. When you start appearing with high capacity and memory, this camp naturally starts to unroll. The results obtained are that the application of this method has expanded rapidly to other areas such as agriculture and medicine. When it begins to work with the processing of digital images it is important to know that it creates a division of an image into a matrix of rectangular elements. each division is known as a pixel, as the next step is to assign a value to the brightness of each pixel. The third step is to change the value of the spectacle of the pixel through the mathematical operations necessary or the transformations to result in the convenient objects of an image. In the last step it is necessary to transfer the representation of the selected pixels to a monitor that can have a high definition television panel to show the processed image.

2.6   Image Segmentation

The segmentation divides the image into parts or components up to the subdivision level, leaving the region or object of interest. The segmentation algorithm is based on the main characteristic of grays as well as the grayscale concoction, the discontinuity or the similarity that can exist over the grayscale level of additional pixels. Intermittent. Divide the base images into repentant cams in the similarity of the image segmentation. According to one predetermined criterion, the image is divided according to the area search with similar values. Segmentation detects or labels a particular section of the image. Is the Separation (the partition) in components of interest (criterion). (connectivity, objects). The criteria that can be used for segmentations are: on boards (corners or fronts) or regions. Binary segmentation by umbrella is eliminated by the level of intensity of the pixels depending on a specific rank.

2.7   Characterization Extraction with LBP

The extraction of characteristics or properties of a binary image is known as the process of converting input information (generally this information is repeated) to realize any condensed representation (denominated conjunction of properties and characteristics). In other words, it is a technology that can reduce the dimensions of an image. This process involves extracting a new set of functions from the original set and traversing a defined function, as long as it retains relevant information as appropriate. the vector determines the efficiency, velocity, complexity and precision of this algorithm. If a descriptive technique is used without effect, it is used to classify objects in the viewer's viewer and to use clear considerations to filter the veins' pixels and obtain representative binary values. The relationship between the intensity of the color of the central pixel of codification and the additional pixels. For its high discrimination, constitute a universal method for resolving many issues. One of its most important characteristics can be its robust invariance to light cages. The characteristics that are considered low level or well basic are those that can be extracted from the pixel information in the original image. High-level information requires pre-processing and is generally based on low-level properties. It is also possible to classify according to the original dimensions or the source of the extracted information. Initially, for the LBP algorithm, only one-color model channel needs to be processed on the color that the image is working on, usually the gray scale or LBP some calculation is performed for the different channels. Elijah a pixel and determine the order of comparison. As long as the user sees the difference in all the analytics related to this technology, you can choose according to the user's requirements.

2.8   Properties Based on the Intensity of the Pixels

These types of features are directly based on the absolute values of pixels in colors and in scale of grays. A low-level property is the intensity as well as constant to scale, which is crucial in the representation of the image. Given its relevance and meaning, it is very useful. Now well, the intensity of a pixel cannot proportion a detailed description of the images. The LBP is not really represented by the color intensity matrix, during the calculation process calculates the histogram of the "local binary mode", which constitutes the value value representative and descriptor of the selected image, the formula 2.1 shows how to calculate the intensity of the pixels in an image.

$$f = \{1, I(x_i, y_i) < I(x_j, y_j)\}$$

When one is going to perform an analysis on some skins or some of the edges one has to come up with a strategy to utilize the correct functioning of the intensity. The missing value can be assigned immediately to its existing circulating entity, or if it ignores the area and does not calculate LBP, depending on which of its applications, transcendental algae will have records during which relevant LBPs are generated, formula calculate the intensity.

$$f = I(x_i, y_i) - I(x_j, y_j)f(x) = \{1, I(x_i, y_i) - I(x_j, y_j) > 0\}$$

To obtain a better result, you can log defining an umbrella, as well as place a bit of value if you center some distortion at the gray level.

Properties based on the texture of an image

A feature that proportions special information based on the most famous texture is the characteristic. for those characteristics based on spectral information, tenor Gabor, filter of Fourier, wavelet and transformed LBP. The local binary mode (LBP) encodes the element of the local map (border, point, area map) in the characteristic histogram.

Properties based on the shape of the object of interest: The techniques for extracting these attributes can be divided into two types: a technique that uses only information provided on the board, the limit of an object, and a technique that uses information in any area. The first ones ignore the information of many objects simply using borders. In addition to being the basis for calculating advanced functions, the gradient detection can also be one of the main ones. Some of the most common gradient-based functions include: Border Orientation Histogram (EOH) and Orientation Gradient Histogram (HOG). The border orientation histogram defines the relationship that exists in two specified directories in an area determined by the image. The EOH and HOG functions are not modified with respect to the general illumination camber, and the HOG is not modified with respect to the geometric transformation (except for rotation). A variation of HOG is PHOG (Oriented Gradient Pyramid Histogram), which proposes to consider the specific characteristics of the local forms. Y DOT (Dominator Direction Template), in order to calculate a complete histogram, only the principal direction is considered. Scale Invariant Feature Transformation (SIFT) [29].

Gradient Histogram (HOG): The orientation histogram is considered as a function that describes the distribution of the gradient directions in each area of the image [9]. The gradient of the image (the derivation of xey) is very useful because its amplitude is very close to the edges and edges (regions where the intensity changes significantly), and it is known that such edges as the edges of an image contain a lot of information object that the planes regions, we assume that will detect a button. We know that a button is redone and can have a lot of money to earn.

Create a gradient histogram for each area and explain the direction and control of these gradients. Recognize the objects presented in an image, the texture of the image can provide valuable information. The human visual system can only distinguish objects according to color, so we can also recognize objects. Also you can distinguish textures. The main characteristic of the texture is the repetition of one of the most patterned patterns in the area. The pattern can be repeated with precision or it can be expressed as a set of small changes, which can be a function of the job. An image obtained using the LBP algorithm has a unique spectrum image. The LBP operator can be considered as a unified method for the statistical model and the structural model of the texture analysis. The value of the central pixel is what serves as the umbrella, lying, multiplying the umbrella by the corresponding weight assigned to the pixel, generates an LBP code for each viewer and finally sums the results. To illustrate the previous content, we will also take the next step: First, place the center pixel, as well as its value; in this example the value that the assignment to the vecindad is the value is 6. Compare the value of the pixel of the vecindad (Pv) with the central pixel (Pc) Computed [29].

KNN Algorithm: KNN is an algorithm that is based on instances for the supervised automatic learning. This algorithm can be used for the classification of discrete values or it can also make such predictions as regression and continuous values. It is considered as a very simple method that allows the insertion of automatic apprenticeships. KNN can therefore predict a time that it will use it and it will be able to practice the learning of the system using this algorithm and also to classify points of interest following the majority of the proportionate dates. As mentioned earlier, it is an algorithm: Vigilance: briefly, it means that we label the conjunction of training dates and give a category or a determined result. Although it is simple, it is used to solve many problems, the most important is that it is easy to learn and implement. His adventure is that we use all the data set to enter "each point", because it requires a lot of memory and resources of the CPU, to agree with these reasons the KNN algorithm works better in small data sets because there is no large number of characteristics (columns) that are related to being evaluated.

- Calculate distance between the element that is classified by the remainder of the joint of training dates.
- Select the "k" element closest (the shortest distance, depending on the function used).
- Perform identification of " in k points: by category, the final classification is determined.

Considering point 3, we see that, in order to determine the category of point, the value of k is very important, because this case may eventually define which group the point will belong to, especially in the "limit" between groups. For example, a priori, choose a value of k to empathize (if we use a number par). It is not possible to determine whether 3 values are mayors greater than 13. It does not mean that obtaining more significant points necessarily increases accuracy.

The most popular form of "bring proximity between points" is the Euclidean distance ("normally") or the similarity of the cosine (the angle of the medicine vector is smaller, more similar to sera). We record that this algorithm and if so all the algorithms in ML can be better used with varying characteristics (columns of the conjunction of dates) of the ones that we obtain dates. The "distance" that we understand in real life will be abstract in many dimensions that we cannot "visualize" [28].

Food security has become a serious issue as COVID-19 has spread over the world, and several major grain-producing countries have taken measures to limit their grain exports. All countries are grappling with the challenge of increasing grain output. As a result, it is critical for farmers to understand the severity of crop illnesses as quickly as possible so that staff may take further precautions to prevent future plant infection. For the detection of tomato leaf disease, a restructured residual dense network was proposed in this paper, which combines the advantages of deep residual networks and dense networks to reduce the number of training process parameters and improve calculation accuracy as well as information flow and gradients. For classification jobs, we need to reorganize the RDN model [29], which was originally designed for image super resolution, by adjusting the input image features and hyper parameters. The Tomato test dataset in AI Challenger 2018 datasets shows that this model can reach a top-1 average identification accuracy of 95%, which confirms its good performance. There are considerable gains in crop leaf recognition with the reconstructed residual dense network model compared to most of the current state of the art.

Tomato leaf disease has a significant impact on the crop's output. Disease detection is critical to the agriculture sector. Because of their inherent limitations, standard data augmentation techniques such as rotation, flipping, and translation cannot provide useful generalizations. A new data augmentation strategy based on generative adversarial networks (GANs) is suggested in this study for the better detection of tomato leaf diseases. This model is capable of achieving a top-1 average identification accuracy of 94.33% when using a combination of generated pictures and original photos as input. A better model for training and evaluating five classes of tomato leaf photos was produced by tweaking hyper-parameters, altering the architecture of convolutional neural networks, and selecting alternative generative adversarial networks [30]. DCGAN pictures, on the other hand, not only increase the amount of the data set, but also contain the properties of variety, which helps the model generalize well. The t-Distributed Stochastic Neighbor Embedding (t-SNE) and Visual Turing Test have both shown that the pictures created by DCGAN are of higher quality and more convincing. An experiment on the detection of tomato leaf disease has demonstrated the ability of DCGAN to generate data that approximates real images, which can be used to

- provide a larger data set for training large neural networks and improve recognition model performance by using highly discriminating image generation technology;
- reduce the cost of data collection;
- increase diversity in data and the generalization ability of the recognition m.

Diseases in tomatoes might be difficult to detect using artificial means. For tomato disease photos, it might be difficult to distinguish between different tomato illnesses because of the difficulty in finding subtle distinguishing traits. Therefore, we have developed a unique model that includes three networks, namely a Location network, a Feedback network, and a Classification network, known as LFC-Net [31], which is a combination of these three networks. In addition, a self-supervision technique is presented in the model that can discover informative portions of tomato picture without the requirement for manual annotation such as bounding boxes/parts. In order to build a new training paradigm, we analyze the coherence between the picture category and the informational value of the image. Using the model's Location network, the Feedback network directs iteration optimization by identifying informative places in the tomato picture. The Classification network then utilizes the Location network's informative areas and the tomato's whole picture to classify it. As a multi-network partnership, our model is able to go forward. Our model outperforms the pre-trained ImageNet model in the tomato dataset, with an accuracy of up to 99.7%. In this study, we show that our model is very accurate and may potentially be extended to different datasets of vegetables and fruits, which can give a reference for the prevention and management of tomato illnesses. of diseases."

This work offers an enhanced Faster RCNN to identify healthy tomato leaves and four diseases: powdery mildew, blight, leaf mold fungus, and ToMV, which may be used to locate damaged leaves. The first step is to replace the VGG16 [32] image feature extraction network with a depth residual network in order to acquire more detailed disease characteristics. Second, the bounding boxes are clustered using the k-means clustering technique. Using clustering findings, we adjust the anchoring. The enhanced anchor frame is closer to the dataset's actual bounding box. There are three kinds of feature extraction networks that we use in our k-means experiment. Compared to Faster RCNN, the findings of the experiments demonstrate that the new approach for detecting agricultural leaf diseases is 2.71% more accurate and has a faster detection speed.

Increased yield and reduced pesticide use can be achieved through the employment of robots in greenhouse disease diagnosis. We describe a robotic detection method for the detection of powdery mildew (PM) and Tomato spotted wilt virus (TSWV) [33] in greenhouse bell peppers (TSWV). A manipulator is at the heart of the device, making it simple to achieve numerous detection postures. PCA and the coefficient of variation have been used to construct a number of detection techniques (CV). However, tests have shown that the system is able to reliably identify and attain the PM detection stance, but has difficulty achieving the TSWV detection pose (above the plant). This problem should be alleviated by increasing the volume of manipulator work. Classification accuracy (90%) was reached using PCA-based classification with leaf vein removal for TSWV, while CV approaches were equally accurate (85% and 87%). The PCA-based pixel-level classification of PM was high (95.2%), but the accuracy of the leaf condition classification was poor (64.3%), since it was based on the top side of the leaf whereas illness symptoms begin on the bottom side of the leaf. Detection of PM conditions is predicted to be improved by exposing the lower leaf sides [34].

3   Methodology and Result

The methodology used for disease detection in tomato generally used.

3.1   Image Acquisition

The DCT transform may be used as an attribute abstraction step in addition to its primary function of reducing redundant information. As a result, the computational complexity of RLDA is kept to a minimum. Additionally, RLDA is an effective strategy for minimizing data dimensionality, as previously stated. The SVD's right and left singular vectors were used instead of those one-of-a-kind values, which saved us time and effort. Additionally, the data demonstrate that when comparing the results of each technique recorded individually, the recognition speed employing SVD+DCT-RLDA [35] produces the best results, in addition to the considerable gain from practise. The Random Woods Tree (RFT) approach, which was invented by Google, is a favourite in computer vision and face recognition. This was the overarching purpose of all of the study and testing that was conducted. Face-recognition software for mobile devices is capable of great levels of accuracy and usefulness when used properly. Every parameter of the kernel may be optimised, including the initialization and termination settings. Each of the 20 different nations is represented with a picture with a resolution of 205×274 pixels. The accuracy of the SVM capture increases from 93.20% to 95.89% when alternative classifications are used in conjunction with it. More than 97% of them, or 97% of the RF, fell within the accuracy range of 97.17%. Images, skin colour recognition, RGB grey histogram, SVM, RF, and classifier performance were all required for the task. They discovered that RF had a high degree of accuracy after experimenting with the 30 trees they created. Specifically, both SVM and RF [36] were effective in extracting feature vectors and computing histograms when applied to this data set. It is necessary to extract the LBP feature distributions and concatenate them into an enhanced feature vector in order to construct a face descriptor for facial recognition. The performance of the proposed solution is assessed in the context of the face recognition problem.... Furthermore, a diverse variety of applications and extensions have been evaluated. In this newspaper, we will suggest a book as well as a fantastic face depiction for your consideration. A descriptive description is created for each position in a decorative picture using local binary patterns, which are constructed using local binary patterns. They are combined to form a feature vector or histogram that has been spatially stretched to include more information. When the All-Region description is used in combination with the region's feel outline, the world geometry of the region's face area is described using the world geometry of the region's face area. A range of domains [37], including texture classification and image recovery, made extensive use of the LBP operator, which has a number of applications. Prior to the research, it was not obvious that the feel operator might be used to represent facial imagery in a sentence. It seems that facial photos are composed of micro patterns such as apartment locations, stains, lines and edges that may be understood by LBP, according to our results. The suggested technique is evaluated on the face recognition issue that is included in the accompanying user's guide. Oriented gradient histograms and random forest techniques are utilised to identify the face of an efficient face recognition system, which may result in faster identification times at a reduced socio-economic cost [38]. Both attribute representation and classification must be precise and time-efficient in order for these systems to function properly. " They're trying all they can to live up to the demands of the HOG description. Despite its rarity, HOG is a less computationally costly electric descriptor in face recognition than other electric descriptors. The results of recent study reveal that categorization processes vary from one another in terms of the accuracy and precision of their classifications. Random Forest surpasses its competitors in both testing and training while requiring less technical expertise. The ORL database has been used to demonstrate the efficacy of this exact combination in an experimental setting (Table 1).

Table 1. Recognition rate and computation cost of different methods. Computation cost includes cost of feature extraction of all the 320 images and time spent to train the classifier.

Method	Recognition rate	Time computation
GABOR + SVM	94.5	47.289
HOG + RF	95.10	5.376
GABOR + RF	95.70	39.345
HOG + SVM	94.4	10.368
LBP + RF	67.7	9.173
HOG + NN	92.2.	14.368

3.2 Pre-processing

Kernel was used to portray their facial expressions and emotions. It was finished during the experimentation [39] to create a trimmed BU-3DFE dataset consisting of 50 participants who presented one lateral and one 60° shot of each

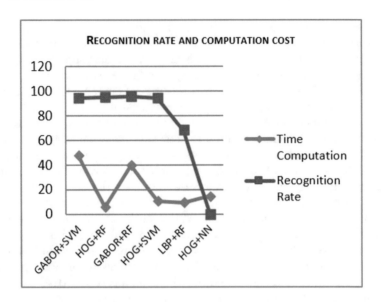

Fig. 1. Comparison of various methods for their recognition rate and computation cost

of the six prototypical facial expressions. A total of ten themes were utilised to teach the device, with the remaining forty issues being used only for data analysis after the training period. The resolution and placement dimensions of this LBP picture were optimised by the use of the practise approach. The best Cable image dimensions were 4060 pixels when a place size of 810 pixels was used as a starting point. The resolution of the rotated picture was determined to be 40 × 50 pixels, which was found to be optimum. Because of the equipment, the accuracy of rectal faces averaged 75%, while the accuracy of spinning faces averaged 70%, according to the manufacturer. As a result of the research. Provided that a person's local texture information about their face was provided, it resulted in a more realistic feature representation of the person's face. If you are dealing with image analysis issues, there are various advantages to adopting LBPs, the most notable of which is improved performance [40]. It is possible to handle notions such as local ternary patterns (LTP), chemical local solar designs (CLBP), and other similar ones. This expansion has a number of downsides, the most significant of which are the increased computation time and the rise in vector measurement. More information about FER-related research that has made use of them may be found in the following sections, which also contain examples. What it takes to recognise and interpret both the usual and the neutral looks on someone's face Fig. 1 shows a sample of photographs from the JAFFE dataset, with the rest of the images to follow. It was chosen to use the first LBP operator [41] on normalised pictures in order to test the hypothesis. The produced LBP image was segmented into 108 pixel-sized areas with no overlap from the beginning to the end. Local histograms of each area of the face were constructed and concatenated in order to build a face emotion feature vector. It was necessary to remove patterns with frequencies below a specific threshold from the feature vectors in order to compensate for the enormous size of the feature vectors. The linear programming approach is used by the classifier of preference to determine its classification. When two disjoint point sets have the same number of points, this system constructs a plane in order to reduce the usual number of points they both have. It was necessary to practise each and every one of the 21 expression pairs for the assignment. The feature vectors from all of the test samples were merged to form the designated class, which was achieved by combining the output of all of those classifiers. Tenfold cross-validation testing was used in the trials on a seven-class JAFFE dataset, and the results were analysed. Following a 20-day trial period, the technique was shown to be successful 93.8% of the time. The LBP algorithm has gone to the top of the pack in recent years when it comes to texture-based feature extraction in computer vision applications. It's getting more popular, particularly in the Specialty. This is due to the fact that simple calculation and resistance to variations in light are required. In order to offer a visual depiction of changes in texture and feel, it is employed in this way. The operator makes a grayscale feel image from a grayscale photo by using a grayscale picture. This authentic LBP operator computes as indicated in Fig. 2 2.6 on every 3 3 pixel grid cell in the grid cell. A threshold function is applied to the centre pixel for each of the three neighbouring pixels in the image. To indicate if the value of the centre pixel is higher than or equal to that of its own neighbour, enter '0,' otherwise, enter '1. The pixel value of your LBP picture is used in order to convert the binary pattern into a usable LBP image for further processing.

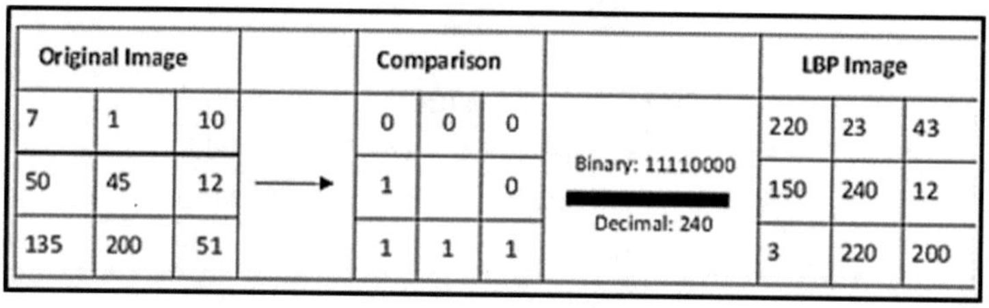

Original Image				Comparison					LBP Image		
7	1	10		0	0	0		Binary: 11110000	220	23	43
50	45	12	▶	1		0			150	240	12
135	200	51		1	1	1		Decimal: 240	3	220	200

Fig. 2. Original LBP operator

3.3   Segmentation

framework was developed for comparing and evaluating machine learning algorithms, which may now be employed to their full capacity. A comparison was made between SVMs and ANNs and RFs in terms of accuracy and the amount of time spent training, optimising, and categorizing [42]. The parameters of the SVM will be represented by the gamma values on the graph. The multi-layer perceptron has been utilised as a classifier for the artificial neural network, or ANN [46]. When creating the ANN, the number of layers and hidden neurons were the two most significant elements to take into consideration. The thickness of the forest and the number of trees were depicted in relation to your RF. The results and conclusions of this inquiry are summarized in the first eating table (Tables 2 and 3).

Table 2. Recognition rate and computation cost of different methods. Computation cost includes cost of feature extraction of all the 320 images and time spent to train the classifier.

Factor	RF	ANN	SVM
Overall accuracy (%)	**87.33**	66.93	84.3
Robust to subjects	high	high	**High**
Robust to hand shapes	high	high	**High**
Classification time (s)	**0.033**	0.061	20.974
Optimisation time (s)	14916	3589	**109**
Training time (s)	101	39	**21**

Table 3. Result for Ep1 and Ep2 by using the At&T database

Evaluation protocol	Approach	Recognition rate
EP1	2DPCA (24)	96.00%
	ICA (24)	85.00%
	Fisher faces (24)	94.50%
	Kernel Eigen faces (24)	94.00%
	ERE (27)	97.00%
	LRC	93.50%
EP2	Fisher faces (24)	98.50%
	Kernel Eigen faces (24)	97.50%
	ICA (24)	93.80%
	B ERE_S (27)	99.25%
	2DPCA (24)	98.30%
	T ERE_S (27)	99.00%
	LRC	98.75

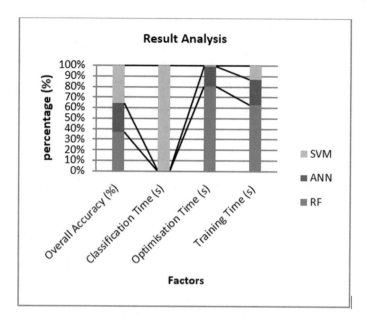

Fig. 3. Performance analysis of SVM, ANN, and RF

3.4   Interpretation

Gunarathna and Rathnayaka [43] it is necessary to be proficient in the LDA in this situation. It is suggested in this article that face recognition algorithms be built on PCA, which employs four metrics to determine a person's facial features. Pre-processing of the input data, as well as PCA loss of measurement LDA is used for feature extraction, while neural networks are utilized for classification in neural networks. In this case, a neurological classifier is used to reduce the amount of photos taken by taking a large number of different samples. Yale University was chosen as the location for testing the proposed technique. According to the findings of the experiments conducted with this database, the proposed facial recognition system is effective and does not make nearly as many mistakes in classification as older approaches did (Fig. 3).

3.5   Classification

The AT&T database has been managed by AT&T. In this collection, there are 10 photographs of 40 different guys. Smiling and not smiling, opening and closing eyes, and eyeglasses or zero glasses are all examples of facial gestures that are contained in the database. This layering of the skin on the face accounts for about 10% of the total amount of skin on the face, with a maximum of 20 layers. The package included the Evaluation Protocol 1 (EP1) and the Evaluation Protocol 2 (EP2), which were both included in the evaluation process (EP2) (Table 4).

Table 4. Relative study of recognition rate using this approach with previous

Technique	PCA + Short distance	PCA + LDA + Short distance	DCT	DCT + LDA + Short distance	DCT + LDA + RB	Proposed method
Recognition Rate	84	94.8	84.5	97.7	98.1	99.5

4   Discussion

For the development and implementation of the project methodology that implements computer vision techniques, the different software is used for the analysis of the information of the processed images. This is how it is observed. In each of these techniques are applied which allow an optimal result to be required.

If different in the titled tests, the results obtained in each of the stages must be applied to the algorithms that most need to be explained and there are methods and techniques that can be applied in each stage, this is what is to be tried prove that the techniques selected are the appropriate ones. For each reason, at each stage it describes the user referring to the computer vision algorithms that best result.

5   Conclusion

To identify and categorize tomato plant leaf diseases, Machine learning/ deep neural network model was presented in the study. Morphological characteristics such as color, texture and leaf margins were also considered. Standard profound learning models were presented in this article, along with their modifications. This article covered fungal and bacterial pathogen-caused biotic illnesses, such as tomato leaf blight, blast, and browning. Accuracy was 98.49% for the model detection rate. The suggested model was compared to the VGG and ResNet versions using the same dataset. It was found that the suggested model outperformed the other models when the results were analyzed. It's a novel idea to identify tomato illness using this method. We want to enhance the model in the future to incorporate abiotic illnesses caused by nutrient deficiencies in the crop leaf. In the long run, our goal is to expand unique data gathering and compile a great quantity of data on various plant diseases. Next-generation technologies will be used to increase accuracy.

References

1. de Luna, R.G., Dadios, E.P., Bandala, A.A.: Automated image capturing system for deep learning-based tomato plant leaf disease detection and recognition. In: TENCON 2018–2018 IEEE Region 10 Conference, pp. 1414–1419. Korea (South), Jeju (2018)

2. Zou, L., Sun, J., Gao, M., et al.: A novel coverless information hiding method based on the average pixel value of the sub-images. Multimed. Tools Appl. **78**, 7965–7980 (2019). https://doi.org/10.1007/s11042-018-6444-0

3. Irmak, G., Saygili, A.: Tomato leaf disease detection and classification using convolutional neural networks. In: Innovations in Intelligent Systems and Applications Conference (ASYU), pp. 1–5. Istanbul, Turkey (2020)

4. Alsmirat, M.A., et al.: Accelerating compute intensive medical imaging segmentation algorithms using hybrid CPU-GPU implementations. Multimed. Tools Appl. **76**(3), 3537–3555 (February 2017). https://doi.org/10.1007/s11042-016-3884-2

5. Gadade, H.D., Kirange, D.K.: Tomato leaf disease diagnosis and severity measurement. In: 2020 Fourth World Conference on Smart Trends in Systems, Security and Sustainability (WorldS4), pp. 318–323. London, UK (2020)

6. Concepcion, R., Lauguico, S., Dadios, E., Bandala, A., Sybingco, E., Alejandrino, J.: Tomato septoria leaf spot necrotic and chlorotic regions computational assessment using artificial bee colony-optimized leaf disease index. In: IEEE REGION 10 CONFERENCE (TENCON). pp. 1243–1248. Osaka, Japan (2020)

7. Al-Ayyoub, M., et al.: Accelerating 3D medical volume segmentation using GPUs. Multimed. Tools Appl. **77**(4), 4939–4958 (2018)

8. Mehedi Masud, M., et al.: Pre-trained convolutional neural networks for breast cancer detection using ultrasound images. ACM Trans. Internet Technol. **21**(4 Article 85), 17 (November 2021). https://doi.org/10.1145/3418355

9. Elhassouny, A., Smarandache, F.: Smart mobile application to recognize tomato leaf diseases using convolutional neural networks. In: International Conference of Computer Science and Renewable Energies (ICCSRE), pp. 1–4. Agadir, Morocco (2019)

10. Widiyanto, S., Fitrianto, R., Wardani, D.T.: Implementation of convolutional neural network method for classification of diseases in tomato leaves. In: 2019 Fourth International Conference on Informatics and Computing (ICIC), pp. 1–5. Semarang, Indonesia (2019)

11. Shijie, J., Peiyi, J., Siping, H., Haibo, S.: Automatic detection of tomato diseases and pests based on leaf images. In: Chinese Automation Congress (CAC), pp. 2537–2510. Jinan, China (2017)

12. Sardogan, M., Tuncer, A., Ozen, Y.: Plant leaf disease detection and classification based on CNN with LVQ algorithm. In: 2018 3rd International Conference on Computer Science and Engineering (UBMK), pp. 382–385. Sarajevo, Bosnia and Herzegovina (2018)

13. Ashok, S., Kishore, G., Rajesh, V., Suchitra, S., Sophia, S.G.G., Pavithra, B.: Tomato leaf disease detection using deep learning techniques. In: 2020 5th International Conference on Communication and Electronics Systems (ICCES), pp. 979–983. Coimbatore, India (2020)

14. Zhou, C., Zhou, S., Xing, J., Song, J.: Tomato leaf disease identification by restructured deep residual dense network. IEEE Access **9**, 28822–28831 (2021)

15. Sabrol, H., Satish, K.: Tomato plant disease classification in digital images using classification tree. In: 2016 International Conference on Communication and Signal Processing (ICCSP), pp. 1242–1246. Melmaruvathur, India (2016)

16. Jiang, D., Li, F., Yang, Y., Yu, S.: A tomato leaf diseases classification method based on deep learning. In: Chinese Control And Decision Conference (CCDC). pp. 1446–1450. Hefei, China (2020)

17. Mamun, M.A.A., Karim, D.Z., Pinku, S.N., Bushra, T.A.: TLNet: a deep CNN model for prediction of tomato leaf diseases. In: 2020 23rd International Conference

on Computer and Information Technology (ICCIT), pp. 1–6. DHAKA, Bangladesh (2020)

18. Kaur, M., Bhatia, R.: Development of an improved tomato leaf disease detection and classification method. In: 2019 IEEE Conference on Information and Communication Technology, pp. 1–5. Allahabad, India (2019)

19. Wu, Q., Chen, Y., Meng, J.: DCGAN-based data augmentation for tomato leaf Ddisease identification. IEEE Access **8**, 98716–98728 (2020)

20. Chakravarthy, A.S., Raman, S.: Early blight identification in tomato leaves using deep learning. In: 2020 International Conference on Contemporary Computing and Applications (IC3A), pp. 154–158. Lucknow, India (2020)

21. Juyal, P., Sharma, S.: Detecting the infectious area along with disease using deep learning in tomato plant leaves. In: 2020 3rd International Conference on Intelligent Sustainable Systems (ICISS), pp. 328–332. Thoothukudi, India (2020)

22. David, H.E., Ramalakshmi, K., Gunasekaran, H., Venkatesan, R.: Literature review of disease detection in tomato leaf using deep learning techniques. In: 2021 7th International Conference on Advanced Computing and Communication Systems (ICACCS), pp. 274–278. Coimbatore, India (2021)

23. Meeradevi, R.V., Mundada, M.R., Sawkar, S.P., Bellad, R.S., Keerthi, P.S.: Design and development of efficient techniques for leaf disease detection using deep convolutional neural networks. In: 2020 IEEE International Conference on Distributed Computing, pp. 153–158. VLSI, Electrical Circuits and Robotics (DISCOVER), Udupi, India (2020)

24. Salonki, V., Baliyan, A., Kukreja, V., Siddiqui, K.M.: Tomato spotted wilt disease severity levels detection: a deep learning methodology. In: 2021 8th International Conference on Signal Processing and Integrated Networks (SPIN), pp. 361–366. Noida, India (2021)

25. Gadade, H.D., Kirange, D.K.: Machine learning based identification of tomato leaf diseases at various stages of development. In: 2021 5th International Conference on Computing Methodologies and Communication (ICCMC), pp. 814–819. Erode, India (2021)

26. Habiba, S.U. Islam, M.K.: Tomato plant diseases classification using deep learning based classifier from leaves images. In: 2021 International Conference on Information and Communication Technology for Sustainable Development (ICICT4SD), pp. 82–86. Dhaka, Bangladesh (2021)

27. Chamli Deshan, L.A., Hans Thisanke, M.K., Herath, D.: Transfer learning for accurate and efficient tomato plant disease classification using leaf images. In: 2021 IEEE 16th International Conference on Industrial and Information Systems (ICIIS), pp. 168–173. Kandy, Sri Lanka (2021)

28. Kibriya, H., Rafique, R., Ahmad, W., Adnan, S.M.: Tomato leaf disease detection using convolution neural network. In: 2021 International Bhurban Conference on Applied Sciences and Technologies (IBCAST), pp. 346–351. Islamabad, Pakistan (2021)

29. Yoren, A.I., Suyanto, S.: Tomato plant disease identification through leaf image using convolutional neural network. In: 2021 9th International Conference on Information and Communication Technology (ICoICT), pp. 320–325. Yogyakarta, Indonesia (2021)

30. Anwar, M.M., Tasneem, Z., Masum, M.A.: An approach to develop a robotic arm for identifying tomato leaf diseases using convolutional neural network. In: 2021 International Conference on Automation, Control and Mechatronics for Industry 4.0 (ACMI), pp. 1–6. Rajshahi, Bangladesh (2021)

31. Hidayatuloh, A., Nursalman, M., Nugraha, E.: Identification of tomato plant diseases by leaf image using squeezenet model. In: 2018 International Conference on Information Technology Systems and Innovation (ICITSI), pp. 199–204. Bandung, Indonesia (2018)
32. Singh, K., Rai, P., Singla, K.: Leveraging deep learning algorithms for classification of tomato leaf diseases. In: 2021 6th International Conference on Signal Processing, Computing and Control (ISPCC), pp. 319–324. Solan, India (2021)
33. Mim, T.T., Sheikh, M.H., Shampa, R.A., Reza, M.S., Islam, M.S.: Leaves diseases detection of tomato using image processing. In: 8th International Conference System Modeling and Advancement in Research Trends (SMART). pp. 244–249. Moradabad, India (2019)
34. N.K.E., Kaushik, M., Prakash, P., Ajay R., Veni, S.: Tomato leaf disease detection using convolutional neural network with data augmentation. In: 2020 5th International Conference on Communication and Electronics Systems (ICCES), pp. 1125–1132. Coimbatore, India (2020)
35. Tm, P., Pranathi, A., SaiAshritha, K., Chittaragi, N.B., Koolagudi, S.G.: Tomato leaf disease detection using convolutional neural networks. In: 2018 Eleventh International Conference on Contemporary Computing (IC3), pp. 1–5. Noida, India (2018)
36. Gibran, M., Wibowo, A.: Convolutional neural network optimization for disease classification tomato plants through leaf image. In: 2021 5th International Conference on Informatics and Computational Sciences (ICICoS), pp. 116–121. Semarang, Indonesia (2021)
37. Hemalatha, A., Vijayakumar, J.: Automatic tomato leaf diseases classification and recognition using transfer learning model with image processing techniques. In: Smart Technologies, Communication and Robotics (STCR), pp. 1–5. Sathyamangalam, India (2021)
38. Kodali, R.K., Gudala, P.: Tomato plant leaf disease detection using CNN. In: IEEE 9th Region 10 Humanitarian Technology Conference (R10-HTC), pp. 1–5. Bangalore, India (2021)
39. Waleed, J., Albawi, S., Flayyih, H.Q., Alkhayyat, A.: An effective and accurate CNN model for detecting tomato leaves diseases. In: 2021 4th International Iraqi Conference on Engineering Technology and Their Applications (IICETA), pp. 33–37. Najaf, Iraq (2021)
40. Lakshmanarao, A., Babu, M.R., Kiran, T.S.R.: Plant disease prediction and classification using deep learning convNets. In: 2021 International Conference on Artificial Intelligence and Machine Vision (AIMV), pp. 1–6. Gandhinagar, India (2021)
41. Tian, Y.-W., Zheng, P.-H., Shi, R.-Y.: The detection system for greenhouse tomato disease degree based on android platform. In: 2016 3rd International Conference on Information Science and Control Engineering (ICISCE), pp. 706–710. Beijing, China (2016)
42. Durmuş, H., Güneş, E.O., Kırcı, M.: Disease detection on the leaves of the tomato plants by using deep learning. In: 2017 6th International Conference on Agro-Geoinformatics, pp. 1–5. Fairfax, VA, USA (2017)
43. Gunarathna, M.M., Rathnayaka, R.M.K.T.: Experimental determination of CNN Hyper-parameters for tomato disease detection using leaf images. In: 2020 2nd International Conference on Advancements in Computing (ICAC), pp. 464–469. Malabe, Sri Lanka (2020)
44. Yang, G., Chen, G., He, Y., Yan, Z., Guo, Y., Ding, J.: Self-supervised collaborative multi-network for fine-grained visual categorization of tomato diseases. IEEE Access 8, 211912–211923 (2020)

45. Kaur, N., Devendran, V.: Ensemble cassification and feature extraction based plant leaf disease recognition. In: 2021 9th International Conference on Reliability, Infocom Technologies and Optimization (Trends and Future Directions) (ICRITO), pp. 1–4. Noida, India (2021)
46. Gupta, N., Sharma, G., Sharma, R.S.: A comparative study of ANFIS membership function to predict ERP user satisfaction using ANN and MLRA. Int. J. Comput. Appl. **105**(5), 11–15 (2014)

Big Data and Deep Learning with Case Study: An Empirical Stock Market Analysis

Divya Kapil$^{(\boxtimes)}$ and Varsha Mittal

CSE, School of Computing, Graphic Era Deemed To Be University,
Graphic Era Hill University, Dehradun, India
divya.k.rksh@gmail.com

Abstract. Big data is a vast produced data quantity that is accumulated at a speed that fast overtakes the edging span. It is difficult for traditional databases and analytical tools to handle the massive quantity of produced data through social sites, Audio/Video Streaming, games etc. Big data requires new technical approaches and architecture to handle such large and complex data. Since big data has myriad attributes, including complexity, variety, value, variability, velocity, and volume and due to these attributes, multifarious challenges emerge. Big data is a new technology with tremendous advantages for business organizations; therefore, it is important to focus on various challenges and issues related to this technology. In this research, we describe big data attributes and big data importance and the employment of deep learning in big data. We also discuss various challenges to adopting big data technologies. We also present stock market analysis as case study. In the Financial market, miscellaneous stakes, bonds, guarantees or banknotes are bartered. Therefore creating many datasets as time series data where the cost is conspired concerning a time series which is a notable property and diverse set of the predictive algorithm. In this research, essential elements of time series data have been examined.

Keywords: Big data · Hadoop distributed file system · MapReduce · Deep learning

1 Introduction

With the arrival of the digital era, the data amount is generated and kept in the data warehouse, streaming of audio/video and webpages, which all are sources of a huge data quantity. Big data [1] describes structured, semi-structured and amorphous data types that are complicated data infrastructures. To handle complex data infrastructure, controlling administrations and scientific solutions are needed. Big data is a comprehensive phrase for datasets that are immense or complicated that conventional data processing tools are insufficient to handle them. Various issues include investigation, capture, visualization, tracking, sharing,

N. Nedjah et al. (Eds.): ICSPN 2021, LNNS 599, pp. 297–308, 2023.
https://doi.org/10.1007/978-3-031-22018-0_27

warehouse, transfer, and data privacy. It merely concerns the service of predictive analytics or other specific cutting-edge techniques to pull value from data and infrequently to a certain dataset measure.

YouTube is an American online video-sharing forum, and its headquarter is in San Bruno, California. The ex-PayPal workers-Jawed Karim, Chad Hurley and Steve Chen, initiated the service in 2005. YouTube was bought by Google in 2006.

Nowadays, Netflix is a leading streaming platform that provides entertainment services. They provide movies, web series, and documentaries in various languages. Members can enjoy the entertainment services from anywhere and anytime. Now big data plays an important role so that a single platform can take on the complete entertainment world. Wall Street Journal states Netflix employs Big Data analytics to maximize the merit and resilience of its video streams and estimate consumer entertainment selections and consider patterns. It presents a wide variety of user-generated and corporate videos. The big data features are discussed below:

Volume: Big data means a enormous quantity of data depicting the volume. The extant data is in petabytes at present and will be increased to zettabytes in the hereafter. The social sites produce data for terabytes on a daily grounds, and managing this amount of data using current traditional systems is a critical challenge. Some other challenges are class imbalance, variance and bias, feature engineering and processing performance.

Variety: Big data is a vast amount of data and has different data types that include conventional data, semi-structured data from several resources such as web pages, emails, papers, data from sensors, web log files, and social sites. These are systematised, semi-systematised and amorphous data that are problematic to be taken care of by the existent conventional analytic approaches. The challenges are data locality, heterogeneity and noisy and dirty data.

Variability: Data flow discrepancies are variability. Data loads are an enormous challenge to manage, particularly with increased social media usage. Challenges are data availability, real-time process, concept drift and distributed random variables.

Velocity: It is a concept that takes into account the rapidity of coming data from different sources. The data which are generated through sensor devices, for example, it would constantly be moving to the database store, which is a huge amount. Our conventional systems are, therefore, not properly capable of analyzing the data that is regularly in movement.

Value: Users can run such queries against stored data and therefore deduce important outcomes from the data, which is filtered and rate them according to the dimensions needed.

Complexity: Hierarchies, relationships and multiple data linkages need to be connected and correlated.

Table 1. Big data characteristics

S. No.	Features	Description
1	Volume	Large amount of data may be in zettabytes in hereafter
2	Variety	Different category of data like structured, semi- structured, un-structured data.
3	Variability	Incongruities of the data flow.
4	Velocity	The rapidity of the data coming from various sources.
5	Value	Huge amounts of data are worthless unless they provide value.
6	Complexity	Big Data has structured, unstructured, and semi-structured types

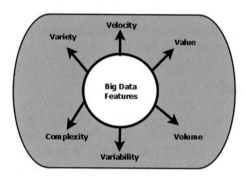

Fig. 1. Features of big data

Big Data Databases: NoSQL Relational Database Management System (RDBMS) needs to build data in stipulated structures that are not permissible in the Big Data environment in which quick, extended magnitudes of amorphous data are created. Numerous NoSQL databases are conceived to supervise amorphous and non-relational data to encounter the issues. They sustain many data models containing document-oriented, key-value pairs like XML, BSON, JSON, and HTML files and graphs created for linked data. The NoSQL databases furnish an inexpensive method than RDMS to manage repository and Big Data administration in a distributed system.

NewSQL: NoSQL databases include multiple limitations that cause the outcome of NewSQL databases that often do not sustain SQL querying and indexing and are usually gradually managing enormous queries. Contrasting RDBMS, they do not guarantee ACID regulations for trustworthy transactions. NewSQL is designed for Big Data applications that comprise an RDBMS based on a distributed architecture to handle these constraints. NewSQL databases combine the most useful prior technologies the RDBMS and the NoSQL.

Searching and Indexing conventional searching techniques are not acclimated to the distributed environment and Big Data's complicatedness. Corporations must execute ample real-time queries via vast magnitudes of amorphous and structured data sets. Big data [2,3] can be generated by web companies that handle the data that is loosely structured or unstructured. Figure 1 shows the big data's characteristics, commonly known as V's of big data. Our research focuses on Hadoop, MapReduce and big data research challenges. The key contributions of this paper:

- To provide the Hadoop system description with its components and MapReduce programming model.
- To present the use of deep learning techniques in big data,
- To present many challenges in the big data.
- To analyze stock market dataset

This research paper is systematised into five sections. Related work is shown in Sect. 2. Section 3 discusses big data technologies Hadoop and MapReduce and the MapReduce programming model. Section 4 discusses research issues, and also discuss the stock market analysis and the conclusion is provided in Sect. 5.

2   Related Work

Information rises swiftly at the speed of 10x every five years [4]. Between 1986 and 2007, the technological abilities of data repository, communication and computation were traced via 60 analog and digital approaches in 2007 [5,6]. For big data, specific architectures and design choices are discussed in the paper [7].

In paper [8], the big data's challenges are presented and also, they discussed big data analysis and design. Madden [9] discusses the conventional databases needed for Big data in the paper and concludes that the databases do not resolve all big data problems. The generic architecture model for Scientific Data Infrastructure (SDI) is proposed in the paper [10]. Available current technologies and best practices, the model provides building interoperable data basics. The authors show that the proposed models can be implemented easily using the infrastructure services provisioning model, which is cloud-based.

Yang et al. [11] presented a new system to store the big graph data in the cloud that is employed to analyse and process the data, in which the big data is squeezed with spatiotemporal features. Najafabadi et al. [12] concentrated on deep learning but remarked the subsequent available barriers for Machine learning when using with Big Data like amorphous data layouts, snappy moving data like streaming, many origin data intakes, noisy and bad differentia data, algorithms scalability, imbalanced allotment of intake data, restricted labelled data and unlabeled data.

Qiu et al. [13] discussed the employment of Machine Learning in Big Data in the area of signal processing. They determined five essential problems extensive scale, dissimilar data types, high data rate, unsure and unfinished data.

An extensive scale network traffic observation and investigation system that is based on Hadoop was presented that is placed in the crux network of an extensive cellular network and has been broadly estimated [14]. Song et al. [15]

perform a comparative analysis between KNN and the various algorithms based on MapReduce and estimate them via a mixture of time and theory.

Financial market index prediction is a very problematic chore for persons who like to invest in the financial market. This issue occurs due to the wavering and volatility of the stock market prices. Since technologies are becoming advanced, so it is possible to predict the stock costs faster and precisely. Machine Learning approaches are widely employed because of the ability to recognise stock tendencies from enormous data that apprehend the underlying stock cost action. Artificial Neural Network technique is primarily executed and plays a vital role in decision making for stock market forecasts. Multi-Layer Perceptron design with backpropagation algorithm can forecast with better accuracy.

3   Hadoop

The big data enhances performance in business model services and products and also provides support for decision-making [6,16]. Big Data's architecture must be compatible with the organization's support infrastructure. We address data management tool in this section, Hadoop, and Mapreduce as well.

Hadoop: It is codded in Java and was released in 2006 as a top-level Apache project [17]. Doug Cutting invented an open-source project, Hadoop, that could be employed in a distributed environment to employ the Google MapReduce programming system. It is used in large quantities of data. MapReduce and Hadoop Distributed File System are the most common components for Big Data. Hadoop consists of HBase, Hive, HCatalog, Pig, Zookeeper, Oozie and Kafka.

HDFS: It is used when a single computer has too much data. HDFS is a very complex system of files. The cluster contains two groups of nodes. The first node serves as a primary node and the second node represents data node.

HBase: HBase is an open source program that has been published and distributed based on Google's Big Table. This approach is based on columns instead of row-based, increasing process efficiency across similar values across large data sets.

Hive: Hive uses Hadoop as a data warehouse and is consummate with the terminology of SQL for querying [18,19]. This technology was developed at Facebook. The working of Hive is done in the form of tables. These are two tables; the first table is managed table. Hive manages the data of this table, and the second table is the external table. The data of this table is managed outside of Hive.

Pig: Yahoo Research developed Pig in 2006, but in 2007 it was transferred to the Apache Software Foundation. Pig has a vocabulary and environment for execution. The language of Pig is PigLatin [20,21]. It's a language of data flow that ties items together.

Yarn: In the latest Hadoop YARN [22,23], Yet Another Resource Negotiator is included. The main goal of YARN is to let the system act as an available data-processing design. Other than MapReduce, it uses a programming model.

Map Reduce: The programming paradigm basically does two distinct jobs: Map and Reduce. Input is taken from the distributed file system and given to the map task. The map task generates a key-value pairs series from the input, which is accomplished pursuant to the map function code written. The master controller collects Such generated values, sorts them according to the key, and divides them into reduced tasks. The sorting ensures that with the same reduction activities, the same key values stop. The Tasks Reduce blends all the related values with a key that deals with only one key simultaneously. Furthermore, the method of combining counts on the code written to relieve the work. We show the algorithm and Mapreduce programming model below.

Algorithm for MapReduce

MapReduce operations (Wordcount)
Map()
- Intake $< file\_name, file\_content >$
- Parses file and exudes ¡word, count¿ pairs
Reduce()
- Adds Values for the same keys and exudes $< word, TotalCount >$

Mapreduce programming model

- Input: This is the file that is to be processed.
- Split: Received data must be split into smaller parts which are called splits.
- Map: It processes any break according to the map function's logic. A mapper is used for each break. Each mapper is called a job and the JobTracker maps and executes multiple tasks through the various Job Trackers.
- Combine: This is an optional step. By reducing the data amount which to be transferred across the network, the performance can be improved.
- Shuffle & Sort: All the mappers generate outputs and after that these are shuffled and sorted to place them in order, and then they are grouped and then send to the next phase.
- Reduce: The mappers' outputs are aggregated for this reduce () function is used. Then this output is sent to the final phase.
- Output: Finally, the reduce phase's output is written into a file in HDFS.

4   Big Data with Deep Learning

We live in the epoch where Machine Learning is keeping a deep impact on a comprehensive scope of applications such as text learning, speech and image recognition [24, 25], and health care [26, 27]. As Machine Learning is employed

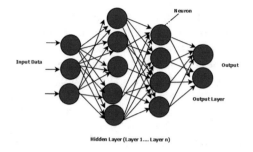

Fig. 2. Deep neural network with hidden layers

in the latest applications, it is generally the subject that there is not sufficient training data. Conventional applications such as entity detection or machine translation relish enormous portions of training data that have been collected for decades. More, current applications have small or no training data. As an illustration, smart factories are progressively growing automated where Machine Learning acquires product grade management. When there is a fresh development or a new fault to glimpse, there is undersized or no training data, to begin with. Machine Learning is a decisive technique that can be employed to make forecasts on the forthcoming data nature established in history.

Machine learning algorithms perform by creating a model from input instances to create data-driven findings for the future. Big Data is the data accumulated from diverse embryones that are employed as the intake for analytics. This analytical approach has a Machine Learning algorithm that learns the provided data and makes itself better from experience. Among the multiple issues in Machine Learning, data gathering is becoming a crucial problem. It is understood that the prevalence of the time for executing machine learning end-to-end is consumed on readying the data, which contains gathering, cleaning, examining, picturing, and characteristic engineering. Whereas these stages are time-consuming, data gathering has become an issue. Though data gathering was conventionally an issue in Machine Learning, since the quantity of training data is growing, data administration research is becoming related.

Deep learning is considerable outstanding Machine Learning approach in numerous applications like text acquaintance, image examination and speech recognition. It employs supervised and unsupervised techniques to understand multilevel models and attributes in hierarchical architecture for pattern recognition and classification. The current growth in communication and sensor network technologies has stimulated the accumulation of big data. It offers incredible chances for a comprehensive of areas, and it has multiple complex problems in data mining and data processing because of its features of volume, velocity, variety, and veracity. Deep learning has a significant position in big data analytical solutions. Deep learning has also been fruitfully involved in enterprise outcomes that carry the benefit of the extensive magnitude of digital data. Corporations such as Facebook and Google that gather and examine immense quantities of data on a day to day basis have been strongly moving to deep learning associated projects. Deep learning is a machine learning approach that comprehends numerous levels of representations in the architecture of deep.

Traditional neural networks [28] have an issue getting entangled in nonconvex objective function's local optima that usually shows unsatisfactory performance [29]. Moreover, they cannot carry the benefit of unmarked data that are usually large and frugal to gather in Big Data. To overcome these issues, a deep belief network employs a deep architecture that is skilled in understanding feature representations from the marked and unmarked data delivered to it [25, 30]. Deep belief network is shown in Fig. 3. Hinton et al. [31] represent a Deep Learning generative model for understanding the binary codes for documents. The Convolutional Neural Network (CNN) is the broadly utilized deep learning model in attribute learning for immense-scale image recognition and categorization [16, 32–34]. It contains three layers that are convolutional layer, pooling or subsampling layer and completely-connected layer.

The convolutional layer employs the convolution function to execute the weight sharing, whereas the subsampling is employed to decrease the dimension. The high magnitudes of data show a tremendously difficult problem for deep learning. Big data usually maintains an enormous number of instances/inputs, wide varieties of class kinds/outputs, and very elevated dimensionality/attributes. These characteristics straight cause execution time complexity and model complicatedness.The most current deep learning system can manage a huge number of instances and parameters and also feasible to rise up with additional GPUs utilised. Another issue related to increased volumes is the noisy labels and data inadequacy. Unlike most traditional datasets employed for machine learning that was noise free and carefully selected, Big Data has usually incomplete outcome from their different sources.

Although big data addresses many current issues related to high data volumes, there are still some problems. The biggest challenge in storage systems architecture. We present some issues that are not solved big data yet in this segment. Privacy and Security and Information Security, Technical Challenges, Analytical challenges, Skill Requirement,and Sharing of Information, Storage and Processing Issues, Computational Complexities and Knowledge Discovery, Visualization and Scalability of Data, IoT for Big Data Analytics, Big Data Analytics with cloud computing, Big Data Analytics with Bio-inspired Computing, Quantum Computing for Big Data Analysis

4.1 Stock Market Prediction Analysis

In this section, we analysis stock market prediction analysis using time series model. This model has generated using weka tool. For this purpose we use Nifty 50 dataset. To forecast the action of the Nifty 50 index of the Indian stock market. The NIFTY 50 is the index that aids in observing the enduring trend of the blue-chip organisations and reminisces the actual scenario of the Indian Stock Market. This study concerns a monthly report on the closing stock indices of NIFTY shrouding the period from 2000 to 2007, including a total number of 3000 instances. The parameters employed in this research contain of Date, Open, High, Low, Close, Volume, Turnover, P/E, P/B, and Div Yield the Nifty 50 Index. We employ 70% dataset for training purpose and 30% for testing and the dataset is trained with time-series classifiers. The usual metrics that are employed for time series data will be Mean Squared Error, Root Mean Squared Error and Mean Absolute Error, and Mean Absolute Percentage Error. The less value of Mean Squared Error and Root Mean Squared Error, the more promising the fit of the model.

Mean Absolute Percentage Error: it estimates of forecast accuracy of a prediction approach in statistics.

Mean Squared Error: it estimates the errors' square average, which means the average squared dissimilarity between the real and forecasted values.

$$MSE = (1/N) \star \sum (real - forecasted)^2$$

where, N = number of total items
real = real value
forecasted = predicted value
Root Mean Squared Error: it estimates the dissimilarity between forecasted values by a model and the values estimated. It denotes the square root of the dissimilarity between forecasted values and the experimental values of these dissimilarities.

$$RMSE = \sqrt{\sum (forecasted - real)^2 / total\, forecasted\, values}$$

Table 2. Result

Mean absolute error	24.0141	38.0528	50.4985	63.134	75.8678
Mean absolute percentage error	1.1717	1.8059	2.3473	2.8798	3.4239
Root mean squared error	35.2889	55.4447	72.206	89.0602	106.3545
Mean squared error	1245.3069	3074.1176	5213.7131	7931.7107	11311.2731

5 Conclusion

Big data has various applications, such as Intelligent Transportation systems. Big data analytics can improve transportation efficiency in many ways. Big data becomes too immense, intricate, and incomprehensible when managed by conventional techniques. As machine learning evolves more broadly employed, it is essential to obtain enormous data volumes and labelled data, particularly for state-of-the-art neural networks. We discuss big data and tools of big data like Hadoop and its components. We also describe the concept of MapReduce from

Fig. 3. Results

Fig. 4. Results

the Hadoop framework. We also discuss the role of deep learning in big data and analysis timeseries model for stock market. Various research problems occur when big data is employed with additional cutting-edge technologies, like the Internet of Things (IoT), cloud computing etc. When big data comes with IoT data, Knowledge acquisition is the biggest challenge. Employing cloud computing for big data, privacy is the main issue in data hosting on public servers. We discuss those issues that will enable those business associations to locomote to this technology.

References

1. Katal, A., Wazid, M., Goudar, R.H.: Big data: issues, challenges, tools and good practices. pp. 404–409, 8–10 Aug 2013, Noida (2013)
2. Borthakur, D.: The hadoop distributed file system: architecture and design. Hadoop Project Website, vol. 11 (2007)
3. Singh, A., et al.: Distributed denial-of-service (DDoS) attacks and defense mechanisms in various web-enabled computing platforms: issues, challenges, and future research directions. Int. J. Semant. Web Inf. Syst. (IJSWIS) **18**(1), 1–43 (2022)
4. Hendrickson, S.: Getting Started with Hadoop with Amazon's Elastic MapReduce, EMR (2010)
5. Hilbert, M., López, P.: The world's technological capacity to store, communicate, and compute information. Science **332**(6025), 60–65 (2011)
6. Manyika, J., Michael, C., Brown, B., et al.: Big data: The Next Frontier for Innovation, Competition, and Productivity. Tech. Rep, Mc Kinsey (2011)
7. Bakshi, K.: Considerations for big data: architecture and approach. In: IEEE, Aerospace Conference (2012)

8. Kaisler, S., Armour, F., Espinosa, J.A., Money, W.: Big data: issues and challenges moving forward. In: IEEE, 46th Hawaii International Conference on System Sciences (2013)
9. Madden, S.: From databases to big data. IEEE Internet Comput. (2012)
10. Demchenko, Y., Zhao, Z., Grosso, P., Wibisono, A., de Laat C.: Addressing big data challenges for scientific data infrastructure. In: IEEE, 4th International Conference on Cloud Computing Technology and Science (2012)
11. Yang, C., Zhang, X., Zhong, C., Liu, C., Pei, J., Ramamohanarao, K., Chen, J.: A spatiotemporal compression based approach for efficient big data processing on cloud. J. Comput. Syst. Sci. **80**(8), 1563–1583 (2014). ISSN: 0022-0000
12. Najafabadi, M.M., Villanustre, F., Khoshgoftaar, T.M., Seliya, N., Wald, R., Muharemagic, E.: Deep learning applications and challenges in big data analytics. Journal of Big Data **2**(1), 1–21 (2015). https://doi.org/10.1186/s40537-014-0007-7
13. Qiu, J., Wu, Q., Ding, G., Xu, Y., Feng, S.: A survey of machine learning for big data processing. EURASIP J. Adv. Signal Process. **67**, 1–16 (2016)
14. Liu, J., Liu, F., Ansari, N.: Monitoring and analyzing big traffic data of a large-scale cellular network with hadoop. Network IEEE **28**(4), 32–39
15. Song, G., Rochas, J., Beze, L., Huet, F., Magoules, F.: K nearest neighbour joins for big data on mapreduce: a theoretical and experimental analysis. IEEE Trans. Knowl. Data Eng. **28**(9), 2376–2392
16. Ling, Z., Hao, Z.J.: An intrusion detection system based on normalized mutual information antibodies feature selection and adaptive quantum artificial immune system. Int. J. Semant. Web Inf. Syst. (IJSWIS) **18**(1), 1–25 (2022)
17. A. Hadoop.: Hadoop (2009). https://hadoop.apache.org/
18. Dean, J., Sanjay, Google, Inc.: MapReduce: simplified data processing on large clusters
19. Cvitić, I., Peraković, D., Periša, M., Gupta, B.: Ensemble machine learning approach for classification of IoT devices in smart home. International Journal of Machine Learning and Cybernetics **12**(11), 3179–3202 (2021). https://doi.org/10.1007/s13042-020-01241-0
20. Thusoo, A., Sarma, J.S., Jain, N., Shao, Z., Chakka, P., Zhang, N., Antony, S., Liu, H., Murthy, R.: Hive—a petabyte scale data warehouse using hadoop By Facebook Data Infrastructure Team
21. Gupta, B.B.: A lightweight mutual authentication approach for RFID tags in IoT devices. Int. J, Networking Virtual Organ (2016)
22. Apache HBase. Available at https://hbase.apache.org
23. Sharma, R., et al.: Detecting and preventing misbehaving intruders in the internet of vehicles. Int. J. Cloud Appl. Comput. (IJCAC) **12**(1), 1–21 (2022)
24. Al-Ayyoub, M., et al.: Accelerating 3D medical volume segmentation using GPUs. Multimedia Tools Appl. **77**(4), 4939–4958 (2018)
25. Chui, K.T., et al.: Handling data heterogeneity in electricity load disaggregation via optimized complete ensemble empirical mode decomposition and wavelet packet transform. Sensors **21**(9), 3133 (2021). https://doi.org/10.3390/s21093133
26. Yu, H.Q., Reiff-Marganiec, S.: Learning disease causality knowledge from the web of health data. Int. J. Semant. Web Inf. Syst. (IJSWIS) **18**(1), 1–19 (2022)
27. Lu, J., Shen, J., et al.: Blockchain-based secure data storage protocol for sensors in the industrial internet of things. IEEE Trans. Ind. Inform. **18**(8), 5422–5431 (2022). https://doi.org/10.1109/TII.2021.3112601

28. Masud, M., et al.: Pre-trained convolutional neural networks for breast cancer detection using ultrasound images. ACM Trans. Internet Technol. **21**(4, Article 85), 17 (2021). https://doi.org/10.1145/3418355
29. Rumelhart, D., Hinton, G., Williams, R.: Learning representations by back-propagating errors. Nature **323**, 533–536 (1986)
30. Hinton, G., Salakhutdinov, R.: Reducing the dimensionality of data with neural networks. Science **313**(5786), 504–507 (2006)
31. Hinton, G., Salakhutdinov, R.: Discovering binary codes for documents by learning deep generative models. Topics Cogn. Sci. **3**(1), 74–91 (2011)
32. Karpathy, A., Toderici, G., Shetty, S., Leung, T., Sukthankar, R., Li, F.: Large-scale video classification with convolutional neural networks. In: Proceedings of IEEE Conference on Computer Vision and Pattern Recognition, pp. 1725–1732. IEEE (2014)
33. Han, Y., Kim, J., Lee, K.: Deep convolutional neural networks for predominant instrument recognition in polyphonic music. IEEE/ACM Trans. Audio Speech Lang. Process. **25**(1), 208–221 (2017)
34. Simonyan, K., Zisserman, A.: Very deep convolutional networks for large-scale image recognition (2014). arXiv:1409.1556

Automated Machine Learning (AutoML): The Future of Computational Intelligence

Gopal Mengi[(✉)], Sunil K. Singh, Sudhakar Kumar, Deepak Mahto,
and Anamika Sharma

Chandigarh College of Engineering and Technology, Chandigarh, India
official.mengi.gopal@gmail.com, sksingh@ccet.ac.in, sudhakar@ccet.ac.in ,
co20306@ccet.ac.in

Abstract. Computer science controls every task in today's environment, and everything in the sector attempts to automate the task. The basic essence of computer science is to minimize human effort and the core of reducing human involvement in any task is to automate. Machine learning (ML) has become an important part of many aspects of our daily life. High-performance machine learning applications, on the other hand, necessitate the use of highly qualified data scientists and domain specialists. Automated machine learning (AutoML) aims to reduce the need for data scientists by allowing domain experts to automatically construct machine learning applications without extensive statistical and machine learning knowledge. This paper provides an overview of existing AutoML methodologies as well as information on related technologies. We present and investigate important AutoML techniques and methodologies along with present challenges and future research directions. We also analyze various security threats that can be posed to the machine learning models and AutoML.

Keywords: AutoML · Machine learning · Automation · Security · Security threats

1 AutoML - Introduction

Automated Machine Learning, or AutoML, is a technique for automating the end-to-end process of applying machine learning to real-world scenarios. End-to-end (E2E) learning is the process of training a potentially complex learning system with a single model that represents the complete target system, without the need for the intermediary layers that are common in traditional pipeline designs [1,2]. Deep neural architectures appear to be breaching the traditional barriers between learning machines and other processing components by casting a potentially complex processing pipeline into the coherent and flexible modelling language of neural networks, which appears to be a natural outcome [3,4].

N. Nedjah et al. (Eds.): ICSPN 2021, LNNS 599, pp. 309–317, 2023.
https://doi.org/10.1007/978-3-031-22018-0_28

Because the AutoML problem is a combinatorial problem, any proposed technique must identify an optimal combination of operations for each section of the ML pipeline to minimize mistakes. Based on a comprehensive assessment of a huge landscape of models and hyperparameters, an automated machine learning method is created to discover the optimum model and feature engineering [5,6].

This software operates on a collection of pre-defined parameters, which include pre-defined operation sets, operations picked by algorithms, a generator function for creating new features, the number of selected features, and the maximum number of features to be selected. Although these standard methods cannot be used to process fully raw data, datasets are normally polished to some level and can work well with them [7–9].

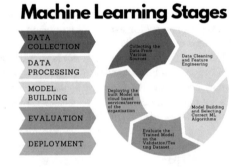

Fig. 1. Machine learning stages

2   Working of AutoML

AutoML works on predefined steps of the Machine Learning Pipelining process. There are eight steps in the AutoML process which are as follows:-

A. **Data Ingestion**. The data intake process includes basic data exploration to ensure that the data can be used for machine learning in the first place, such as checking that there aren't too many missing values. It's important to remember that most AutoML tools won't work until the model has enough labelled data to work with. As a result, this stage guarantees that sufficient data is available to train a feasible model [10].

B. **Data Preparation**. Deduplication, filling in missing values, scaling, and normalization are all examples of data preparation, or data preprocessing. This stage guarantees that the data quality is sufficient enough to be used for modelling, as machine learning algorithms might be finicky about the data they take as input [11–13].

C. **Data Engineering.** Data engineering is the third phase in the AutoML process, and it entails deciding how features are retrieved and processed, as well as data sampling and shuffling. Data engineering, also known as feature engineering, can be done manually or automatically using machine learning techniques such as deep learning, which extracts and selects features from data [14–16].

D. **Model Selection.** Choosing a good and appropriate model for the data you have processed is the key feature of the AutoML process. AutoML tools automatically identify the right model for your data which would give the best accuracy [17–19].

E. **Model Training.** Several models are often trained on subsets of data, and the most accurate one is chosen for further tuning and deployment. A sequence of validation steps with held-out data is performed on the final model.

F. **Hyperparameter Tuning.** AutoML must be able to tune hyperparameters, or meta parameters, to maximize performance to work effectively. Hyperparameter optimization is a term used to describe this process. This means AutoML systems must be able to make a series of predictions for various hyperparameter combinations and then choose the best one based on performance. Initial weights, learning rate, momentum, maximum tree depth, and other hyperparameters are all common [20, 21].

G. **Model Deployment.** Deploying a trained ML model can be very difficult, especially for large-scale systems that normally require intensive data engineering efforts. By exploiting in-built information about how to distribute the model to multiple systems and contexts, an AutoML system may make developing a machine learning pipeline much easier [22–24].

H. **Model Updates.** Finally, as fresh data becomes available, AutoML systems are capable of updating models over time. This ensures that models are always up to date with fresh data, which is critical in fast-paced corporate situations.

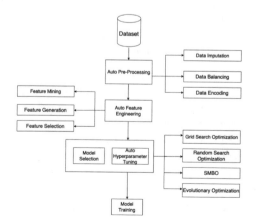

Fig. 2. Basic taxonomy of AutoML

3   Advantages of AutoML

One advantage of AutoML is that it eliminates the need for specialists and data scientists to conduct the time-consuming process of manually training and evaluating models. AutoML helps data scientists in choosing the best model and architecture of the model to obtain the best accuracy. If you don't accomplish the feature selection correctly, your score will suffer, and your process will be slowed. Instead, your company may rely on AutoML's automatic evaluation process to choose features for you [25]. Using AutoML can help organizations, data scientists, and ML engineers to scale and deploy models easily. Reduce human bias with AutoML and be confident that your algorithms will benefit your company. These tools will measure the effectiveness of your model and compare it to evaluation metrics automatically, eliminating the need for your team to do it manually [26].

4   Security Threats to AutoML

Based on the training status of the learning model, security risks to machine learning systems may be classified into two classes. Threats that arise before or during model training are distinguished from threats that occur after the machine learning model has been trained [27].

4.1   Threats to Machine Learning Models Before the Model training

Stealthy Channel Attack. Data quality is basic for building a powerful AI model. Subsequently, it is basic to assemble precise and applicable information. Data is collected from various sources to make a real-world application. An attacker might compromise the machine learning system by giving it phoney and erroneous data. As a result, an attacker could endanger the entire system by entering large amounts of incorrect data, even before building the model. This is known as a secret channel attack. Therefore, data collectors need to be very careful when collecting data from machine learning systems [28].

Data Poisoning. Data is so important to machine learning that even the slightest changes to the data can make the algorithm worthless. Attackers attempt to exploit flaws and vulnerabilities in machine learning systems to modify the data used for training. Data poisoning directly affects two important data qualities: confidentiality and reliability. The data required to train a system typically contains sensitive or sensitive information. Data confidentiality is endangered by poison attacks.

4.2   Threats to Machine Learning Systems After the ML Model has been Trained

1. **Evasion Attack**. Another notable and very much read-up security risk for AI frameworks is attack avoidance. Inside these types of attacks, the input

information or testing information is messed with, driving the AI framework to foresee mistaken information. This endangers the framework's trustworthiness and the public's confidence in it.

2. **Manipulation of the System.** An authentic AI framework learns constantly; it continually learns and moves along. Constant criticism of the environmental factors is used to improve what is happening. This is similar to support models, which highlight ceaseless contributions from a critic. A powerful AI framework ought to offer ordinary criticism and consistently endeavour to move along. An assailant can exploit this part of a skilful AI framework by giving sham information as input, making the framework move in the wrong heading. Accordingly, rather than working on it over the long run, the framework's presentation debases, causing a change in the framework's way of behaving and, in the end, delivering it useless.

3. **Transfer Learning Attack.** Pre-trained machine learning models are often used for high-speed output. One of the main reasons for using a pre-trained model is that training time is exponentially long for some applications that require large amounts of training data. Computing resources are still scarce, so it makes sense to use a pre-trained model. These pre-trained models are used after being optimized and refined to meet your requirements. However, because the model is pre-trained, it is difficult to know if it is trained on a particular dataset. An attacker could exploit this by manipulating a real model or replacing it with a malicious model [29].

4. **Output Integrity Attack.** If an attacker has access to the model and the interface that presents the findings, he or she can display manipulated results. This is known as an output integrity attack. Because we have a theoretical lack of understanding of the fundamental internal workings of a machine learning system, we cannot anticipate the actual outcome. As a result, when the output of the system is shown, it is taken at face value. By jeopardizing the output's integrity, the attacker can take advantage of the attacker's lack of expertise.

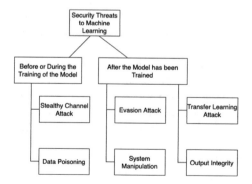

Fig. 3. Security threats to machine learning

5   Challenges of AutoML

Data and model application are two issues that AutoML has to deal with. High-quality labelled data is insufficient, and data discrepancies during offline data processing will have negative consequences. In addition, teams must do tough technical tasks such as automatic machine learning processing of unstructured and semi-structured data. People have limited means to appropriately judge before the outcomes are obtained with this type of multi-objective investigation.

Another thing worth mentioning is that performing AutoML in a dynamic environment is more challenging than in a static environment since the environments are always changing. Companies will need to react to changes in data faster, detect distribution changes, and automatically adjust to new types of models, among other things, to use dynamic feature learning.

While automated robots are capable of finding solutions, they may not always be what the user desires. An explainable model may be desired by the user. Explainability, experts say, has a lot of uncertainty because everyone's understanding is different, and it has a lot to do with personal judgment. Making the model understandable is considerably more difficult. Organizations must work to advance the creation of explainable machine learning standards. Experts decide whether the outputs fit their own interpretable and explainable standards and consistency, which AutoML can provide [30,31].

When we look at how much time a data scientist spends solving an ML problem, we can see that the majority of the time is spent thinking about and understanding the problem, which cannot be automated. The stress of repetitious labour can only be alleviated using AutoML.

6   Future Research Direction

AutoML will shift its focus in the future to no-code AI platforms that employ drag-and-drop interfaces to allow individuals without coding skills to submit data and immediately begin training machine learning models. AutoML may eliminate the need for skilled data scientists with extensive data science expertise, as well as programming languages such as Python and training hardware such as CPU or GPU.

AutoML-Zero is an AutoML strategy that reduces human bias by employing relatively simple mathematical operations (such as cos, sin, mean, and std) as the search space's core operations and developing two-layer neural networks on top of them. Even though the AutoML-Zero network is significantly simpler than both human-designed and Neural Architecture Search (NAS)-designed networks, the experimental results suggest that a new model design paradigm with little human design may be developed, allowing for a more broad and flexible design [32].

Automated machine learning (AutoML) attempts to automate the configuration of machine learning algorithms and their integration into a larger (software) solution but at a significant resource cost. To decrease this use, Green AutoML, a sustainable AutoML research program, has been proposed. It categorizes efforts into four categories: AutoML system design, benchmarking, transparency, and research incentives. Furthermore, AutoML has been presented as a potential method for detecting invasive ductal carcinoma (IDC) incomplete slide images (WSI). A large public IDC picture dataset is used and upgraded for model evaluation, as is an experimental IDC identification model constructed with Google Cloud AutoML Vision [33].

References

1. Chauhan, K., Jani, S., Thakkar, D., Dave, R., Bhatia, J., Tanwar, S., Obaidat, M.S.: Automated machine learning: the new wave of machine learning. In: 2nd International Conference on Innovative Mechanisms for Industry Applications (ICIMIA), 2020, pp. 205–212 (2020). https://doi.org/10.1109/ICIMIA48430.2020.9074859

2. Liu, Y., et al.: Survey on atrial fibrillation detection from a single-lead ECG wave for internet of medical things. Comput. Commun. **178**, 245–258 (2021). ISSN: 0140-3664

3. Karmaker, S.K., ("Santu"), Hassan, M.M., Smith, M.J., Xu, L., Zhai, C., Veeramachaneni, K.: AutoML to date and beyond: challenges and opportunities. ACM Comput. Surv. **54**(8), Article 175, 36 (2021) (2022). https://doi.org/10.1145/3470918

4. Yu, H.Q., Reiff-Marganiec, S.: Learning disease causality knowledge from the web of health data. Int. J. Semant. Web Inf. Syst. (IJSWIS) **18**(1), 1–19 (2022)

5. Zhang, L., Shen, W., Li, P., Chi, X., Liu, D., He, W., Zhimeng, X., Wang, D., Zhang, C., Jiang, H., Zheng, M., Qiao, N.: AutoGGN: a gene graph network AutoML tool for multi-omics research. Artif. Intell. Life Sci. **1**, 100019 (2021). ISSN: 2667-3185. https://doi.org/10.1016/j.ailsci.2021.100019

6. Gupta, B.B., Badve, O.P.: GARCH and ANN-based DDoS detection and filtering in cloud computing environment. Int. J. Embed. Syst. **9**(5), 391–400 (2017)

7. Liu, D., Xu, C., He, W., Xu, Z., Fu, W., Zhang, L., Yang, J., Wang, Z., Liu, B., Peng, G., Han, D., Bai, X., Qiao, N.: AutoGenome: an AutoML tool for genomic research. Artif. Intell. Life Sci. **1**, 100017 (2021). ISSN: 2667-3185. https://www.sciencedirect.com/science/article/pii/S2667318521000179

8. He, X., Zhao, K., Chu, X.: Automl: a survey of the state-of-the-art. Knowl.-Based Syst. **212**, 106622 (2021). ISSN: 0950-7051. https://www.sciencedirect.com/science/article/abs/pii/S0950705120307516

9. Chui, K.T., et al.: Handling data heterogeneity in electricity load disaggregation via optimized complete ensemble empirical mode decomposition and wavelet packet transform. Sensors **21**(9), 3133 (2021). https://doi.org/10.3390/s21093133

10. Singh, I., Singh, S.K., Singh, R., Kumar, S.: Efficient loop unrolling factor prediction algorithm using machine learning models. In: 3rd International Conference for Emerging Technology (INCET), 2022, pp. 1–8 (2022). https://doi.org/10.1109/INCET54531.2022.9825092

11. Pang, R., Xi, Z., Ji, S., Luo, X., Wang, T.: On the security risks of AutoML. arXiv.org. In: USENIX Security '22, Knowledge-Based Systems, 2021. Elsevier (2021). https://doi.org/10.48550/arxiv.2110.06018

12. Truong, A., Walters, A., Goodsitt, J., Hines, K., Bruss, C.B., Farivar, R.: Towards automated machine learning: evaluation and comparison of AutoML approaches and tools. In: IEEE 31st International Conference on Tools with Artificial Intelligence (ICTAI), 2019, pp. 1471–1479 (2019). https://doi.org/10.1109/ICTAI.2019.00209
13. Sharma, R., Sharma, T.P., Sharma, A.K.: Detecting and preventing misbehaving intruders in the internet of vehicles. Int. J. Cloud Appl. Comput. (IJCAC) **12**(1), 1–21 (2022)
14. Li, Y., Wang, Z., Xie, Y., Ding, B., Zeng, K., Zhang, C.: AutoML: from methodology to application. In: Proceedings of the 30th ACM International Conference on Information and Knowledge Management. Association for Computing Machinery, New York, NY, USA, pp. 4853–4856 (2021). https://doi.org/10.1145/3459637.3483279. Industry Applications (ICIMIA), pp. 205-212 (2020). https://doi.org/10.1109/ICIMIA48430.2020.9074859
15. Chopra, M., et al.: Assess and analysis Covid-19 immunization process: a data science approach to make India self-reliant and safe. In: International Conference on Smart Systems and Advanced Computing (SysCom 2021) (2022). http://ceur-ws.org/Vol-3080/10.pdf
16. Gupta, B.B.: A lightweight mutual authentication approach for RFID tags in IoT devices. Int. J. Networking Virtual Organ. (2016)
17. Chopra, M., Singh, S. K., Sharma, S., Mahto, D.: Impact and usability of artificial intelligence in manufacturing workflow to empower Industry 4.0. In: International Conference on Smart Systems and Advanced Computing (SysCom 2021) (2022). http://ceur-ws.org/Vol-3080/10.pdf
18. Chopra, M., et al.: Predicting catastrophic events using machine learning models for natural language processing. In: Data Mining Approaches for Big Data and Sentiment Analysis in Social Media, pp. 223–243. IGI Global (2022). https://doi.org/10.4018/978-1-7998-8413-2.ch010
19. Ling, Z., Hao, Z.J.: An intrusion detection system based on normalized mutual information antibodies feature selection and adaptive quantum artificial immune system. Int. J. Semant. Web Inf. Syst. (IJSWIS) **18**(1), 1–25 (2022)
20. Singh, I., Singh, S.K., Kumar, S., Aggarwal, K.: Dropout-VGG based convolutional neural network for traffic sign categorization. In: Congress on Intelligent Systems. Lecture Notes on Data Engineering and Communications Technologies, vol. 114. Springer, Singapore (2022). https://doi.org/10.1007/978-981-16-9416-5_18
21. Lu, J., Shen, J., et al.: Blockchain-based secure data storage protocol for sensors in the industrial internet of things. IEEE Trans. Ind. Inform. **18**(8), 5422–5431 (2022). https://doi.org/10.1109/TII.2021.3112601
22. Xin, D., Wu, E.Y., Lee, D.J.-L., Salehi, N., Parameswaran, A.: Whither AutoML? Understanding the role of automation in machine learning workflows. In: Proceedings of the 2021 CHI Conference on Human Factors in Computing Systems (CHI '21). Association for Computing Machinery, New York, NY, USA, Article 83, pp. 1–16 (2021). https://doi.org/10.1145/3411764.3445306
23. Li, Y., Wang, Z., Ding, B., Zhang., C.: AutoML: a perspective where industry meets academy. In: Proceedings of the 27th ACM SIGKDD Conference on Knowledge Discovery and Data Mining (KDD '21). Association for Computing Machinery, New York, NY, USA, pp. 4048–4049 (2021). https://doi.org/10.1145/3447548.3470827
24. Cvitić, I., Peraković, D., Periša, M., et al.: Ensemble machine learning approach for classification of IoT devices in smart home. Int. J. Mach. Learn. Cyber. **12**, 3179–3202 (2021). https://doi.org/10.1007/s13042-020-01241-0

25. Drozdal, J., Weisz, J., Wang, D., Dass, G., Yao, B., Zhao, C., Muller, M., Ju, L., Su, H.: Trust in AutoML: exploring information needs for establishing trust in automated machine learning systems. In: Proceedings of the 25th International Conference on Intelligent User Interfaces (IUI '20). Association for Computing Machinery, New York, NY, USA, pp. 297–307 (2020). https://doi.org/10.1145/3377325.3377501

26. Crisan, A., Fiore-Gartland, B.: Fits and starts: enterprise use of AutoML and the role of humans in the loop. In: Proceedings of the 2021 CHI Conference on Human Factors in Computing Systems (CHI '21). Association for Computing Machinery, New York, NY, USA, Article 601, pp. 1–15 (2021). https://doi.org/10.1145/3411764.3445775

27. Zeng, Y., Zhang, J.: A machine learning model for detecting invasive ductal carcinoma with Google Cloud AutoML vision. Comput. Biol. Med. **122**, 103861 (2020). ISSN: 0010-4825. https://doi.org/10.1016/j.compbiomed.2020.103861

28. Chopra, M., Singh, et al.: Analysis and prognosis of sustainable development goals using big data-based approach during COVID-19 pandemic. Sustain. Technol. Entrepreneurship (2022). https://www.sciencedirect.com/science/article/pii/S2773032822000128

29. Chopra, M., Singh, S.K., Sharma, A., Gill, S.S.: A comparative study of generative adversarial networks for text-to-image synthesis. Int. J. Softw. Sci. Comput. Intell. (IJSSCI) **14**(1), 1–12 (2022). https://doi.org/10.4018/IJSSCI.300364

30. Guyon, I., Chaabane, I., Escalante, H.J., Escalera, S., Jajetic, D., Lloyd, J.R., Macià, N., Ray, B., Romaszko, L., Sebag, M., Statnikov, A., Treguer, S., Viegas, E.: A brief review of the ChaLearn AutoML challenge: any-time any-dataset learning without human intervention. In: Proceedings of the Workshop on Automatic Machine Learning. Proceedings of Machine Learning Research (2016). https://proceedings.mlr.press/v64/guyon_review_2016.html

31. Singh, A., Singh, S.K., Mittal, A.: A review on dataset acquisition techniques in gesture recognition from Indian sign language. In: Advances in Data Computing, Communication and Security. Lecture Notes on Data Engineering and Communications Technologies, vol. 106. Springer, Singapore (2022). https://doi.org/10.1007/978-981-16-8403-6_27

32. Gijsbers, P., LeDell, E., Thomas, J., Poirier, S., Bischl, B., Vanschoren, J.: An open source AutoML benchmark. arXiv (2019). https://doi.org/10.48550/arxiv.1907.00909

33. Tornede, T., Tornede, A., Hanselle, J., Wever, M., Mohr, F., Hüllermeier, E.: Towards green automated machine learning: status quo and future directions (2022). arXiv.org. https://doi.org/10.48550/arxiv.2111.05850

34. Aggarwal, K., Singh, S.K., Chopra, M., Kumar, S.: Role of social media in the COVID-19 pandemic: a literature review. In: Data Mining Approaches for Big Data and Sentiment Analysis in Social Media, pp. 91–115 (2022). https://doi.org/10.4018/978-1-7998-8413-2.ch004

35. Aggarwal, K., Singh, S.K., Chopra, M., Kumar, S., Colace, F.: Deep learning in robotics for strengthening Industry 4.0.: opportunities, challenges and future directions. In: Robotics and AI for Cybersecurity and Critical Infrastructure in Smart Cities. Studies in Computational Intelligence, vol. 1030. Springer, Cham (2022). https://doi.org/10.1007/978-3-030-96737-6_1

Semi-supervised Federated Learning Based Sentiment Analysis Technique Across Geographical Region

Aarushi Sethi[1(✉)], Himashree Deka[1], Justin Zhang[2], and Wadee Alhalabi[3]

[1] National Institute of Technology, Kurukshetra, India
`sethi.aarushi30@gmail.com`
[2] University of North Florida, Jacksonville, USA
`justin.zhang@unf.edu`
[3] King Abdulaziz University, Jeddah, Saudi Arabia
`wsalhalabi@kau.edu`

Abstract. With fast globalization, the need for developing models that are suited to global data and not simply localized data is the need of the hour. However, collecting sensitive data globally gives rise to privacy restrictions and concerns. Privacy-preserving machine learning techniques like federated learning can be very effective in tackling the problem of privacy yet building a promising model with comparable accuracy which we have tried to depict in the paper. Additionally, we have discussed an algorithm that automates the labelling of the unlabelled mined tweets from seven different locations. In conclusion, we have provided a pipeline to develop a global model while ensuring the privacy of the data thus ensuring that sensitive and restricted information can be used to build machine learning models as well.

Keywords: Federated learning · Unlabelled dataset · Semi-supervised learning · Privacy

1 Introduction

With social media attaining a dominant position as a method of communication over the past few years, sentiment analysis has increasingly gained importance in the world of Natural Language Processing [1,2]. Sentiment Analysis is the act of determining whether a given piece of information is positive, negative or neutral in nature. This determination can be on a document level, sentence level or aspect based. These days the engagement and sentiment towards an event, brand or organization plays an important role in public perception and reception of the same. Twitter is one of the most popular social media platforms where people resort to tweeting about current prevalent topics. Hence, tweets are a valuable source of information for sentiment analysis. With globalization and social media networks not confined to a particular region for functioning, the

© The Author(s), under exclusive license to Springer Nature Switzerland AG 2023
N. Nedjah et al. (Eds.): ICSPN 2021, LNNS 599, pp. 318–328, 2023.
https://doi.org/10.1007/978-3-031-22018-0_29

scope of data for sentiment analysis is endless. However, in the real world, mining of copious amounts of data is bound by several strict government policies [3,4]. Over the years, governments in various countries have imposed arduous policies that make it difficult to perform cross-country data mining so as to prevent cyber threats and leakage of confidential data. Hence, in such a scenario, the data extracted for a centralized model will either be of low value, or would not be sufficient and diverse enough to build a good global model. Furthermore, even if one does acquire a large dataset, the computational cost for transferring this data is huge. Given such a condition, privacy preservation must be given high importance. One possible solution to this problem is federated learning [5,6]. Federated Learning relies on the concept of decentralization of the model. Instead of one centralized model collecting data and thereafter training it, it uses several smaller client side models to train local data. Once the client side models have been trained, the weights, instead of the data, are transferred to a global model where they are averaged. The weights of these global models are then sent to the clients again in order for their local models to be updated. Although federated learning may still attract cyber attacks, it is far more safe and secure than a centralized model that makes use of transfer of raw data.

In federated learning, we don't directly interact with the client data, and hence client data exists in a raw form. This raw form of data might be labeled or unlabeled in nature. When the data is unlabeled and we are using a supervised classifier, we must change this unlabeled data into a labeled form. Directly providing labels by interacting with the data violates the principle of federated learning. Additionally, providing human crafted labels to the large amount of data is a tedious task and may lead to biases and erroneous labels. To solve this problem, we have used a semi-supervised-based labeling technique [7–9] which is robust and autonomous in nature. Semi-supervised labeling techniques learn the mapping of the labels from small labeled dataset and use it to predict the sentiments of large unlabeled dataset.

In this paper, we have tried to recreate a similar situation where-in tweets are extracted based on their location. These locations serve as clients for the decentralized model. We have also discussed a semi-supervised approach to labeling the raw tweets for training. Once we have created a dataset out of the tweets that we scraped based on the location, we compare the performance of a centralized model with a homogenous federated learning model.

Section 2 describes similar work in the area of federated learning and sentiment analysis. Section 3 describes the problem statement and Sect. 4 explains our approach to tackling the suggested problem. Section 5 displays the results and observations of our experiments and finally, Sect. 6 concludes the paper.

2   Related Work

2.1   Federated Learning

Federated Learning was a concept proposed by Google in 2016 [10] as a solution to the privacy concerns that arise from transferring data for the development of a machine learning architecture. It is a concept developed on the transfer of weights between a central server and several other client servers and not the raw data. Most federated learning experiments have used variations of LSTM, RNN and other smaller deep learning architectures as their base client model [11, 12] owing to lesser number of trainable parameters, computational cost and time. Federated Learning can be broadly divided into two types based on the type of data that is fed to the clients-IID and non-IID. IID datasets assume that the training and testing data both are independently and identically distributed. Most of the developments in federated learning have been made assuming the data is independent and identical in distribution however, recent developments have attempted to tackle the problem of non-IID datasets as well [13]. Various algorithms can be used to calculate the weights for the centralized server [14] however, in this paper we have used the federated average function. While federated learning is a leap forward when it comes to privacy preservation of user data, the gradients that are being transferred to the global model can still be attacked upon and privacy can be compromised. A recommendation to prevent such a security breach is to use encryption techniques such as differential privacy, secure multiparty computation and homomorphic encryption [15, 16]. However, a downside to implementing some of these techniques along with federated learning is the massive loss of information. Some of the fields where federated learning has made researchers privy to information which was earlier difficult to access are the health-care sector [17, 18], pharmaceutical sector [19, 20] and financial sector [21]. In our paper, we have used data which is extracted from various locations and is unlabelled. We first label this data and train a federated learning model to compare the effect of a centrally trained model on one general dataset and a decentralized model with data that belongs to a specific geographical region. Also, integration of a transfer learning model with a federated learning framework is performed.

2.2 Transfer Learning

Over the past few years the field of Natural Language Processing has seen a drastic improvement and appreciation towards transfer learning. These pre-trained models provide the user with the ability to get a robust and generalized model even with lower amounts of data available. Also, these pre-trained models like BERT [22, 23], GPT [24] can be fine-tuned to the user's preferred extent. Such abilities are favorable when it comes to integrating it with federated learning. However, it has been duly noted that transfer learning models have high numbers of parameters that hinder them to be applied in real world scenarios such as keyboard learning from mobiles. Efforts have been made to make use of conditional or selective training to reduce the number of trainable parameters and make the fine-tuning of transfer learning models more efficient [25].

3   Problem Formation

3.1   Labeling a Dataset

Let $N = \{n_1, n_2, n_3 \ldots n_n\}$ be the number of devices located across different geographical regions. Let $D = \{d_1, d_2, d_3 \ldots d_n\}$ be the number of dataset such that $d_k \in n_k$. When we are collecting data from twitter for a dataset $d_k = \{x_i, y_i\}_{i=1}^{N}$, we have $Y = \{y_i\}_{i=1}^{N} = \{\phi\}$.

To find y_k, we propose a majority-voting based semi-supervised algorithm which is defined in Sect. 4.1, such that if A is an algorithm, then for a data sample x_i, we get its desired label from Eq. 1.

$$f(x_i, A) = y_i \ni y_i \in Y \tag{1}$$

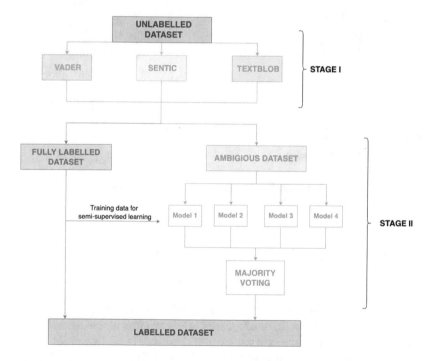

Fig. 1. Multi-stage labeling

3.2   Privacy Preserving Model

Let p_k be the privacy parameters associated with the dataset d_k, then the total privacy across all the datasets is given by

$$P = \sum_{i=1}^{N} p_k \tag{2}$$

In Eq. 2, the value of p_k ranges between 0 and 1. In a centralized machine learning model $P \longrightarrow 1$, whereas, in a federated model $P \longrightarrow 0$.

Let $J = \{j_1, j_2, j_3 \ldots j_n\}$ be the parameters associated with the global model located across each client in a federated training, then the total parameters for all devices will be

$$min_J = \sum_{i=1}^{N} j_k \tag{3}$$

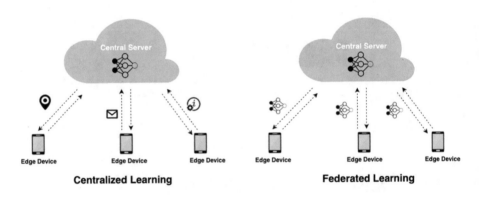

Fig. 2. Centralized learning and federated learning

4   Proposed Approach

4.1   Multi Stage Semi-supervised Labeling

Our multi-stage labeling process is presented in Fig. 1. Let $X = \{x_1, x_2, x_3 \ldots x_n\}$ be the number of records presented inside the dataset d_k stored at node n_k . Initially the labels of each record $Y = \{\phi\}$ is an empty set. As shown in Fig. 1, we pass each record present in X through three APIs, i.e., Vader [26], Textblob [27], and Sentic [28]. Vader, Textblob, and Sentic are prebuilt APIs that provide sentiments to the input tweets. Each API processes the tweets and provides its own sentiments. For a dataset x_k, let $y_{pred} = \{y_1, y_2, y_3 \ldots y_n\}$ be the sentiments resulted by three APIs. We further define two different dataset, i.e., fully labeled dataset, $D_f = \{\}$ and ambiguous dataset, $D_a = \{\}$ such that $D_f \cup D_a = d_k$. If the predictions resulted by the APIs are the same, i,e.,

$y_1 = y_2 = y_3$ then we append that record into the fully labeled dataset, i.e., D_f with its corresponding label y_k otherwise we append the corresponding record into D_a.

D_a as of stage 1 labeling procedure does not contain any labels y_k to its respective records x_k. This is because the given three APIs were not able to provide the same labels to any particular record x_k. In a stage 2 of labeling process, our aim is to provide labels to the records present in D_a. To achieve this, we have used a semi-supervised learning approach. As a part of semi-supervised labeling, we first need a labeled dataset which can be further split into training and testing dataset. The classifier is first trained on a training dataset and the trained classifier is used to predict the labels of an unlabeled dataset. Once we change an unlabeled dataset into labeled dataset, we combine training dataset and labeled dataset together and retrain the classifier and predict the labels of a test dataset. Using this method we can predict the label of an unlabeled dataset and evaluate the performance of a semi-supervised classifier in predicting labels with the help of a test dataset.

In our approach, we consider D_f as our labeled dataset which can be split into D_f^{train} and D_f^{test}. The unlabeled dataset is D_a for which we want to predict the sentiment. Here, we have trained four base models, i.e., Random Forest, KNN, Naive Bayes, and Decision Tree on a dataset. D_f^{train} where $D_f^{train} = \{x_i, y_i\}$. The four base models discover an ability to map the corresponding records x_i to its label y_i. We use this mapping ability further to generate the labels of the records present in D_a. Let $Y_{predict} = \{y_1, y_2, y_3 \ldots y_n\}$ be the labels predicted by four base classifiers. The predicted labels by different classifiers may be the same or different, so we choose the correct labels with the majority voting process. During the majority voting, some predicted sentiments may show ambiguous nature, i.e., two of the classifiers predict the sentiment positive, whereas, the other two of the classifiers predict the sentiments as negative. In such a scenario, we discard those tweets. Discarding tweets due to ambiguity in labels may decrease the size of the dataset and hinder the performance in building a robust sentiment analysis model; however, our experimental result shows that the discarded sentiments are very low in number and discarding such tweets have very little effect on model performance.

4.2 Federated Learning

In a general deep learning scenario, we have a central dataset D and only one processing node N; however in a federated learning setup, instead of a centralized dataset D, we have multiple dataset $D = \{d_1, d_2, d_3 \ldots d_n\}$ distributed among nodes $N = \{n_1, n_2, n_3 \ldots n_n\}$ where each dataset $d_k \in n_k$. Federated learning is a process to train a global model by extracting knowledge from the private training data of each node without exporting the private data to the server. When federated training starts, the global model has a DNN architecture that generates the initial parameter θ_0. The initial parameter is sent to the subset of participating clients P with $P \in N$. Each participant client p_k after receiving the parameter θ_0 starts its local SGD training on the dataset d_k such that the global parameter $\theta_1 \longleftarrow \theta_0$. For the updated parameter θ_1, along with corresponding dataset d_k the loss function at each node is given by

$$l = \frac{1}{n} \sum_{i=1}^{N} l(\theta_1, x_i) \tag{4}$$

All the nodes present inside P tries to minimize the given loss function. Once all the nodes update their parameters, all the parameters are aggregated and averaged. If θ_i represent the list of parameters updated by each node then the averaged parameters is

$$\theta_{avg} = \frac{i}{k} \sum_{i=1}^{N} \theta_i \tag{5}$$

where k is the number of nodes present inside the subset of clients P.

In a general federated learning scenario, for a record $x_i \in D_k$, we have its corresponding label y_i; however, in our scenario, we don't have any such labels and the labels need to be generated. To generate such labels, along with the DNN model M_1 that gets trained on the local dataset, we also have a model M_2 that generates the labels of the corresponding records x_i. During the first round of federated training, the model M_2 is transferred at the client side. Here the model M_2 collectively refers to all the base models discussed in Sect. 4.1 to generate the labels. Before the first federated training round is initiated, the model M_2 is required to do all the processing at the client side to generate the labels from the unlabeled dataset. Although we have additional model M_2, we make sure that all the processing is done at the client side to generate labels and no dataset or knowledge transfer is there between the node n_k and $\{N - n_k\}$ or with the server.

5   Experiment and Results

In this section we have discussed the experimental details and the corresponding results that we achieved by performing the experiments. We have divided the experiments into two major sub-sections. The first section discusses the details surrounding the formulation of the dataset. The second subsection describes the results that have been obtained using BERT and federated learning.

5.1   Dataset

To form a dataset, we extracted tweets from different geographical regions. The locations that were chosen to extract the tweets are - New York, London, Mumbai, Singapore, Paris, Sydney and Johannesburg. Figure 2 represents the number of tweets extracted from different places along with the numerical count of sentiment across each region. Extracting tweets from different geographical regions was important so that we can have a variation in the language of different sentiments such that our federated learning model can jointly learn this variation. The extracted tweets were then preprocessed. Preprocessing of tweets includes tokenization, stemming and removal of stop words. Initially, we extracted 105,000 number of tweets; however, some of the tweets were discarded which had insufficient length, i.e. sentences having a length of less than three words. Discarding these kinds of tweets is necessary as they do not contribute much in pattern extraction and are usually considered noise in the dataset. We also removed some duplicate tweets in the dataset and the resulting dataset after preprocessing contained 102,500 tweets. We further used this dataset to provide labels to the tweets.

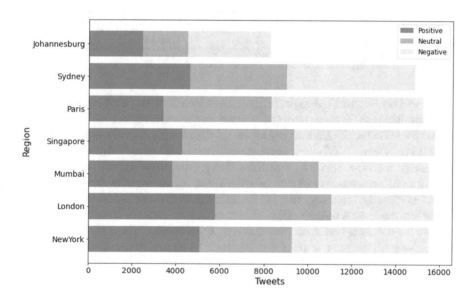

Fig. 3. Number of tweets across each geographical region

5.2 Implementation and Result

We performed required experimentation on a machine having a core i9 processor, 32 GB of RAM, and 12 GB of graphics card. We used two versions of BERT, i.e., BERT-base and BERT-multilingual to perform our experiment on a centralized and federated environment. Both the versions of BERT are pretrained models where BERT-base is pretrained on only English language, whereas, BERT-multilingual is trained on 104 language texts. We used both the models to perform a comparative analysis of accuracy on the federated learning dataset since the primary language for our dataset is English but it is sourced from various locations that use varied forms of English. During the experimentation, we imported both the models of BERT and fine tuned it using Huggingface on the formed dataset with a learning rate of $5e^{-5}$ and weight decay of 0.01 using learning rate scheduler. This means that the pre- trained model imported from huggingface was trained on our custom dataset for a specific task, in our case sentiment analysis. We kept the transformer layers frozen and trained a feed forward neural network to our custom dataset. The centralized models were trained for 6 epochs whereas the decentralized models were trained for 4 rounds where clients in each round were trained for 2 epochs to achieve a nearly stagnant accuracy. The accuracy obtained from both these models on a centralized and federated setting is presented in Table 1 (Fig. 3).

Table 1. Bert-base-uncased and bert-multilingual-uncased accuracy

	BERT-base-uncased	M-BERT-uncased
Centralised accuracy	88.93	88.51
Federated accuracy	86.92	83.41

It can be observed through Table 1 that BERT-base performs better than BERT-multilingual even though the dataset was based on tweets accounting from varying locations with different use of the English language, although the samples were all in english. BERT-base also outperforms BERT-multilingual significantly in federated learning. It can be theorized that a generalized BERT model is better suited for such a scenario as using M-BERT may cause negative interference and lead to decline is accuracy. Hence, BERT-multilingual does not improve upon the performance on a dataset with regional connotations. A basic model is more suitable for a region-based English dataset. Furthermore, the accuracy obtained from federated learning is on par with that of centralized learning. Hence, it can be concluded that a reasonable performance can be seen while building a privacy-preserving machine learning model with federated learning.

Although we extracted an equal number of tweets from each location, i.e. 15,000, after preprocessing all the locations don't contain equal number of tweets. The limited amount of sample across some locations may hinder the accuracy of a federated learning model. So, we varied the number of tweets used per ocation to compare the effect of the number of datapoints on the accuracy of the BERT model. For this, we trained BERT- base-uncased on 3000 tweets per location and 10,000 tweets per location to compare the accuracy while keeping all other hyperparameters the same. Table 2 compares the performance of BERT with variations in the number of samples. This illustrates that an increase in accuracy is observed when the sample size of the training dataset is increased. The increase in sample size has a greater impact on federated learning than centralized learning.

6   Conclusion

In this paper, we have built an end-to-end privacy preserving model with the help of federated learning. We gained comparable accuracy in a federated setting

Table 2. Model accuracy with varying data samples

	3000/location	10000/location
Centralized accuracy	88.90	90.23
Federated accuracy	86.96	90.11

to that of a centralized setting while significantly decreasing the vulnerability of the data. Additionally, we have proposed a novel semi-supervised way of labeling unlabeled tweets obtained from different geographical regions. This technique ensures that the data used at the client end is completely devoid of any human intervention and maximum privacy is achieved. The result obtained from feed forward neural network classifiers shows that our semi-supervised labeling methods perform very well. In the future, we would like to measure time required by a BERT and a normal deep learning classifier in a federated setting. We would like to compare the trade-off between number of trainable parameters, training time and accuracy. We would like to compare lighter models like LSTM to BERT based relevance in difference scenarios.

References

1. Sanders, A.C., et al.: Unmasking the conversation on masks: natural language processing for topical sentiment analysis of COVID-19 Twitter discourse. In: AMIA Annual Symposium Proceedings, vol. 2021. American Medical Informatics Association (2021)
2. Al-Ayyoub, M., et al.: Accelerating 3D medical volume segmentation using GPUs. Multimedia Tools Appl. **77**(4), 4939–4958 (2018)
3. Attard, J., et al.: A systematic review of open government data initiatives. Gov. Inf. Q. **32**(4), 399–418 (2015)
4. Gupta, B.B., Badve, O.P.: GARCH and ANN-based DDoS detection and filtering in cloud computing environment. Int. J. Embed. Syst. **9**(5), 391–400 (2017)
5. Yu, H.Q., Reiff-Marganiec, S.: Learning disease causality knowledge from the web of health data. Int. J. Semant. Web Inf. Syst. (IJSWIS) **18**(1), 1–19 (2022)
6. Kairouz, P., et al.: Advances and open problems in federated learning. Found. Trends® Mach. Learn. **14**(1–2), 1–210 (2021)
7. da Silva, et al.: Using unsupervised information to improve semi-supervised tweet sentiment classification. Inf. Sci. **355**, 348–365 (2016)
8. Chui, K.T., et al.: Handling data heterogeneity in electricity load disaggregation via optimized complete ensemble empirical mode decomposition and wavelet packet transform. Sensors **21**(9), 3133 (2021). https://doi.org/10.3390/s21093133
9. Lu, J., et al.: Blockchain-based secure data storage protocol for sensors in the industrial internet of things. IEEE Trans. Ind. Inform. **18**(8), 5422–5431 (2022). https://doi.org/10.1109/TII.2021.3112601
10. Brendan McMahan, H., Moore, E., Ramage, D., Hampson, S., Aguera y Arcas, B.: Communication-efficient learning of deep networks from decentralized data. In: Proceedings of the 20th International Conference on Artificial Intelligence and Statistics, pp. 1273–1282 (2017) (original version on arxiv Feb 2016)

11. Yang, T., Andrew, G., Eichner, H., Sun, H., Li, W., Kong, N., Ramage, D., Beaufays, F.: Applied federated learning: improving google keyboard query suggestions. ArXiv abs/1812.02903 (2018)
12. Hard, A., Rao, K., Mathews, R., Beaufays, F., Augenstein, S., Eichner, H., Kiddon, C., Ramage, D.: Federated learning for mobile keyboard prediction. ArXiv abs/1811.03604 (2018)
13. Hsieh, K., Phanishayee, A., Mutlu, O., Gibbons, P.: The non-iid data quagmire of decentralized machine learning. In: International Conference on Machine Learning, pp. 4387–4398. PMLR (2020)
14. Hu, S., et al.: The oarf benchmark suite: characterization and implications for federated learning systems. ACM Trans. Intell. Syst. Technol. (TIST) **13**(4), 1–32 (2022)
15. Deng, J., Wang, C., Meng, X., Wang, Y., Li, J., Lin, S., Han, S., Miao, F., Rajasekaran, S., Ding, C.: A secure and efficient federated learning framework for NLP. arXiv Preprint (2022). arXiv:2201.11934
16. Ling, Z., Hao, Z.J.: An intrusion detection system based on normalized mutual information antibodies feature selection and adaptive quantum artificial immune system. Int. J. Semant. Web Inf. Syst. (IJSWIS) **18**(1), 1–25 (2022)
17. EU CORDIS. Machine learning ledger orchestration for drug discovery, 2019. https: cordis.europa.eu/project/rcn/223634/factsheet/en?WT.mc id=RSS-FeedWT.rssf=projectWT.rssa=223634WT.rssev=a. Retrieved Aug 2019
18. Gupta, B.B.: A lightweight mutual authentication approach for RFID tags in IoT devices. Int. J. Networking Virtual Organ. (2016)
19. Sharma, R., Sharma, T.P., Sharma, A.K.: Detecting and preventing misbehaving intruders in the internet of vehicles. Int. J. Cloud Appl. Comput. (IJCAC) **12**(1), 1–21 (2022)
20. Courtiol, P., Maussion, C., Moarii, M., Pronier, E., Pilcer, S., Sefta, M., Manceron, P., Toldo, S., Zaslavskiy, M., Le Stang, N., et al.: Deep learning-based classification of mesothelioma improves prediction of patient outcome. Nat. Med. 1–7 (2019)
21. WeBank. WeBank and Swiss re-signed cooperation MOU, 2019. https:// finance.yahoo.com/news/webank-swiss-signed-cooperation-mou-112300218.html. Retrieved Aug 2019
22. Cvitić, I., Peraković, D., Periša, M., et al.: Ensemble machine learning approach for classification of IoT devices in smart home. Int. J. Mach. Learn. Cyber. **12**, 3179–3202 (2021). https://doi.org/10.1007/s13042-020-01241-0
23. Tenney, I., Das, D., Pavlick, E.: BERT rediscovers the classical NLP pipeline. arXiv Preprint (2019). arXiv:1905.05950
24. Singh, S., Mahmood, A.: The NLP cookbook: modern recipes for transformer based deep learning architectures. IEEE Access **9**, 68675–68702 (2021)
25. Pilault, J., Elhattami, A., Pal, C.: Conditionally adaptive multi-task learning: improving transfer learning in nlp using fewer parameters less data. arXiv Preprint (2020). arXiv:2009.09139
26. Pano, T., Kashef, R.: A complete VADER-based sentiment analysis of bitcoin (BTC) tweets during the era of COVID-19. Big Data Cogn. Comput. **4**(4), 33 (2020)
27. Steven, L.: textblob Documentation. Release 0.15 2 (2018): 269
28. Cambria, E., et al.: Sentic API: a common-sense based API for concept-level sentiment analysis. In: CEUR Workshop Proceedings, vol. 144 (2014)

Sustainable Framework for Metaverse Security and Privacy: Opportunities and Challenges

Manraj Singh[✉], Sunil K. Singh, Sudhakar Kumar, Uday Madan, and Tamanna Maan

Chandigarh College of Engineering and Technology,
Sector-26, Chandigarh 160019, India
co20332@ccet.ac.in, sksingh@ccet.ac.in, sudhakar@ccet.ac.in ,
co18351@ccet.ac.in

Abstract. A metaverse is a network of 3D virtual worlds focused on social connection. In futurism and science fiction, it is often described as a hypothetical iteration of the Internet as a single, universal virtual world that is facilitated by the use of virtual reality (VR) and augmented reality (AR) headsets. Many of the technologies that make up the metaverse's building pieces are already in advanced stages of development. Many others, on the other hand, are still some years away from being practical. We start by looking at its establishments, then, at that point, continue on toward the clever protection and security issues raised by this new worldview. At last, we feature a portion of the sweeping yet consistent ramifications on an assortment of spaces. The discussed topics are privacy user profiling in the metaverse, user privacy, and countermeasures and security perspective with humans in and out of the loop, integrity, and authentication in the metaverse, Polarization and radicalization in a singleton world, distributed denial of service, device vulnerability, and data explosion and exploitation.

Keywords: Augmented reality · Metaverse · Security · Virtual reality · Meta media · Multiverse · Singularity · Security · Privacy · Facebook · Meta

1 Introduction

The Metaverse opens up the possibility of a totally new specialized world. The possibility of computer-generated reality coinciding with this present reality is a feasible one. There has been practically no examination of how this new worldview would tackle current and future security concerns. In this paper, we analyze a portion of the indicated security weaknesses [1].

In this review, we return to the beginnings of the metaverse thought. We underscore contemporary security and protection concerns, featured as at no other time. We contend that software engineering and designing are basic, but not similar to different teaches like ways of thinking, regulation, and sociologies.

© The Author(s), under exclusive license to Springer Nature Switzerland AG 2023
N. Nedjah et al. (Eds.): ICSPN 2021, LNNS 599, pp. 329–340, 2023.
https://doi.org/10.1007/978-3-031-22018-0_30

2   Related Work

The metaverse is a computerized universe open through a virtual environment. It is laid out through the converging of essentially worked on physical and computerized reality. Metaverse offers enhanced immersive experiences and a more interactive learning experience for students in learning and educational settings. Big data are collected and analyzed in various fields [2,3]. Considering such a trend, this study aims to predict user satisfaction using big data from online reviews. Particularly, the authors use various machine learning models and embedding methods to find the optimal model that predicts user satisfaction. Furthermore, to remove biased text data, Valence Aware Dictionary and sentiment Reasoner (VADER) are applied [4–6]. Considering online review data, the user satisfaction classifier model (e.g., Logistic Regression) was suggested by Nwakanma. Moreover, several machine learning models were proposed as approaches to the prediction of user satisfaction [7–9]. Furthermore, accurate results have been obtained by employing machine learning technology when investigating user satisfaction [10].

Metaverse is characterized as an assortment of innovation devices and metaverse associated with IoT, Blockchain, Artificial Intelligence, and the wide range of various tech ventures including the clinical region [11]. IoT and Metaverse are computerized twins, Metaverse is involving the greatest IoT gadgets in their virtual workstations [12]. This information has a special recognizing tag and is utilized as detectable information in the blockchain-based Metaverse. In the Metaverse, such information is turning into an important re-hotspot for man-made consciousness [13]. Metaverse utilizes computerized reasoning and blockchain innovation to construct an advanced virtual existence where you can securely and uninhibitedly participate in friendly and monetary exercises that rise above the restrictions of this present reality, and the utilization of these most recent advancements will be sped up. Worldwide organizations, for example, Samsung or Facebook are presently making a phenomenally tremendous monetary interest in developing metaverse stages or related promoting. Inside the Metaverse, individuals can live "unique" or reproduced public activities (e.g., playing with another symbol, for example, making companions on a social metaverse stage like Zepeto, VRChat, Roblox, or Second Life. Practically movements of every sort in reality can be mimicked inside the metaverse like playing sports, promoting, and trading items. Of different exercises inside the metaverse universes, the proposed study will zero in on the people's utilization of media contents inside the metaverse [14,15].

The utilization of vivid 3D virtual universes for training keeps on becoming because of their true capacity for establishing imaginative learning conditions and their gigantic fame with understudies. Nonetheless, versatility stays a significant test. Though MOOCs adapt to a huge number of clients downloading or streaming learning materials, the exceptionally intuitive, multi-client nature of virtual universes is undeniably seriously requesting - supporting even 100 clients in a similar district simultaneously is viewed as an accomplishment. Be that as it may, assuming the ordinary number of simultaneous clients is somewhat low and there is just a periodic need to have an enormous gathering in-world simultaneously for a unique occasion might the Cloud at any point be utilized for supporting these high, yet fleeting, tops popular [16–18].

Likewise, Metaverse advances will change the transportation framework as far as we might be concerned. Arrangements for the progress of the transportation frameworks into the universe of metaverse are in progress [19,20]. The metaverse matches the actual world and is a virtual world precisely like the actual world, consequently laying out a super-enormous space that coordinates both the actual world and virtual world; this can likewise be perceived as an equal universe made in the virtual world [21–23]. MetaGraspNet is an enormous scope benchmark dataset for vision-driven mechanical handling through physical science-based metaverse combination. The proposed dataset contains 100,000 pictures and 25 different article types and is parted into 5 hardships to assess object identification and division model execution in various getting a handle on situations [24]. A metaverse will utilize virtual portrayal (e.g., computerized twin), computerized symbols, and intelligent experience innovations (e.g., expanded reality) to help examinations, improvements, and tasks of different remote applications. In particular, the metaverse can offer virtual remote framework activities through the computerized twin that permits network architects, portable designers, and media communications engineers to screen, notice, examine, and mimic their answers cooperatively and practically. Presented an overall design for metaverse-based remote frameworks. We talk about key driving applications, plan patterns, and key empowering influences of metaverse-based remote frameworks [25,26]. Finally, we present several open challenges and their potential solutions.

3 Privacy Issues in Metaverse Found by Analysis

Indeed, even with the present innovation, the computerized shards we leave behind uncover a ton about our characters, inclinations, and directions (e.g., political and sexual) This has been obvious from the earliest examinations about 10 years prior, and such expectations capacities have developed decisively from that point forward.

3.1 User Profiling in the Metaverse

Clients are as of now attracted to long-range informal communication stages as strong magnets. Additionally, the metaverse will draw in (significantly more) individuals, as well as satisfied makers, business people, and ventures [27]. At the end of the day, it will be a uniform meta-stage for clients—no matter what their interests and most loved applications (e.g., peruses, gamers, understudies, and so forth)—as well as designers and organizations that host such projects. Clients may not know that accounts and examinations are occurring, and subsequently, their protection might be endangered. Current information assortment methods and going with investigations will be considered awkward, best case scenario, in the metaverse [28]. Advertisers might examine where clients move their mouse, where they look on a screen, and how long they spend on a specific portrayed thing.

3.2   User Privacy

Three regions are especially vital as far as client protection in the metaverse:

1. personal information;
2. behavior;
3. communications

Every one of these spaces will give undeniably a greater number of information to stages than they presently have, bringing about new and increased perils. Individual and delicate information, for instance, will spill through the metaverse, remembering a large number of genuine information for client ways of behaving and physiological attributes. How might we continue doxing under control.

Worries about protection in metaverse correspondences are not restricted to the undeniable dangers of corporate information breaks [29]. They additionally incorporate various sorts of client correspondence, for example, sexting and different types of physically arranged messages and communications. Representatives who are disappointed have not many choices for freely harming their organization's image.

3.3   Countermeasures

Given the assortment of security dangers that metaverse clients face, a few specialists have proactively started to consider how to manage client protection in 3D social metaverses. Among them, a couple of arrangements have been introduced, every one of which depends on a blend of three fundamental methodologies:

1. making a life-sized model or different clones of one's symbol to shadow one's own exercises;
2. making a private duplicate (e.g., an occasion) of a public space for the client's select use, or briefly keeping different clients out of public space; and
3. permitting client instant transportation, imperceptibility, or different types of masks.

Clients can pick their ideal degree of security, additionally relying upon their exercises, and the manner by which to apply the chosen protection highlights. Better than ever arrangements ought to be contrived in such a manner.

4   Security Issues in the Metaverse Found by Analysis

4.1   Humans In and Out of the Loop

We are starting to see the risks of re-appropriating culturally significant positions to calculations that, in spite of their unrivaled exhibition, are tormented by various difficulties. These incorporate inclination, absence of straightforwardness, weakness to attacks and controls, and the gigantic computational and energy requests of complicated AI models (e.g., profound learning) [27]. To be subdued and addressed, every one of these difficulties comprises an open logical test that will require numerous cooperative endeavors from a few logical fields.

4.2   Trustworthiness and Verification in the Metaverse

The security difficulties of content honesty and client validation can be used as a functional and pertinent illustration of the issues delivered by calculations and robotization. In the metaverse, human-machine communications will turn out to be more normal, on the off chance that not needed, for specific occupations. How simple will it be to imitate one's own age, orientation, occupation, or some other trademark. Can we recognize individuals and machines

4.3   Polarization and Radicalization in the Singleton World

Security difficulties will emerge because of the metaverse's status as a singleton. A metaverse is basically a huge aggregator of applications, administrations, and things, as well as clients. Its prosperity is not entirely set in stone by its capacity to act as a focal passage for such content. The greater part of the present Web stages will be displaced by a modest bunch of gigantic metaverses. A singleton metaverse will propel the synchronous presence, concurrence, and connection, everything being equal. A few clients might use their metaverse "conduct" to savage, disturb, or in any case exploit different clients.

4.4   Distributed Denial of Service

DDoS assaults have been around for some time, so it's nothing unexpected they're something to know about in the metaverse. Consider a specialist who depends on AR glasses for fundamental data during a medical procedure and is denied of it because of an assault.

4.5   Device Vulnerability

Programmers can catch information passing through head-mounted gadgets associated with the metaverse by means of the web. This can be seen as a social assault assuming the assailant posts realistic or explicit substance. It can likewise be viewed as a political instrument in the event that they post disdain discourse or other misleading publicity.

5   Proposed Mechanism

5.1   Confusion – Making a Cloud of Duplicates

A convoluted metaverse will ultimately give convincing explanations for clients to want to confound different symbols in their detectable zone. Almost certainly, the sheer irritation brought about by certain symbols (e-malicious outsiders or bots) will make confusion strategies engaging at times. "Cloud of clones" protection plan intends to make a haze of disarray in the environmental elements to darken client area, exercises, convictions, wants, as well as expectations [30]. The framework makes at least one symbol "clones" with the equivalent or comparative appearance as the client's symbol in this plan.

Whenever clones are begun, the client can demonstrate which ways of behaving they need for which gathering of clones by utilizing order semantics like "vivify" or "cloners eyes".

1. "Dole out all clones a way of behaving with an elevated degree of haphazardness and intuitiveness".
2. "Relegate half of the clones the conduct 'stroll around a house,' and the other a portion of the conduct 'stroll around and around."

Figure 1 portrays an adapted picture of the metaverse, showing our hero (symbol B) close to her virtual home. B chooses, arranges, and dispatches the "haze of clones" security procedure in Fig. 1 (centerboard). The quick appearance of a gathering of for all intents and purposes indistinguishable symbols to the client will confound an in- metaverse observer [31]. The goal is for observers to disregard the first "copy" due to the unexpected appearance and dispersal of these clones [32]. Bunch An integrated client B, who hopes to continue with her exercises without being bothered.

The plan ought to have the option to give secrecy as long as the spectator doesn't get visual access again before the clones are made. Whether another specialist could identify a clone by noticing a pseudo-arbitrary way of behaving is as yet an inquiry. Making an enormous number of clones has suggestions for execution and sending that we will not get into here.

5.2   Towards a Framework of Privacy Plans

Table 1 frameworks (at a significant level) some suggested security plan essentials that we examined in before areas. It makes reasonable to envision this course of action of plans as an insurance construction to be given to the client in a controlled manner. The difference makes it reasonable to envision this course of action of plans as a security construction to be given to the client in a controlled manner. The different security plans might be figured out into an "Assurance Options Menu" that gives fast induction to the different capabilities.t security plans might be figured out into an "Insurance Options Menu" that gives rapid permission to the various abilities. Figure 2(a) portrays a client's point of view

Fig. 1. A privacy plan (including clone)

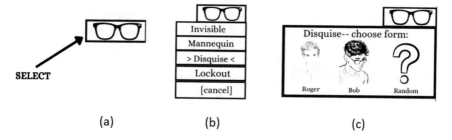

Fig. 2. (a)–(c) A client chooses one of various accessible security plans from a protection choices

Table 1. Frameworks

Name	Summary
1. Clones	The surface makes a "swarm" of new image clones, each unclear in appearance to the client's image, in such a way as observers are perplexed and may disregard the client's image.
2. Private copy	Allows basically a piece of the VR world to be delivered as a "Private Copy" which the client is exclusively involved and speaks with, inconspicuous to other people.
3. Mannequin	The texture replaces the client's symbol with a solitary clone of the client which displays a credible way of behaving, while the client's actual symbol is shipped to somewhere else. A life-sized model is a kind of clone that regularly subs for a client's symbol while the client's consideration is somewhere else.
4. Lockout	Licenses a piece of the VR world to be 'walled off' momentarily for private use; various images are momentarily locked out and limited from entering.
5. Disguise	Permits the symbol to remain in the neighborhood in a new (for example camouflaged) structure. The hidden appearance might be run domly produced by the texture, or the client might utilize a symbol appearance proofreader to make at least one camouflaged structure for use with this arrangement. Spectators don't effortlessly see the change and hence become befuddled.
6. Teleport	The client's symbol is moved to another area in the virtual world. The goal may be picked by the client, for instance from a summary of complaints or by using a "map" interface.
7. Invisibility	The client's symbol takes an undetectable structure so symbols and additionally bots can't identify the presence or activities of the client

of a virtual world's neighborhood. The framework permits the client to choose the camouflage's appearance, as found in Fig. 2. After the client designs the camouflage, the client's symbol takes on a new masked appearance. It becomes possible to empower more extravagant associations if various protection-saving innovations are coordinated into a structure.

1. At the point when a client consolidates Disguise or Invisibility with Teleport, the client's symbol disappears from its unique spot and returns in a subsequent area. The symbol is camouflaged or undetectable at the subsequent area (as recently picked by the client).When magically transport is initiated, any symbols or bots noticing the client at the first area will forget about the client.
2. Whenever a client decides to consolidate the Clones plan with Invisibility, the client's actual symbol becomes undetectable simultaneously as clones of the client show up. Any symbols or bots checking the client before the execution of this blend of plans will be confounded by the presence of clones.
3. At the point when a client joins Private Copy and Teleport, a private duplicate of a piece of the virtual world that the client chooses is made, and the client is then magically transported into that private duplicate. Surveilling symbols or bots will just see the client briefly before they enter or move toward where the private duplicate will be put away.

The framework might give a guide interface and request the client to choose the part from the virtual world on which the private duplicate is based. At the point when the client's protection settings are finished, the framework makes a private clone of the gambling club and magically transports them into it. The framework might have the option to help the client in using the accessible security plans. Framework screens the virtual world for situations in which a client's security might be imperiled. In these cases, the framework advises the client and suggests that they use protection apparatuses. This data could be founded on the client's area or sort of commitment, as well as the area of different symbols or bots.

6   Conclusion

The Metaverse is another computerized climate that consolidates virtual and increased reality. This examination gives an early glance at the security worries that the metaverse will go up against. It's conceivable that security and protection issues won't be among the most genuine. The best way to explore this strange universe we've started to research is to analyze it utilizing a genuinely multidisciplinary approach that joins innovative spaces with humanities and sociologies. Taken individually, the (limited) list of described technologies and concepts may represent a step forward—possibly even a huge one—but only along a previously traveled path. Instead, our argument is that the interaction of the aforementioned forces would result in an effect that is amplified both numerically and qualitatively, drastically and irreversibly altering the technological and cognitive landscapes. Traditional systemic risks introduced by technology, such as security and, particularly, privacy, are crucial in this new world that we are beginning to refer to as the metaverse.

References

1. Wang, Y., Su, Z., Zhang, N., Liu, D., Xing, R., Luan, T. H., Shen, X.: A survey on metaverse: fundamentals, security, and privacy. arXiv Preprint (2022). arXiv:2203.02662
2. Gupta, S., et al.: PHP-sensor: a prototype method to discover workflow violation and XSS vulnerabilities in PHP web applications. In: Proceedings of the 12th ACM International Conference on Computing Frontiers (CF '15). Association for Computing Machinery, New York, NY, USA, Article 59, pp. 1–8 (2015). https://doi.org/10.1145/2742854.2745719
3. Chopra, M., Singh, et al.: Analysis and prognosis of sustainable development goals using big data-based approach during COVID-19 pandemic. Sustain. Technol. Entrepreneurship **1**(2), 100012 (2022)
4. Liu, Y., et al.: Survey on atrial fibrillation detection from a single-lead ECG wave for internet of medical things. Comput. Commun. **178**, 245–258 (2021). ISSN: 0140-3664
5. Chopra, M., Singh, S.K., Sharma, A., Gill, S.S.: A comparative study of generative adversarial networks for text-to-image synthesis. Int. J. Softw. Sci. Comput. Intell. (IJSSCI) **14**(1), 1–12 (2022)
6. Sahoo, S.R., et al.: Security issues and challenges in online social networks (OSNs) based on user perspective. Comput. Cyber Secur. 591–606 (2018)
7. Singh, A., Singh, S.K., Mittal, A.: A review on dataset acquisition techniques in gesture recognition from Indian sign language, pp. 305–313. In: Advances in Data Computing, Communication and Security (2022)
8. Chopra, M., et al.: Predicting catastrophic events using machine learning models for natural language processing. In: Data Mining Approaches for Big Data and Sentiment Analysis in Social Media, pp. 223–243. IGI Global (2022)
9. Singh, I., Singh, S.K., Kumar, S., Aggarwal, K.: Dropout-VGG based convolutional neural network for traffic sign categorization. In: Congress on Intelligent Systems, pp. 247–261. Springer, Singapore (2022)
10. Singh, R., Rana, R., Singh, S.K.: Performance evaluation of VGG models in detection of wheat rust. Asian J. Comput. Sci. Technol. **7**(3), 76–81 (2018)
11. Singh, S.K., Kaur, K., Aggrawal, A.: Emerging trends and limitations in technology and system of ubiquitous computing. Int. J. Adv. Res. Comput. Sci. **5**(7) (2014)
12. Gupta, S., Singh, S.K., Jain, R.: Analysis and optimisation of various transmission issues in video streaming over Bluetooth. Int. J. Comput. Appl. **11**(7), 44–48 (2010)
13. Chopra, M., Singh, S.K., Sharma, S., Mahto, D.: Impact and usability of artificial intelligence in manufacturing workflow to empower Industry 4.0 (2020)
14. Aggarwal, K., Singh, S.K., Chopra, M., Kumar, S.: Role of social media in the COVID- 19 pandemic: a literature review. In: Data Mining Approaches for Big Data and Sentiment Analysis in Social Media, pp. 91–115 (2022)

15. Gupta, A., Singh, S.K., Chopra, M., Gill, S.S.: An inquisitive prospect on the shift toward online media, before, during, and after the COVID-19 pandemic: a techno-logical analysis. In: Advances in Data Computing. Communication and Security, pp. 229–238. Springer, Singapore (2022)
16. Singla, D., et al.: Evolving requirements of smart healthcare in cloud computing and MIoT. In: International Conference on Smart Systems and Advanced Computing (Syscom-2021) (2021)
17. Sharma, S.K., Singh, S.K., Panja, S.C.: Human factors of vehicle automation. In: Autonomous Driving and Advanced Driver-Assistance Systems (ADAS), pp. 335–358. CRC Press (2021)
18. Singh, S.K., Kaur, K., Aggarwal, A., Verma, D.: Achieving High Performance Distributed System: Using Grid Cluster and Cloud Computing. Int. J. Eng. Res. Appl. **5**(2), 59–67 (2015)
19. Nedjah, N., et al.: Robotics and AI for Cybersecurity and Critical Infrastructure in Smart Cities. In: Robotics and AI for Cybersecurity and Critical Infrastructure in Smart Cities 1030 (2022)
20. Gupta, A., Singh, S.K., Gupta, A.: A novel smart transportation based framework interlinking the advancements in technology and system engineering. In: International Conference on Smart Systems and Advanced Computing (Syscom-2021) (2021)
21. Kumar, S., et al.: Evaluation of automatic parallelization algorithms to minimize speculative parallelism overheads: an experiment. J. Discrete Math. Sci. Crypt. **24**(5), 1517–1528 (2021)
22. Singh, Singh, S.K., Singh R., Kumar, S.: Efficient loop unrolling factor prediction algorithm using machine learning models. In: 2022 3rd International Conference for Emerging Technology (INCET), pp. 1–8 2022. https://doi.org/10.1109/INCET54531.2022.9825092
23. Singh, S.K., Singh, R.K., Bhatia, M.S.: System level architectural synthesis and compilation technique in reconfigurable computing system. In: ESA 2010: Proceedings of the 2010 International Conference on Embedded Systems and Applications (Las Vegas NV, July 12–15, 2010), pp. 109–115 (2010)
24. Aggarwal, K., et al.: Deep learning in robotics for strengthening Industry 4.0.: opportunities, challenges and future directions. In: Robotics and AI for Cybersecurity and Critical Infrastructure in Smart Cities, pp. 1–19 (2022)
25. Singh, S.K., Singh, R.K., Bhatia, M.P.S.: Design flow of reconfigurable embedded system architecture using LUTs/PLAs. In: 2012 2nd IEEE International Conference on Parallel, Distributed and Grid Computing, pp. 385–390. IEEE (2012)
26. Singh, S.K., Singh, R.K., Bhatia, M.P.S., Singh, S.P.: CAD for delay optimization of symmetrical FPGA architecture through hybrid LUTs/PLAs. In: Advances in Computing and Information Technology, pp. 581–591. Springer, Berlin, Heidelberg (2013)
27. Su, Z., Zhang, N., Liu, D., Luan, T.H., Shen, X.: A survey on metaverse: funda-mentals, security, and privacy (2022)
28. Di Pietro, R., Cresci, S.: Metaverse: security and privacy issues. In 2021 Third IEEE International Conference on Trust, Privacy and Security in Intelligent Systems and Applications (TPS-ISA), pp. 281–288. IEEE (2021)
29. Zhao, R., Zhang, Y., Zhu, Y., Lan, R., Hua, Z.: Metaverse: security and privacy concerns. arXiv Preprint (2022). arXiv:2203.03854
30. Dionisio, J.D.N., III., W.G.B., Gilbert, R.: 3D virtual worlds and the metaverse: current status and future possibilities. ACM Comput. Surv. (CSUR) **45**(3), 1–38 (2013)

31. Wang, F.Y., Qin, R., Wang, X., Hu, B.: Metasocieties in metaverse: metaeconomics and metamanagement for metaenterprises and metacities. IEEE Trans. Comput. Soc. Syst. **9**(1), 2–7 (2022)
32. Kemp, J., Livingstone, D.: Putting a second life "metaverse" skin on learning management systems. In: Proceedings of the Second Life education workshop at the Second Life community convention, vol. 20. The University of Paisle, CA, San Francisco (2006)

Email Spam Detection Using Naive Bayes and Random Forest Classifiers

Varsha Arya[1], Ammar Ali Deeb Almomani[2], Anupama Mishra[3(✉)],
Dragan Peraković[4], and Marjan Kuchaki Rafsanjani[5]

[1] Insights2Techinfo, Chennai, India
varsha.arya@insights2techinfo.com
[2] Skyline University College, Sharjah, United Arab Emirates
[3] Swami Rama Himalayan University, Dehradun, India
tiwari.anupama@gmail.com
[4] University of Zagreb, Zagreb, Croatia
dperakovic@fpz.unizg.hr
[5] Shahid Bahonar University of Kerman, Kerman, Iran
kuchaki@uk.ac.ir

Abstract. Email is used in business and education and almost everywhere for communications. Email is categorized into many categories based on its content like primary, social, promotional, and spam. So spams are also a sub-category under all the categories. Spam may be in the form of text messages,web messages, images and others also. In this paper, we are discussing Email spam. Email spam, sometimes termed junk emails or undesired emails that consumes computing resources, users time and information. These emails are also involved in many cyber crimes like phishing, vishing, identity stolen, data stolen and more. Spam detection and filtration are major issues for email providers and users. Email filtering is a key tool for detecting and combating spam. In our research work, Machine learning algorithms Naive Bayes and random forest are utilized to achieve the objective of spam detection. To evaluate performance, we use accuracy, precision, recall, and f score metrics and discussed the results. We found that these classifiers performed well. In the future, we can utilize deep learning to achieve better results in both text and email categories.

Keywords: Cyber security · Cyber crime · SPAM · Naive Bayes · Random Forest

1   Introduction

Information sharing is easy and rapid because to technology. Users can exchange information globally on several sites. Email is most convenient and professional way for information-sharing medium worldwide. On the internet, attackers are always seeking ways to attack and it also include spam [1–3]. Unwanted emails consume time and resources. Malicious attachments or URLs in these emails

N. Nedjah et al. (Eds.): ICSPN 2021, LNNS 599, pp. 341–348, 2023.
https://doi.org/10.1007/978-3-031-22018-0_31

can lead to security vulnerabilities [4–6]. Spam is any useless and unwelcome email sent by an attacker to a huge number of recipients [7–9]. Email security is crucial. Spam can contain viruses, worms and Trojans. Attackers exploit this to lure users to online services. They may send spam emails with multiple-file attachments and packed URLs that lead to dangerous and spamming websites [10–12]. Many email providers let customers set keyword-based filtering criteria. This strategy is difficult, and consumers don't want to alter their emails, thus spammers attack their email accounts. The technology like Internet of things (IoT) has significantly grown in recent decades. Smart cities need IoT, IoT-based social networking platforms and apps abound. IoT is causing a rise in spamming. Researchers presented approaches to detect and filter spam and spammers. Existing spam detection systems are mostly behavior-pattern-based and semantic-pattern-based. These methods have flaws. Spam emails have grown with the Internet and global communication [6,13–15]. The spam rate is high despite many anti-spam tools and strategies. Harmful emails with links to malicious websites are the most deadly spams. Spam emails can slow server response by overflowing memory or capacity. Every organization assesses spam technologies to accurately detect spam emails and avoid escalating email spam difficulties. Whitelist/Blacklist [7,16–18], mail header analysis, keyword checking, etc. are popular spam detection strategies. Figure 1 shows the statistics of spam.

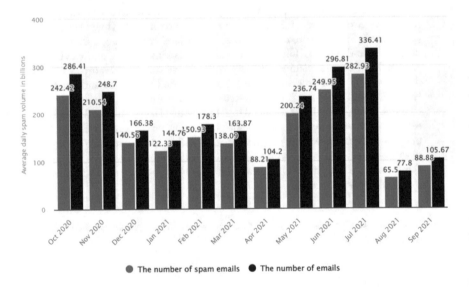

Fig. 1. Spam statistics

Form the social media users, as per researchers, 40% accounts are used for spam [8,19,20]. Spammers target certain areas, review pages, or fan pages to deliver concealed links to pornographic or other product sites from bogus accounts. Malicious emails sent to the same people or groups exhibit common features. Investigating these highlights can enhance email detection. AI [9] can categorise emails as spam or non spam. This method uses header, subject, and body feature extraction. After extracting the data, we can classify it as spam or ham. Spam detection today uses learning-based classifiers [10,21]. In learning-based classification, the detection procedure assumes spam emails have specified qualities [11]. Many factors complicate learning-based spam identification. Subjectivity in spam, concept drift, language issues, processing overhead, and text latency.

NLP (Natural Language Processing) analyses and represents naturally occurring texts at one or more linguistic levels to achieve human-like language processing for a range of activities or applications. "Natural texts" can be any language, manner, genre, etc. Oral or written writings must be in a human language. Significantly, the text being evaluated should be acquired from actual usage. NLP is the use of computers to process written and spoken language for practical purposes, such as translating languages, retrieving information from the web to answer inquiries, and conversing with machines. Symbolic, statistical, connectionist, and hybrid are natural language processing techniques. This study proposes a solution using NLP and big text corpora to approximate linguistic models.

The other sections are organized as we have discussed related work in Sect. 2, proposed methodology in Sect. 3, experimental details in Sect. 4, then evaluated results and performance in Sect. 5 and last concluded in Sect. 6.

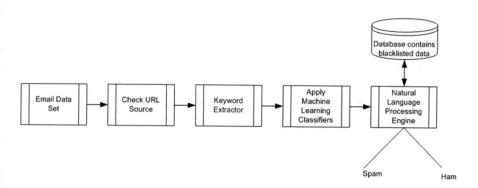

Fig. 2. Proposed methodology

2   Related Work

The solution for the detection of a spammer is different based on the collected dataset and predictive user analysis. This survey includes (1) Various detection mechanisms including the way of spreading spam. (2) Analysis of user contents based on their account information and posts including a comparative analysis of various methodologies for the detection of spammers. (3) Open issues and challenges in social network spam detection techniques. We propose a novel approach that can analyze message content and find features using the TF-IDF techniques to efficiently detect Spam Messages and Ham messages using different Machine Learning Classifiers. The Classifiers going to use in proposed work can be measured with the help of metrics such as Accuracy, Precision and Recall. In our proposed approach accuracy rate will be increased by using the Voting Classifier [Ubale, G., & Gaikwad, S. (2022, March). SMS Spam Detection Using TFIDF and Voting Classifier. In 2022 International Mobile and Embedded Technology Conference (MECON) (pp. 363–366). IEEE.] We have checked cross-validation as the correlation between features. By using ExtraTreesClassifier we have encountered the feature importances. Finally, we have tested XGBoost Classifier, Random Forest Classifier, Decision Tree Classifier, KNN Model, SVM Classifier, Logistic Regression Model, AdaBoost Classifier algorithms for better accuracy and we found out XGBoost Classifier and Random Forest Classifier has Better accuracy. Again we have applied SMOTE and PCA Techniques on Dataset for accuracy deviations on XGBoost Classifier and Random Forest Classifier but we got the same results as in normal state as XGBoost Classifier accuracy is 96.8312% and Random Forest Classifier accuracy is 96.7853% [Chaudhari, M. S. S., Gujar, S. N., & Jummani, F. Detection of Phishing Web as an Attack: A Comprehensive Analysis of Machine Learning Algorithms on Phishing Dataset.] This chapter highlights the scope of OMSA strategies and forms of implementing OMSA principles. Besides technological issues of OMSA, this chapter also outlined both technical problems regarding its production and non-technical issues regarding its use.

3   Proposed Methodology

The processing begins with the input of an email that is not classified. The URL Source checking block examines the URL of the incoming email and looks for a match in the URL Blacklist Database. If a match is found, the email is marked as spam. If the search is successful, the system classifies the email as spam or proceeds as though it were ham, depending on the results, and processes it further. The Fig. 2 presents the proposed methodology. In addition to this, there is something called a Threshold Counter that monitors the amount of time that has passed and keeps a tally of how many emails have originated from this particular source. This counter will directly classify the email as spam once it hits its threshold value and will also update the black list simultaneously. This happens when emails sent from a given URL or IP address go beyond a specific large number. The email is then forwarded to a Keyword Extractor, which will parse the message and turn it into a list of keywords. The Classifier is given these keywords, and it places the email into the appropriate category based on the keywords. The email is then sent to the NLP Engine, along with the category that it belongs to. Here is where the meat of the Content Analysis process takes place. Using a variety of algorithms and methods, the email is analysed to determine whether it is spam or real.

4 Experimental Details

4.1 Experiment Setup and Data Sets

The experimental setup includes Python 3.10.4 64 bit on Jupyter Notebook in Visual Studio Code with CPU: 11th Gen Intel(R) Core(TM) i5-1135G7 with Clock Speed @ 2.40GHz-3.32GHz , GPU RAM: 16GB DDR4, Storage: 1TB HDD with 256GB SSD. The scikit-learn, Matplot, Numpy, nltk and Panda libraries were used. Dataset was taken from kaggle plateform.

4.2 Machine Learning Classifiers

Random Forest Random forest (RF) is a technique for classification and regression that is based on ensemble learning. It works particularly effectively for problems that require data to be sorted into classes. The algorithm was conceived of by Breiman and Cutler, who are credited as its creators (Breiman and Cutler, 2007). The use of decision trees as a means of making predictions is central to the RF paradigm. Multiple decision trees are constructed and used for class prediction after the training phase; this is accomplished by taking into account the classes that were voted on for each individual tree during the training phase (Fig. 4). The class that has the highest margin function (MF) is the one that is deemed to be the output (Liu et al., 2012).

Multinomial Naive Bayes In the field of natural language processing (NLP), probabilistic learning is frequently accomplished through the use of the Multinomial Naive Bayes technique, which is a strategy for probabilistic learning. The Bayes theorem must be utilised in order to make a prediction regarding the tag in regard to the independent variable. It does this by computing the probability of each title appearing in a sample and then providing the label with the highest probability of being accurate.

5 Result and Performance Evaluation

For the evaluation of the performance in classification machine learning, we have the following metrics: RECALL: how many spam emails recalled from all spam emails.
PRECISION: what is the ratio of email correctly classified as spam (Fig. 3).
ACCURACY: it measures how many observations, both positive and negative, were correctly classified.
F1-Score: it combines precision and recall into one metric. The higher the score the better our model is.

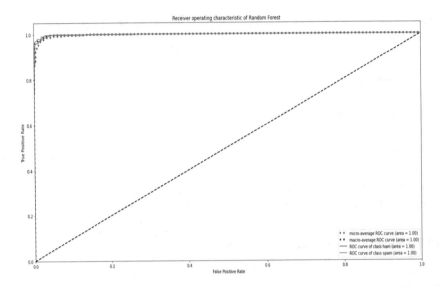

Fig. 3. RoC curve of Random Forest

ROC Curve: It is a chart that visualizes the tradeoff between true positive rate (TPR) and false positive rate (FPR). Basically, for every threshold, we calculate TPR and FPR and plot it on one chart. Of course, the higher TPR and the lower FPR is for each threshold the better and so classifiers that have curves that are more top-left side are better.

Table 1. Confusion matrix

Random Forest		Naive Bayes	
842	1	834	9
25	271	2	294

Table 2. Classification report

Classifiers	Precision	Recall	F1 score	Accuracy
Random Forest	1.00	0.90	0.94	97.27
Naive Bayes	0.97	0.99	0.98	99.03

From the Tables 1 and 2 it can be seen that the Naive Bayes performed well over detection of spam emails. All the metrics of random forest produced results lower than the naive bayes but still able to achieve a good accuracy.

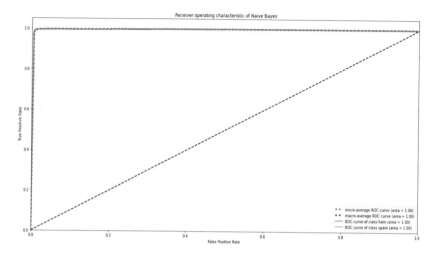

Fig. 4. RoC curve of Naive Bayes

6 Conclusion

For many people who use the internet, receiving spam emails is both a significant source of anxiety and a significant source of frustration. In our research work, we have used two supervised classifiers and detected spam emails from the dataset. From the roc curve also it can be seen that the results are good and performed better. The mode that was suggested as a solution in this paper is extremely helpful because it introduces a threshold counter, which helps overcome congestion on the web server and also maintains the spam filter's efficiency. However, at the same time, it requires overhead storage space for the databases. While this mode has a number of advantages, the main one is that it has a number of benefits. Since Natural Language Processing is still a research topic that is relatively undeveloped, future improvements in the field of spam detection for online security can be made utilising Natural Language Processing.

References

1. Srivastava, A., et al.: A recent survey on DDoS attacks and defense mechanisms. In: International Conference on Parallel Distributed Computing Technologies and Applications, pp. 570–580. Springer, Berlin, Heidelberg (2011)
2. Major-websites-across-east-coast-knocked-out-in-apparent-ddos-attack. https://www.cnbc.com/2016/10/21/major-websites-across-east-coast-knocked-out-in-apparent-ddos-attack.html
3. Network 2020. Worldwide infrastructure security report. Network. https://www.arbornetworks.com/images/documents/WISR2020/EN/Web.pdf
4. Patil, A., Laturkar, A., Athawale, S.V., Takale, R., Tathawade, P.: A multilevel system to mitigate DDOS, brute force and SQL injection attack for cloud security. In: 2017 International Conference on Information, Communication, Instrumentation and Control (ICICIC), pp. 1–7. IEEE (2017)

5. Dinh, H.T., Lee, C., Niyato, D., Wang, P.: A survey of mobile cloud computing: architecture, applications, and approaches. Wireless Commun. Mobile Comput. **13**(18), 1587–1611 (2013)
6. Miao, R., Potharaju, R., Yu, M., Jain, N.: The dark menace: characterizing network-based attacks in the cloud. In: Proceedings of the 2015 Internet Measurement Conference, pp. 169–182 (2015)
7. McAfee 2020. McAfee Labs Threats Report Statistics. https://www.mcafee.com/enterprise/en-us/assets/reports/rp-quarterly-threats-july-2020.pdf
8. He, Z., Zhang, T., Lee, R.B.: Machine learning based DDoS attack detection from source side in cloud. In: 2017 IEEE 4th International Conference on Cyber Security and Cloud Computing (CSCloud), pp. 114–120. IEEE (2017)
9. Garg, S., Kaur, K., Kumar, N., Kaddoum, G., Zomaya, A.Y., Ranjan, R.: A hybrid deep learning-based model for anomaly detection in cloud datacenter networks. IEEE Trans. Netw. Serv. Manag. **16**(3), 924–935 (2019)
10. Murtaza, U., Aslan, Z.: Performance analysis of machine learning techniques used in intrusion detection systems
11. Agarwal, S., Tyagi, A., Usha, G.: A deep neural network strategy to distinguish and avoid cyber-attacks. In: Artificial Intelligence and Evolutionary Computations in Engineering Systems, pp. 673–681. Springer, Singapore (2020)
12. Santos, R., Souza, D., Santo, W., Ribeiro, A., Moreno, E.: Machine learning algorithms to detect DDoS attacks in SDN. Concurrency Comput. Pract. Experience **32**(16), e5402 (2020)
13. Al-Duwairi, B., OOzkasap, O., Uysal, A., Kocaoggullar, C., Yildrim, K.: LogDos: a novel logging-based DDoS prevention mechanism in path identifier-based information centric networks. arXiv Preprint (2020). arXiv:2006.01540
14. Verma, P., Tapaswi, S., Godfrey, W.W.: An adaptive threshold-based attribute selection to classify requests under DDoS attack in cloud-based systems. Arab. J. Sci. Eng. **45**(4), 2813–2834 (2020)
15. Gupta, B.B., Misra, M., Joshi, R.C.: An ISP level solution to combat DDoS attacks using combined statistical based approach. arXiv Preprint (2012). arXiv:1203.2400
16. Gupta, B.B., Badve, O.P.: Taxonomy of DoS and DDoS attacks and desirable defense mechanism in a cloud computing environment. Neural Comput. Appl. **28**(12), 3655–3682 (2017)
17. Gupta, B.B., Dahiya, A., Upneja, C., Garg, A., Choudhary, R.: A comprehensive survey on DDoS attacks and recent defense mechanisms. In: Handbook of Research on Intrusion Detection Systems, pp. 186–218. IGI Global (2020)
18. Mishra, A., et al.: Security threats and recent countermeasures in cloud computing. In: Modern Principles, Practices, and Algorithms for Cloud Security, pp. 145–161. IGI Global (2020)
19. Mishra, A., Gupta, N.: Analysis of cloud computing vulnerability against DDoS. In: 2019 International Conference on Innovative Sustainable Computational Technologies (CISCT), pp. 1–6. IEEE (2019)
20. Mirkovic, J., Reiher, P.: A taxonomy of DDoS attack and DDoS defense mechanisms. ACM SIGCOMM Comput. Commun. Rev. **34**(2), 39–53 (2004)
21. Alzahrani, R.J., Alzahrani, A.: Security analysis of DDoS attacks using machine learning algorithms in networks traffic. Electronics **10**(23), 2919 (2021)
22. Sahoo, S. R., Gupta, B.B., Peraković, D., Peñalvo, F.J.G., Cvitić, I.: Spammer detection approaches in online social network (OSNs): a survey. In: Sustainable Management of Manufacturing Systems in Industry 4.0, pp. 159–180. Springer, Cham (2022)

A Proposed Darknet Traffic Classification System Based on Max Voting Algorithms

Ammar Almomani[1,2(✉)], Mohammad Alauthman[3], Mouhammad Alkasassbeh[4], Ghassan Samara[5], and Ryan Wen Liu[6]

[1] School of Information Technology, Skyline University College, P.O. Box 1797 Sharjah, UAE
[2] IT-department- Al-Huson University College, Al-Balqa Applied University, P. O. Box 50 Irbid, Jordan
ammarnav6@bau.edu.jo
[3] Department of Information Security, Faculty of Information Technology, University of Petra, Amman, Jordan
mohammad.alauthman@uop.edu.jo
[4] Department of Computer Science, Princess Sumaya University for Technology, Amman 11941, Jordan
m.alkasassbeh@psut.edu.jo
[5] Computer Science Department, Zarqa University, Zarqa, Jordan
gsamara@zu.edu.jo
[6] Hubei Key Laboratory of Inland Shipping Technology, School of Navigation, Wuhan University of Technology, Wuhan, China
wenliu@whut.edu.cn

Abstract. The darknet refers to an anonymous address space on the Internet where people do not expect to interact with their computers so that it is considered malicious and provides them with a safe place where people can express themselves anonymously. Tools such as "tor" are also used in illegal activities such as disseminating child abuse, selling Drugs and Weapons, or distributing malicious software In addition to cyber-attacks. This proposed proposes a darknet traffic detection and classification system by using the Max Voting classifier Ensemble technique to produce the final predictions in a manner based on three algorithms: Random Forest and KNN and Gradient Boosting combined to get accurate results in case the client gets access to the internet using the darknet or not. As an expected contribution, the system will be got a higher Accuracy compared with the current systems, The proposed suggested working on a dataset of more than 100,465 records analyzed from the 2020 CIC-Dark collected from the Darknet.

Keywords: Darknet · Machine learning techniques · Deep learning · Classification · Max voting

© The Author(s), under exclusive license to Springer Nature Switzerland AG 2023
N. Nedjah et al. (Eds.): ICSPN 2021, LNNS 599, pp. 349–355, 2023.
https://doi.org/10.1007/978-3-031-22018-0_32

1   Introduction

The Internet connects billions of devices around the world through a network called the World Wide Web. The Internet is a means that connects the whole world, and it is the copper wire that connects devices to be connected to transmit information and benefit from Connecting computers to each other, the purpose of which is to share between these devices to exchange and obtain necessary information through this method.

It also aids in the detection, prevention, and recovery of data from malicious actions. Analyzing darknet traffic is one of the Cyber security strategies [1, 2]. Darknet and darknet traffic is also known as black hole monitors, network telescopes, unwanted network traffic, and Internet Background Radiation (IBR) [3–6], spurious traffic, most the hackers use the darknet to do their malicious activities.

The main idea of this proposed system is data traffic detection and analysis, This proposes an analysis and classification system for dark data traffic that is based on producing final predictions in the max voting ensemble method, It is beneficial to study this traffic to identify attack trends in the real network.

Each attack follows a similar pattern of stealing network data [7–9]. The linked study featured Ensemble learning and max voting in section No. 2, which explores machine learning utilizing multiple ways to construct hybrid systems. Section 3 offered a framework for detailing the new darknet Traffic analysis and classification system using the max voting classifiers approach in the latter section. Finally, Sect. 4 concludes future work research.

2   Related Work

In this section we will discuss previous researchers' work Using a variety of algorithms, researchers have devised many approaches for detecting and classifying Darknet traffic or Tor users. This section provides an overview of the combined techniques that are aimed at improving overall performance. A new approach to intrusion sensing with an ensemble design was introduced using the Max voting ensemble method.

Habibi Lashkari et al. [10] The authors of the paper proposed using the Deep Image approach, which employs feature selection to build a grayscale image and feed it into a 2D convolutional neural network to detect and classify dark traffic. They tested their approach on the CXVPN2016 and ISCXTor2017 datasets, and they achieved an accuracy of 86% for the multi-class classification.

Sarwar et al. [11] The authors of the paper proposed a general approach to detect and classify dark web traffic using deep learning. Then study the most recent complex dataset, which has a lot of information about the movement of dark data, and they preprocess the data. The results show that the suggested approach surpasses existing approaches, with a maximum accuracy of 96% in Darknet traffic detection and 89% in Darknet traffic classification using XGB as a feature selection strategy and CNN-LSTM as a recognition model.

Ozawa et al. [12] The authors of the paper were successful in identifying the activities of attackers connected with well-known malware classes. As an example, they give a fascinating analysis of attack activities before and after the first source code release of the well-known IoT virus Mirai. Experiments show that the proposed technique is successful and efficient in detecting and tracking new malware activity on the Internet, signaling a potential strategy for automating and accelerating the detection and mitigation of new cyber threats.

In our research, the max voting model was built using three different machines learning from various categories. We utilized Random Forest (RF) from the tree categories, K-Nearest Neighbor(KNN) Algorithm, and Gradient Boosting in Machine Learning, which is discussed in section three as the proposed methodology in detail.

3 The Proposed System

In this part, the researchers discuss a Darknet detection and Classifier System based on Max Voting utilizing a darknet dataset. To handle classification difficulties, the maximum voting approach is often utilized. This technique uses many models to produce predictions for each data point. The forecasts of each model are viewed as a 'vote.' The majority of the forecasts from the models are used to make the final projection. This part will go over the System Design phases as well as the Dataset utilized to figure out that System. In this part, the researchers discuss a Darknet detection and Classifier System based on Max Voting utilizing a darknet dataset. To handle classification difficulties, the maximum voting approach is often utilized. This technique uses many models to produce predictions for each data point. The forecasts of each model are viewed as a 'vote.' The majority of the forecasts from the models are used to make the final projection. This part will go over the System Design phases as well as the Dataset utilized to figure out that System.

3.1 Dataset Description: CIC-Darknet2020 Dataset [13, 14]

The CIC-Darknet2020 is the Dataset will used in our work, to figure out the Classifier System, This dataset will collected by the Canadian Institute for Cyber-Security at the University of New Brunswick to test novel methods for classifying darknet traffic. A two-layered technique is utilized to produce the first layer of the CICDarknet2020 dataset contains benign and darknet traffic. The second layer of darknet traffic consists of Audio-Stream, Browsing, Chat, Email, P2P, Transfer, Video-Stream, and VOIP. We combined our previously developed datasets, ISCXTor2016 and ISCXVPN2016, as well as the relevant VPN and Tor traffic in the corresponding Darknet categories, to create the representative dataset.

3.2 Data Pre-processing

The dataset in actuality had no issues and doesn't need any type of filtering or pre-process, because it didn't include any null values and All values inside the dataset are numeric and compatible with Python

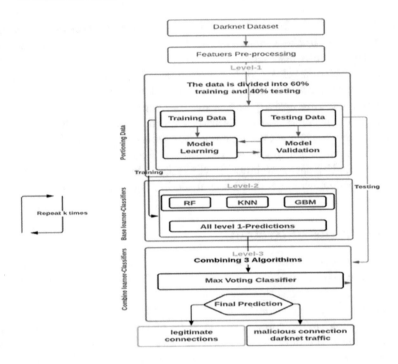

Fig. 1. Darknet traffic detection and classification using Max Voting

3.3 Algorithms List

To solve the main problem which is detection and classification for Darknet data traffic and define people who use tor to get access to the dark web by using Max Voting we did a list of experiments discussed in Chap. 4, Based on the results of our experiments, We decided to use Max Voting within the algorithms KNN, Random Forest and Gradient Boosting.

Random Forest Algorithm [15] Several decision trees may correctly predict a dataset's class using the random forest, but not all of them will. However, when all of the trees are linked together, they correctly predict the result. In other words, the following are the key assumptions for a more accurate Random Forest classifier: For the classifier to accurately predict outcomes, the dataset's feature variable must have some actual values. Each tree's forecasts must have a low correlation with each other (Fig. 1).

K-Nearest Neighbor(KNN) Algorithm for Machine Learning [16] Machine learning algorithm K-Nearest Neighbor uses the Supervised Learning method. In the K-NN method, the new case/data is assigned to the most similar category to the existing categories based on similarity. Keeps all the current data and classifies incoming ones according to their similarity, which is how the K-NN algorithm works. New data can be quickly classified using the K-NN approach as soon as it is received into the appropriate category.

Gradient Boosting in Machine Learning [17] Gradient boosting is a machine learning technique for solving regression and classification issues. As a result, a prediction model that is an ensemble of weak prediction models is produced. The accuracy of a predictive model can be increased in two ways: a. by using feature engineering or b. by immediately employing boosting approaches.

Max Voting [18] One of the easiest ways of merging predictions from different machine learning algorithms is by max-voting, which is commonly used for classification tasks. Each base model predicts and votes for each sample in max-voting. The final prediction class contains only the sample class with the most votes. The maximum voting strategy is often used for classification problems. This technique uses multiple models to generate predictions for each data point. The forecasts of each model are viewed as a 'vote.' The final projection is based on the majority of the models' predictions. The models discussed in this section are merged in an ensemble technique, and categorization is based on the majority vote of the models (hard voting). When each model forecasts, voting begins for each occurrence, with the projected output receiving more than half of the votes (Fig. 2).

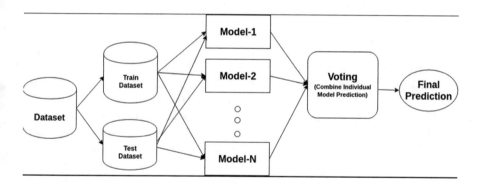

Fig. 2. The basic architecture of Max Voting

4 Designing Phase and Experimental Results

First, the dataset will be split into two parts: 60% for train and 40% for tests. the label is denoted with Y and the other features with X.

Secondly, We have done several experiments to select the Machine Learning Model from several lists, and after reviewing the previous work we decide to make a new classifier never used before in this kind of problem, so we selected the higher three Models' accuracy between the experiment results which discussed in Chap. 4 in details and we make the decision to combine these models to gain the advantage.

The RandomForest, KNeighbors, and GradientBoosting have the higher accuracy results in the experiment. so we combined them to apply a Max Voting classifier to it because we found that the Max Voting method is the ideal method to deal with classification problems, especially to classify between two things. the RandomForest algorithm denoted with model1, KNeighbors algorithm denoted with model2 and GradientBoosting algorithm with model3. also, we denoted the max voting classifier with Model and we added the last 3 algorithms as model1,model2, and model3 as parameters for VotingClassifier inside the Model(as a Max Voting Model).

Finally, We suggest training the final_pred and getting a result, then we print the Accuracy for the result by using a max voting classifier, During working on this proposed and having the output result, we will do several experiments and the expected contributions in the results will show that our proposed can be higher accuracy compared with other proposed

5   Conclusion

concerning the darknet, users who use Tor or Vpn to be in anonymity mood may have illegal actions, and the darknet is a section of allocated IP routed space where none of the servers or active services are located. This work proposes the Darknet traffic analysis and classification by using the Max voting ensemble method. Max voting it's a method from the Ensemble techniques which allows the workers to combine their predictions for having more advantages, which divides the data into several points, and the models selected to be combined by Max Voting predict each data point. the Max Voting Model takes the Majority of the vote for each point and makes the final prediction decision.

We will use to combine three algorithms (Random Forest, KNN, GradientBoosting). Algorithms were selected based on several experiments depending on common measurements such as the accuracy of the model, the recall and the precision, and Model running time so the highest three models were selected to be combined in the Max Voting Classifier.

Based on the related work, the Max Voting method hasn't been used before in this problem. This type of research problem requires a huge amount of data to make the optimal solution. In this study, the dataset was split into a training dataset and a testing dataset. The expected contribution will be high accuracy, this proposal can be extended by improving the model to deal with online data.

Acknowledgments. The research reported in this publication was supported the Research and Innovation Department, Skyline University College, University City of Sharjah – P.O. Box 1797 - Sharjah, UAE. and by Al-Balqa Applied University in Jordan.

References

1. Cvitić, I., Peraković, D., Periša, M., et al.: Ensemble machine learning approach for classification of IoT devices in smart home. Int. J. Mach. Learn. Cyber. **12**, 3179–3202 (2021). https://doi.org/10.1007/s13042-020-01241-0
2. Dar, A.W., Farooq, S.U.: A survey of different approaches for the class imbalance problem in software defect prediction. Int. J. Softw. Sci. Comput. Intell. (IJSSCI) **14**(1), 1–26 (2022)
3. Wang, X., et al.: You are your photographs: Detecting multiple identities of vendors in the darknet marketplaces. In: Proceedings of the 2018 on Asia Conference on Computer and Communications Security (2018)
4. Almomani, A.: Classification of virtual private networks encrypted traffic using ensemble learning algorithms. Egypt. Inform. J. (2022)
5. Almomani, A., et al.: Phishing website detection with semantic features based on machine learning classifiers: a comparative study. Int. J. Semant. Web Inf. Syst. (IJSWIS) **18**(1), 1–24 (2022)
6. Almomani, A.: Darknet traffic analysis and classification system based on stacking ensemble learning/in press. Inf. Syst. e-Business Manag. 1–24 (2022)
7. Gaurav, A., et al.: A comprehensive survey on machine learning approaches for malware detection in IoT-based enterprise information system. Enterprise Inf. Syst. 1–25 (2022)
8. Pan, X., Yamaguchi, S., Kageyama, T., Kamilin, M.H.B.: Machine-learning-based White-Hat worm launcher in botnet defense system. Int. J. Softw. Sci. Comput. Intell. (IJSSCI) **14**(1), 1–14 (2022)
9. Ling, Z., Hao, Z.J.: An intrusion detection system based on normalized mutual information antibodies feature selection and adaptive quantum artificial immune system. Int. J. Semant. Web Inf. Syst. (IJSWIS) **18**(1), 1–25 (2022)
10. Cambiaso, E., et al.: Darknet security: a categorization of attacks to the Tor network. In: ITASEC (2019)
11. Singh, D., Shukla, A., Sajwan, M.: Deep transfer learning framework for the identification of malicious activities to combat cyberattack. Future Gener. Comput. Syst. **125**, 687–697 (2021)
12. Johnson, C., et al.: Application of deep learning on the characterization of Tor traffic using time based features. J. Internet Serv. Inf. Secur. **11**(1), 44–63 (2021)
13. Habibi Lashkari, A., Kaur, G., Rahali, A.: DIDarknet: a contemporary approach to detect and characterize the darknet traffic using deep image learning. In: 2020 the 10th International Conference on Communication and Network Security (2020)
14. Habibi Lashkari, A., Kaur, G., Rahali, A.: CIC-Darknet2020. 2020, Cited 10-5-2021. Available from: https://www.unb.ca/cic/datasets/darknet2020.html
15. Lin, W., et al.: An ensemble random forest algorithm for insurance big data analysis. IEEE Access **5**, 16568–16575 (2017)
16. K-Nearest Neighbor(KNN) Algorithm for Machine Learning. Cited 2021 20-10-2021. Available from: https://www.javatpoint.com/k-nearest-neighbor-algorithm-for-machine-learning
17. Lawrence, R., et al.: Classification of remotely sensed imagery using stochastic gradient boosting as a refinement of classification tree analysis. Remote Sens. Environ. **90**(3), 331–336 (2004)
18. Sarkar, D.. Natarajan, V.: Ensemble Machine Learning Cookbook: Over 35 Practical Recipes to Explore Ensemble Machine Learning Techniques Using Python. Packt Publishing Ltd (2019)

Role of Artificial Intelligence in Agriculture—A Paradigm Shift

Avadhesh Kumar Gupta[1], N. R. N. V. Gowripathi Rao[2(✉)], Purti Bilgaiyan[1],
N. Kavya Shruthi[3], and Raju Shanmugam[1]

[1] Unitedworld School of Computational Intelligence, Karnavati University,
Gandhinagar, Gujarat, India
[2] Karnavati School of Research, Karnavati University, Gandhinagar, Gujarat, India
gowripathiraofmpe@gmail.com
[3] Rashtriya Raksha University, Gandhinagar, Gujarat, India

Abstract. Technological interventions are happening at a very rapid race. It has affected and impacted Agricultural Industry positively as well, as new advances are being reported to facilitate better agricultural outcomes across the globe. Despite several inventions and technological interventions, the main occupation in several countries worldwide is agriculture. Constantly rising population is bound to create additional pressure and critical challenges in order to meet the demand of agri-products. Artificial Intelligence is becoming a driving force and technological support to enhance the global food production which will be enhanced by 50% till 2050 for feeding additional people. Use of Artificial intelligence is the needs of the hour for different agricultural parameters identification. In this paper a preliminary study is made for identifying the role of artificial intelligence in agriculture practices..

Keywords: Machine learning · Artificial intelligence · Drones · Robots in agriculture · Iot in agriculture

1 Introduction

Due to its rapid technological development and broad range of applications, artificial intelligence is one of the most important areas of research in computer science. Agriculture is one of the key industries where Artificial Intelligence (AI) is absolutely important. Generally speaking, agriculture is a primary occupation that requires a lot of effort, tenacity, and persistence in the face of poor income and a difficult way of life. Farmers must accept agriculture as their primary source of income because it requires a lot of time and effort to grow suitable crops. However, due to low income and occasionally no gain from land due to weather conditions or resource scarcity, farmers must deal with loss and decline in financial conditions, which ultimately leads to suicide due to depression [1,2].

Artificial Intelligence (AI) is the term used to describe systems or machines that simulate human intellect. AI seeks to create machines that can emulate

N. Nedjah et al. (Eds.): ICSPN 2021, LNNS 599, pp. 356–360, 2023.
https://doi.org/10.1007/978-3-031-22018-0_33

human thought processes and actions, such as perception, reasoning, learning, planning, and prediction. One of the key traits that set humans apart from other animals is intelligence. A growing number of machine kinds are continuously replacing human work in all spheres of life as a result of the endless cycle of industrial revolutions, and the impending. The next major obstacle is replacing human resources with artificial intelligence. The research in the field of AI is rich and diversified because so many scientists are concentrating on it. Search algorithms, knowledge graphs, natural language processing, expert systems, evolutionary algorithms, machine learning (ML), deep learning (DL), and other areas of AI research are just a few examples [3,4,9].

Artificial intelligence (AI) can be used in a variety of fields and change the way we think about things today's farming. Solutions powered by AI will not only farmers to use fewer resources while still producing more, but it will also a higher yield in light of the rising use of high technologies used in everyday life, including in healthcare and education as well as government. The most widely discussed topic is agriculture. all, as artificial intelligence is centred on simplicity and intelligence working. The use of AI in agricultural sectors should be improved reduced prices and simple processing via use of artificial intelligence Different agricultural issues are swiftly resolved in time. In Artificial intelligence various strategies like increase the quality of harvests and implement indoor farming higher agricultural output rates [5–7].

Crop yield production and price forecasts are supposed to be monitored effectively with the usage of Artificial Intelligence to help the farmer for maximization of profit. Use of IoT based sensors to detect weed affected zones identifying right regions through intelligent spraying is one of the prominent game changer. Actionable predictive insights so as to ensure minimum impact of weather conditions and evaluate right time to sow the seeds are few of the other AI based developments and creating right eco-system in the area of agriculture. Machine Learning models are leveraging data with huge potential in real time for various data points on temperature, water usage, weather conditions and soil etc. Real time AI applications are being designed with the use of ML algorithms to choose the right time to sow seeds, crop choices, hybrid seed choices for generating more yields and improve overall harvest quality and accuracy. Advances in computer vision, mechatronics, image processing provides unique opportunities to develop intelligent precision fertilization, pesting and precision farming thereafter. Even disruptive technology and cognitive computing are playing an important role to understand and interact with different environments, situations and circumstances to enrich agricultural production worldwide.

AI is also offering strong aid in reference with development of seasonal forecast models. Remote sensing, proximity sensing and various other IoT based innovations are being utilized for intelligent agricultural data integration related to meteorology, soil testing, insect infections, rain forecast along with drone imagery for in-depth analysis. Advanced analytics, cloud computing, satellite imagery, deep learning etc. have created strong eco-system for smart agriculture. Deep learning based techniques like principal component analysis (PCA) detection system is being used to help farmers make timely management decisions and control the spread of disease. Robots driven AI techniques are increasing the efficiency of labor as agricultural operations are getting hampered now a days due to mass migration of labor to cities. Chatbots can be used by farmers for resolving various techniques to provide guidance on critical agricultural issues including labor shortages. Several visualization problems are being addressed through AI techniques like lighting conditions, improper cropping and poor alignment in integration with mobile devices to provide an efficient platform for pest and disease detection and pesticide mapping.

The crop selection can be simplified by using AI can simplify crop selection and assess what produce will be most profitable and predictive analytics equipped with forecasting can be used to reduce errors in business processes by minimizing the risk of crop failures. Artificial intelligence mechanisms and Machine learning based algorithms can be used to collect data on plant growth and evaluating crops that are less prone comparatively to disease and adaption to weather conditions. AI based advancements are useful in prediction of optimal mix of agronomic products, chemical soil analysis, accurate estimation of missing nutrients in soil etc. Innovative farming practices like vertical agriculture may help increase food production while minimizing the use of resources and ensuring cost savings. Adoption of AI and machine learning automation with the access and availability of driverless tractors, smart irrigation, fertilizing systems, smart spraying, vertical farming, AI based robots are few of the game changers in the domain of Agri-Tech developments.

By combining AI with big data, farmers can get valid recommendations based on well-sorted real-time information on crop needs. Indian farm sector and farmers are understanding efficiently that Artificial Intelligence is only an advanced part of simpler technologies for processing, gathering and monitoring data for sustainable farming, though some serious constraints exist. Truly AI requires a proper technology infrastructure for it to work. Privacy and security issues in the vertical of agriculture raise various legal issues, unfortunately cyber threats and network intrusion etc. are vulnerable. This research paper is focused on identifying and implementing Artificial Intelligence techniques for the inclusive welfare of all stakeholders in the domain of agricultural developments.

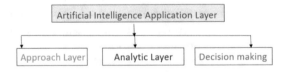

Fig. 1. General Architecture of Artificial Intelligence [1, 7]

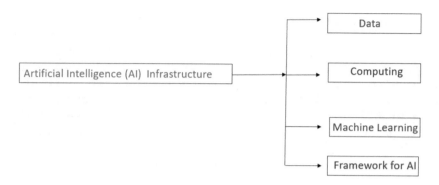

Fig. 2. Artificial Intelligence (AI) Infrastructure [1,7]

2   General Framework of Artificial Intelligence

Agriculture is undergoing a fourth revolution, known as agriculture 4.0 or smart agriculture, which is fueled by the advent of the big data era and the rapid advancement of many cutting-edge technologies, particularly machine learning, contemporary information technology, and communication technology. There are different stages and phases of agricultural activities starting from production to harvesting that can be simplified through the use of Artificial intelligence (Figs. 1 and 2).

Cross-disciplinary applications of artificial intelligence (AI) are possible, and they have the potential to change the way we currently think about farming. In addition to helping farmers work more efficiently with fewer resources, AI-powered solutions will also help them increase their production as a result of the rising use of high-tech machinery in other areas of life, like governance, healthcare, and even education. The impact of artificial intelligence on agriculture is the greatest of all since it is geared toward efficiency and easy labour. AI should be used to improve agricultural areas at low cost and with simple processing. A variety of agricultural issues are quickly resolved thanks to artificial intelligence. Artificial intelligence uses a variety of strategies to increase food production rates, such as increasing harvest quality and introducing indoor farming.

Many AI applications can actually benefit farmers, including those that analyze farm data to improve crop quality and accuracy, identify target weeds with the use of sensors, and identify pests, diseases, and other issues with plants. AI addresses labour issues. Since fewer people are entering this field, farmers are experiencing a labour shortage. One solution to this is the use of agriculture bots, which will collaborate with farmers. These bots harvest crops more quickly and in greater quantity. Blue River Technology employs agricultural robots for the control of weeds. Robotics had created a robot for farmers that will pick and pack crops as part of Harvest CROO Robotics [8].

3   Conclusion

This paper introduces the concept of Artificial Intelligence in agriculture and has the goal of providing as much information as possible, including specifics on the various AI techniques used in agriculture. Different neural network models and fuzzy systems have taken the role and contributed as the expert systems since a long time. Many respondents to the study expressed the opinion that AI will be very helpful to farmers both now and in the future. More in-depth studies are being carried out using more cutting-edge equipment so that conventional agricultural can transition to precision, low-cost agriculture.

References

1. Xu, Y., Liu, X., Cao, X., Huang, C., Liu, E., Qian, S., Zhang, J.: Artificial intelligence: a powerful paradigm for scientific research. Innov. **2**(4), 100179 (2021)
2. Al-Ayyoub, M., et al.: Accelerating 3D medical volume segmentation using GPUs. Multimedia Tools Appl. **77**(4), 4939–4958 (2018)
3. Jaipong, P., Sriboonruang, P., Siripipattanakul, S., Sitthipon, T., Kaewpuang, P., Auttawechasakoon, P.: A review of intentions to use artificial intelligence in big data analytics for Thailand agriculture. Rev. Adv. Multi. Sci. Eng. Innov. **1**(2), 1–8 (2022)
4. Yu, H.Q., Reiff-Marganiec, S.: Learning disease causality knowledge from the web of health data. Int. J. Seman. Web Inf. Syst. (IJSWIS) **18**(1), 1–19 (2022)
5. Sarma, S.K., Singh, K.R., Singh, A.: An expert system for diagnosis of diseases in rice plant. Int. J. Artif. Intell. **1**(1), 26–31 (2010)
6. Agrawal, P.K., et al.: Estimating strength of a DDoS attack in real time using ANN based scheme. In: International Conference on Information Processing, pp. 301–310. Springer, Berlin, Heidelberg (2011)
7. Report on Artificial Intelligence for Agriculture Innovation: World Economic Forum. Niti Aayog, India (2021)
8. Khandelwal, P.M., Chavhan, H.: Artificial intelligence in agriculture: an emerging era of research. Research Gate Publication (2019)
9. Gupta, B.B., Badve, O.P.: GARCH and ANN-based DDoS detection and filtering in cloud computing environment. Int. J. Embedd. Syst. **9**(5), 391–400 (2017)

A Novel Attack Detection Technique to Protect AR-Based IoT Devices from DDoS Attacks

Kwok Tai Chui[1]([✉]), Varsha Arya[2,3], Dragan Peraković[4], and Wadee Alhalabi[5]

[1] Hong Kong Metropolitan University (HKMU), Ho Man Tin, Hong Kong
jktchui@hkmu.edu.hk

[2] Insights2Techinfo, New Delhi, India
varsha.arya@insights2techinfo.com

[3] Lebanese American University, Beirut 1102, Lebanon

[4] University of Zagreb, Zagreb, Croatia
dperakovic@fpz.unizg.hr

[5] Department of Computer Science, King Abdulaziz University, Jeddah, Saudi Arabia
wsalhalabi@kau.edu.sa

Abstract. Argumented reality (AR) has become increasingly important tools for transferring data between machines and people as a result of advances in smart gadgets and artificial intelligence (AI). Traditional rigid sensors have outpaced flexible and IoT devices that can be easily incorporated into AR systems. In recent years, adaptable IoT-based AR have been extensively studied for smart skins in the areas of physiological monitoring, motion detection, robotics, healthcare, and other applications. Researchers are focusing on developing the idea of AI in all domains. There are, of course, certain limits to AR ideas and concepts at this stage in their development. To protect AR enabled smart devices from various types of cyber threat, it is necessary to provide security and authentication mechanisms. In this context, in this paper we present a novel attack detection technology that is used to detect DDoS attacks AR-based IoT devices. The proposed approach detects malicious operations with an accuracy of 92.84%. Also works efficiently as compared to other machine learning techniques.

Keywords: Augmented reality · Smart devices · IoT · Virtual reality · DDoS · Mixed reality · Edge computing

1 Introduction

The term "augmented reality" refers to a method of interacting with a real-world environment in which specific parts are "enhanced" by computer-generated perceptual data, such as smart glasses or tablets. There are many other variations of AR like virtual reality that is an interactive, multisensory, computer-generated

N. Nedjah et al. (Eds.): ICSPN 2021, LNNS 599, pp. 361–373, 2023.
https://doi.org/10.1007/978-3-031-22018-0_34

experience in which the user is immersed in a world that has been created from scratch [9,22], and Mixed Reality (MR), which blends AR and VR technology to create new settings and representations where real and digital items may coexist [1,19].

AR has its application in all major domains and after the COVID-19 pendemic, research in the field of AR is gaining momentum [26.8]. The most common application of AR nowadays is to provide user interfaces for Internet of Things (IoT) devices. Current IoT devices are controlled by their owners using typical user interfaces. In comparison to conventional non-IoT gadgets, they are widely available and allow remote control. However, this presents certain difficulties for users since they are physically removed from the gadgets they control. Furthermore, when additional devices are added to the network, these interfaces become less obvious and more difficult as the relationships between icons, custom names and real devices must be tracked [31] (Fig. 1).

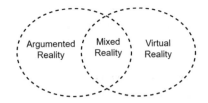

Fig. 1. Relation between AR, VR, and MR

In smart cities, there are many smart devices based on the Internet of Things. Currently, there are many services in smart cities that use augmented reality concepts, such as Ikea Place, SmartSantanderRA, and Google Glass. However, there is still a need to develop new AR technologies with respect to smart cities [2]. Currently, many researchers are developing technologies that can act as a proper interface for AR [9,25]. In addition to smart cities, AR can be applicable in other sectors, such as maintenance work on production lines [5], the development of smart buildings [33], industry 4.0 [20]. Controlling smart devices through AR-based control panels. The main benefit of these types of AR-based panels is that they increase the user experience [27,31]. In order to develop the user interface, mixed reality (MR), a variation of AR, is very useful. MR used the concepts of AR and VR to develop the user interfaces [22]. AR can also be used in the health care sector for the development of VR devices and portable EEG sets [11], for AR-based surgical navigation systems [23], to improve user food habits [3]. Some researchers also propose the use of AR in the education sector and 3D shape modeling [7].

The latest devices that use AR technology generate a large amount of data. Therefore, there is a need for technology that can process data in real time efficiently. Hence, the researchers proposed the concept of edge computing, which processes the IoT data near the edge devices [28], as represented in Fig. 2. Furthermore, researchers are working on the integration of the latest technologies, such as 5G / 6G with AR [42]. However, these techniques are not capable of being susceptible to different cyber attacks. In this context, we propose the DDoS attack detection technique for IoT devices that are used for AR-based applications. Our proposed approach uses the concepts of edge computing and machine learning technique to find malicious users who want to access the IoT devices using the AR-application.

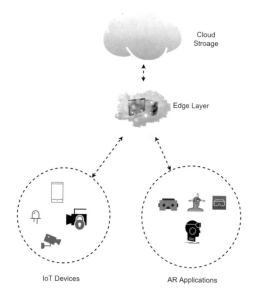

Fig. 2. Edge computing and AR technology

According to this structure, the rest of the paper will be laid down. The literature review appears in Sect. 2, and the recommended technique is presented in Sect. 3. Section 4 then summarises the simulation findings, and Sect. 5 concludes the discussion.

2   Literature Survey

The author reviews different AR technologies with respect to smart cities [2]. The author in [4] proposed a technique to use the smart mobile camera as an interface to AR.

One of the tasks of AR is to provide a smooth user experience for smart devices. In this context, the authors in [27] have developed prototypes that can seamlessly interact with sound and lighting systems by using hand-controlled increased interaction panels. Some researchers proposed the use of AR to improve the habits of users.

The first DDoS attack was detected in 1999, and by then it had been considered as the biggest cybersecurity threat to all major communication platforms like IoT, cloud, etc. Researchers had been proposing different solutions for its detection. At first, some researchers proposed a sliding window-based detection module, in which two sliding windows were used to detect attack traffic. Later, this method is improved by adding a clustering method. As hackers develop different variants of DDoS attacks, these classic techniques were not able to detect the attack traffic. However, recently, the author [17] improved the sliding window technique for the detection of the source IP address entropy. The proposed method quickly calculates the entropy value of incoming traffic; hence, it is an efficient method for the detection of volumetric DDoS attacks.

With the development of new techniques such as machine learning, deep learning, and fuzzy logic, new models for DDoS detection were developed. The author [34] proposed a DDoS detection module based on fuzzy logic, which uses the combination of entropy and fuzzy logic to identify and filter the attack. Some researchers used thread intelligence information shearing and command-line intelligence for the detection of attack packets. Some researchers suggest that machine learning techniques are the most efficient way to detect anomalies in network traffic. In this context, the author [24] proposed an SVM-based technique for the detection of HTTP flooding, SYN flooding, and amplification attacks.

Sometimes the use of honeypots is effective for the detection of DDoS attack, the author [30] proposed a honeypot-based defense model. In the proposed approach, the honeypot technique is used to store the unauthorized attempt of malicious software installation, and then a machine learning approach is used to predict the types of attacks. However, this technique cannot detect zero-day attacks. The author [6] proposed DDoS attack detection based on the Online Discrepancy Test (ODIT) [32] method. In the proposed approach, the author uses asymptotic optimality to handle a large number of training datasets. The proposed approach uses the concept of a dynamic scenario in which the number of devices continues to change. However, all of these detection techniques need a large amount of training data and work only in a specific test scenario. The author [21] also uses the SVM technique for DDoS detection, firstly the author extracted useful features from incoming traffic and then applied the SVM algorithm. The author [15] proposed an anomaly-based detection method using the dimensionality reduction technique; this approach can detect malicious traffic in a higher-dimensionality setting. In [18] auto-encoders are used for the detection of the DDoS attack traffic. However, the proposed approach is not acceptable for large networks because auto-encoders are trained for every device present in the network.

In addition to DL and ML techniques, statistical methods are also used for the detection of DDoS attack traffic. In these methods, the statistical properties of incoming traffic, such as probability or entropy, are analyzed. These techniques are useful for the detection of volumetric DDoS attacks because the statistical properties of the network traffic change at the time of attack. Mostly, the variation of the source's IP address entropy is the factor used by the authors to differentiate normal traffic from DDoS attack traffic. Some author calculates the score of the packet from the IP addresses and attributes of the incoming packets, and then combines the score with the destination IP address entropy for the identification of DDoS attack traffic. In all statistical-based DDoS attack detection methods, there is a need to select a threshold value beforehand. This is the main limitation of most of these statistically-based techniques. To overcome this limitation, researchers proposed a technique for dynamically selecting the threshold value. In addition to entropy, different authors also use other statistical parameters such as traffic flow variation for the preparation of models for DDoS attack detection. Also, some authors proposed a load-balancing technique for the mitigation of DNS flooding. The author also proposed a model through which the victim's site is reachable even if the DNS server is down. However, these techniques are not efficient for large networks.

As no single approach can efficiently detect DDoS attack traffic, some authors proposed a hybrid approach. In this context, the author [14] proposed a DDoS detection approach that is based on the combination of entropy and deep learning. In the proposed approach, the deep lea ring algorithm is used to find the threshold value for entropy-based filtering of incoming packets. In another approach, the author [12] proposed an approach that is a combination of flow control and CNN algorithm. Flow maintenance is managed by the statistical method, and filtering is done by the CNN algorithm.

3   Methodology

This section contains details of a detection approach that uses entropy to locate DDoS attacks for AR-based IoT devices. Our detection method has two phases. The arrival packet entropy is calculated at first pahse. The entropy value of a packet is used in machine learning models in the second phase to assess whether it is malicious or legitimate.

3.1   First Phase

In this phase, we calculate the entropy of the incoming traffic of the IoT devices.

Definition 1. *Probability* *For an independent random variable X, that can take 'n' distinct values, the probability is defined by Eq. 1.*

$$P(X) = \frac{P(x)}{\sum_{i=1}^{n} x_i} \tag{1}$$

Definition 2. *Entropy* *Entropy is the measure of the randomness of the random variable, andvariable, and it is measured by using Shannon's entropy formula.*

$$E(X) = -\sum_{i=1}^{n} P_i \times \log(P_i) \tag{2}$$

– Feature extraction: It is the most important step of our proposed approach, in this step the source and destination IP address of the incoming packets are extracted and placed in the appropriate clusters. We used the IP masking process to find the appropriate group for the IP address.

$$Mask_{IP} = 255.255.255.255 \tag{3}$$

$$Group_i = IP_i \times Mask_{IP} \tag{4}$$

Definition 3. Group Probability and Group Entropy *For every time window δt seconds according to the incoming IP addresses, different clusters are created. The probability and group entropy of each of the groups is given by Eq. 5 and Eq. 6*

$$P(G_i) = \frac{N_{ni}}{\sum\limits_{i=1}^{i=n} N_{ni}} \tag{5}$$

$$H(G) = -\sum_{i=1}^{i=n} P(G_i) \log(P(G_i)) \tag{6}$$

where N_{ni} and 'n' are the number of packets in i^{th} group and the total number of groups formed.

Theorem 1. *When a DDoS attack occurs, the group's entropy is greater than when regular traffic occurs.*

Proof. Let us consider two different scenarios, in the first scenario there is no DDoS attack occurring,occurring, and in the second scenario, a flooding type DDoS attack occurs. Let 'n' be theoccurs. Let 'n' be the number of groups formed during theduring the DDoS attack time and the groups are represented by Eq. 7

$$G_D = \{G_{D1}, G_{D2}, ..., G_{Dn}\} \tag{7}$$

Then using Eq. 5 and 6, we can calculate the probability and entropy of the group. Entropy is represented by Eq. 8

$$H(G_D) = -\sum_{i=1}^{i=n} P(G_{Di}) \log(P(G_{Di})) \tag{8}$$

However, as we know, during a volumetric DDoS attack, the malicious user generates a large amount of fake traffic from spoofed IP addresses to flood the victim's bandwidth. Due to this reason, during a volumetric DDoS attack, the number of groups formed in our proposed approach is increased and the overall entropy increases and the distribution of entropy can be represented as a monotonically increasing concave function. Therefore, according to Jensen's inequality [10], we can represent the entropy during DDoS attack by Eq. 9

$$H_D\{P(x)\} \geq P(H_D(x)) \tag{9}$$

where H_D is the entropy of the group and P (x) is the probability of occurrence of the group.

Similar to Eq. 9, we can calculate the probability and entropy of the groups during no attack period. The entropy during no attack is given by Eq. 10

$$H_N(G_N) = -\sum_{i=1}^{i=n} \overline{P}(G_{Ni}) \log(\overline{P}(G_{Ni})) \tag{10}$$

However, during no attack period, the probability distribution of the groups is normal and the entropy is represented by the convex function because the network traffic is random in nature. Therefore, according to Jensen's inequality, we can represent the entropy by Eq. 11

$$\overline{H}_N\{\overline{P}(x)\} \le \overline{P}(\overline{H}_N(x)) \tag{11}$$

where $\overline{H}_N$ is the entropy of the group and $\overline{P}$ (x) is the probability of occurrence of the group. As we know, the group probability at the time of the DDoS attack is greater than the probability of the group at the time of no attack. (Eq. 12).

$$P(x) > (\overline{P})(x) \tag{12}$$

Then by using, Eqs. 9, 10, and 12 we can compare the group entropy during DDoS attack and no DDoS attack time in Eq. 13. That will prove the theorem.

$$H(x) < \overline{H}(x) \tag{13}$$

3.2 Second Phase

In this phase, the linear regression (LR) technique is used to differentiate malicious packets from legitimate packets. LR is a machine learning technique that used a liner equation to predict the dependent variables, as represented in Fig. 3.

$$y = mx + c \tag{14}$$

Equation 14 represents the basic LR equation; in which x represents the independent variable, m represents the slope, c represents the intercept, and y represents the dependent variable.

However, there are many best-fit lines that can represent the LR curve. Hence, to find the optimal best-fit line, we use the cost function, which is calculated by Eq. 15.

$$Cs = \frac{1}{2N} \sum_{i=1}^{N} (y - \hat{y})^2 \tag{15}$$

Figure 4 represents the cost function curve. From the cost function curve, we find the global min. Value that represents the optimal cost function value.

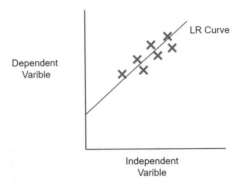

Fig. 3. Linear Regression

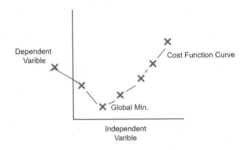

Fig. 4. Cost Function Curve

4   Results and Discussion

The Internet of Things (IoT) network was modelled and simulated using OMNET++. It is an event multi-multior that creates traffic at regular periods of time. OMNET++ The data set collected from OMNET++ was analyzed by Tensorflow on a 64-bit Intel i5 CPU with 16 GB of RAM in a Windows 11 environment.

In a stimulation model, the most important part is finding the answer. optimal threshold to differentiate malicious nodes from legitimate nodes. For this, we plot the ROC curve and find the optimal threshold value for our simulation model. The ROC curve is represented in Fig. 5. From Fig. 5 it is clear that our proposed model works efficiently. From ROC curve, we can find two types of values:

- Higher true negatives are indicated by lower x-coordinate values.
- The larger the y-axis of the figure, the more true positives and fewer false negatives there are in the data.

On the whole, a skilled model will provide a greater probability for a randomly selected genuine positive event than it does for a randomly chosen real negative occurrence. It is this kind of competence that we are referring to when

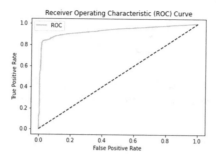

Fig. 5. ROC Curve

we say that the model has talent. Skilled models are usually shown by curved lines that rise from the left side of the plot. When a classifier has no ability to distinguish between classes, it will always provide a random or constant classification. A model with minimal talent is shown at this particular juncture (0.5, 0.5). At each threshold, the ROC of a model without skill is shown by a diagonal line from the bottom left of the figure to the top right. At a specific location, a skilled model may be seen in action (0, 1). Using a line from the bottom left to the top left, and then across the top, a model with flawless talent is shown. Therefore, from Fig. 5, we can say that our model is skilled.

It's possible to draw the ROC curve for the final model and set a threshold that balances the false positives with the true ones, and this can be done manually. Apart from the ROC curve, undderfitting and overfitting are important factors that can affect the simulation results. In an underfitting situation, we get a high error at the time of training and testing the proposed model. However, in the overfitting situation, an error occurs for testing datasets. Overfitting is a more serious problem than an underfitting condition. Therefore, to reduce the overfitting condition, we used the Rigit regression cost function (Eq. 16). The exact value of λ is presented in Table 1.

Table 1. Logical Regression Representation

Parameters	Values
Cost function	–0.054059696277611834
λ	0.01
Precision	0.93
Reall	0.99
F-1 Score	0.96
Accuracy	0.9284

$$R_{cs} = \sum_{i=1}^{N} (y - \hat{y})^2 + \lambda(slop)^2 \tag{16}$$

- **Precision** - This value indicates how many packets are properly filtered as a result of using our suggested strategy. It is calculated by Eq. 17.

$$P = \frac{\phi_{tp}}{\phi_{tp} + \phi_{fp}} \tag{17}$$

- **Recall** - Measures the percentage of legitimate packets from the total undetected packets by our algorithm. It is calculated by Eq. 18

$$R = \frac{\phi_{tp}}{\phi_{tp} + \phi_{fp}} \tag{18}$$

- **True negative rate** - represents the total of filtered packets sent by malicious users. It is calculated by Eq. 19

$$TNR = \frac{\phi_{tn}}{\phi_{tn} + \phi_{fp}} \tag{19}$$

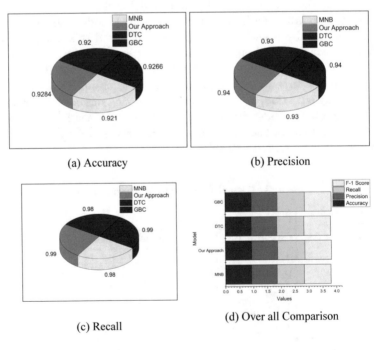

(a) Accuracy

(b) Precision

(c) Recall

(d) Over all Comparison

Fig. 6. Comparison of Proposed Approach With other Machine learning Algorithms

- **Negative predictive value** - The number of packets delivered by a malicious user may be calculated on the basis of the filtered data.

$$NPV = \frac{TN}{TN + FN} \tag{20}$$

Now, we compare our proposed approach with other machine learning techniques. We used accuracy, precision, and recall for the comparison. WE select MNB, decision tree classifier, and gradient boosting algorithms for comparison. This comparison is presented in Fig. 6. From Fig. 6a, it is clear that the accuracy of our proposed approach is greater than that of the other three machine learning techniques. Thus, we can say that our proposed approach detects malicious packets more accurately. From Fig. 6b and c it is clear that our proposed approach has the highest precision value and recall value. Therefore, from Fig. 6 it is clear that our technique is efficient compared to other machine learning techniques.

5 Conclusion

Augmented Reality (AR) has reemerged as a dominant medium of information visualization and interaction as a result of recent advances in smart devices. By spatially supplementing relevant information to essential interaction items in everyday tasks, AR not only has several potential uses, but also gives users a new interface to operate smart devices.But AR applications are still in the developing phase, and as they depend on IoT and other smart devices, they are vulnerable to different types of cyber attack such as DDoS. In this context, we develop a machine learning approach that can detect DDoS attacks efficiently and effectively in an AR-based IoT environment. Our proposed approach uses the linear regression technique to distinguish malicious traffic from legitimate traffic. We tested the performance of our proposed approach using standard statistical parameters (accuracy, precision, and recall) and then compared the performance with other machine learning approaches. From the comparison, it is clear that our proposed algorithm can efficiently detect the significant traffic from the AR-applications. For future work, we would like to test the model on real-world datasets.

References

1. Andrade, T., Bastos, D.: Extended reality in iot scenarios: concepts, applications and future trends, pp. 107–112 (2019). https://doi.org/10.1109/EXPAT.2019.8876559
2. Badouch, A., Krit, S.D., Kabrane, M., Karimi, K.: Augmented reality services implemented within smart cities, based on an internet of things infrastructure, concepts and challenges: an overview (2018). https://doi.org/10.1145/3234698.3234751

3. Bordegoni, M., Carulli, M., Spadoni, E.: Support users towards more conscious food consumption habits: a case study. vol. 1, pp. 2801–2810 (2021). https://doi.org/10.1017/pds.2021.541
4. Chang, I.Y.: Augmented reality interfaces for the internet of things. In: ACM SIGGRAPH 2018 Appy Hour, pp. 1–2 (2018)
5. Dey, S., Sarkar, P.: Augmented reality based integrated intelligent maintenance system for production line. vol. 07–09-December-2016, pp. 126–131 (2016). https://doi.org/10.1145/3014362.3014377
6. Doshi, K., Yilmaz, Y., Uludag, S.: Timely detection and mitigation of stealthy DDos attacks via IoT networks. IEEE Trans. Dependable Secure Comput. (2021)
7. Guo, L., Wang, P.: Art product design and vr user experience based on iot technology and visualization system. J. Sens. **2021** (2021). https://doi.org/10.1155/2021/6412703
8. Gupta, B.B., Misra, M., Joshi, R.C.: An ISP level solution to combat DDos attacks using combined statistical based approach. arXiv preprint arXiv:1203.2400 (2012)
9. Gupta, S., et al.: PHP-sensor: a prototype method to discover workflow violation and XSS vulnerabilities in PHP web applications. In: Proceedings of the 12th ACM International Conference on Computing Frontiers, pp. 1–8 (2015)
10. Jaafari, E., Asgari, M.S., Hosseini, M., Moosavi, B.: On the Jensen's inequality and its variants. AIMS Math. **5**, 1177–1185 (2020). https://doi.org/10.3934/math.2020081
11. Jagarlapudi, A., Patil, A., Rathod, D.: A proposed model on merging iot applications and portable eegs for migraine detection and prevention, pp. 265–269 (2021). https://doi.org/10.1109/DISCOVER52564.2021.9663615
12. Jia, Y., Zhong, F., Alrawais, A., Gong, B., Cheng, X.: Flowguard: an intelligent edge defense mechanism against IoT DDos attacks. IEEE Internet of Things J. **7**(10), 9552–9562 (2020)
13. Kimura, N., Okita, Y., Goka, R., Yamazaki, T., Satake, T., Igo, N.: Development of IoT educational materials for engineering students, pp. 449–454 (2021). https://doi.org/10.1109/IEEECONF49454.2021.9382682
14. Koay, A., Chen, A., Welch, I., Seah, W.K.: A new multi classifier system using entropy based features in DDos attack detection. In: 2018 International Conference on Information Networking (ICOIN), pp. 162–167. IEEE (2018)
15. Kurt, M.N., Yilmaz, Y., Wang, X.: Real-time nonparametric anomaly detection in high dimensional settings. IEEE Trans. Pattern Anal. Mach. Intell. (2020)
16. Kusuma, H., Shukla, V., Gupta, S.: Enabling vr/ar and tactile through 5g network (2021). https://doi.org/10.1109/ICCICT50803.2021.9510181
17. Li, J., Liu, M., Xue, Z., Fan, X., He, X.: Rtvd: a real-time volumetric detection scheme for DDos in the Internet of Things. IEEE Access **8**, 36191–36201 (2020)
18. Meidan, Y., Bohadana, M., Mathov, Y., Mirsky, Y., Shabtai, A., Breitenbacher, D., Elovici, Y.: N-baiot-network-based detection of IoT botnet attacks using deep autoencoders. IEEE Pervasive Comput. **17**(3), 12–22 (2018)
19. Mishra, A., et al.: A comparative study of distributed denial of service attacks, intrusion tolerance and mitigation techniques. In: 2011 European Intelligence and Security Informatics Conference, pp. 286–289. IEEE (2011)
20. Montalvo, W., Bonilla-Vasconez, P., Altamirano, S., Garcia, C., Garcia, M.: Industrial control robot based on augmented reality and IoT protocol. Lecture Notes in Computer Science (including subseries Lecture Notes in Artificial Intelligence and Lecture Notes in Bioinformatics) 12243 LNCS, pp. 345–363 (2020). https://doi.org/10.1007/978-3-030-58468-925

21. Nomm, S., Bahi, H.: Unsupervised anomaly based botnet detection in IoT networks. In: 2018 17th IEEE International Conference on Machine Learning and Applications (ICMLA), pp. 1048–1053. IEEE (2018)

22. Pfeiffer, T., Pfeiffer-Lemann, N.: Virtual prototyping of mixed reality interfaces with Internet of Things (IoT) connectivity. i-com **17**(2), 179–186 (2018). https://doi.org/10.1515/icom-2018-0025

23. Rong, F., Juan, Z., ShuoFeng, Z.: Surgical navigation technology based on computer vision and vr towards IoT. Int. J. Comput. Appl. **43**(2), 142–146 (2021). https://doi.org/10.1080/1206212X.2018.1534371

24. She, C., Wen, W., Lin, Z., Zheng, K.: Application-layer DDos detection based on a one class support vector machine. Int. J. Netw. Secur. Appl. (IJNSA) **9**(1), 13–24 (2017)

25. Soedji, B., Lacoche, J., Villain, E.: Creating ar applications for the IoT : a new pipeline (2020). https://doi.org/10.1145/3385956.3422088

26. Subramanian, M., Shanmuga Vadivel, K., Hatamleh, W., Alnuaim, A., Abdelhady, M., Sathishkumar, V.: The role of contemporary digital tools and technologies in covid-19 crisis: an exploratory analysis. Expert Syst. (2021). https://doi.org/10.1111/exsy.12834

27. Sun, Y., Armengol-Urpi, A., Reddy Kantareddy, S., Siegel, J., Sarma, S.: Magic-hand: Interact with IoT devices in augmented reality environment, pp. 1738–1743 (2019). https://doi.org/10.1109/VR.2019.8798053

28. Tang, G., Shi, Q., Zhang, Z., He, T., Sun, Z., Lee, C.: Hybridized wearable patch as a multi-parameter and multi-functional human-machine interface. Nano Energy **81** (2021). https://doi.org/10.1016/j.nanoen.2020.105582

29. Tewari, A., et al.: Secure timestamp-based mutual authentication protocol for IoT devices using rfid tags. Int. J. Seman. Web Inf. Syst. (IJSWIS) **16**(3), 20–34 (2020)

30. Vishwakarma, R., Jain, A.K.: A honeypot with machine learning based detection framework for defending IoT based botnet DDos attacks. In: 2019 3rd International Conference on Trends in Electronics and Informatics (ICOEI), pp. 1019–1024. IEEE (2019)

31. Xu, Z., Lympouridis, V.: Virtual control interface: Discover and control IoT devices intuitively through ar glasses with multi-model interactions, pp. 763–764 (2021). https://doi.org/10.1109/VRW52623.2021.00264

32. Yilmaz, Y.: Online nonparametric anomaly detection based on geometric entropy minimization. In: 2017 IEEE International Symposium on Information Theory (ISIT), pp. 3010–3014. IEEE (2017)

33. Zhao, Y., Xie, C., Lu, M.: Design of smart building operations and maintenance management service system. Lect. Notes Netw. Syst. **261**, 183–190 (2021). https://doi.org/10.1007/978-3-030-79760-723

34. Zhao, Y., Zhang, W., Feng, Y., Yu, B.: A classification detection algorithm based on joint entropy vector against application-layer DDos attack. Secur. Commun. Netw. **2018** (2018)

Application of Artificial Neural Network (ANN) in the Estimation of Financial Ratios: A Model

Karamath Ateeq[1(✉)], Jesus Cuauhtemoc Tellez Gaytan[2], Aqila Rafiuddin[3], and Chien-wen Shen[4]

[1] Senior Faculty, School of Information Technology, Skyline University, Sharjah, UAE
Karamath.ateeq@skylineuniversity.ac.ae
[2] Business School, Tecnologico de Monterrey, San Pedro Garza Garca, Mexico
cuauhtemoc.tellez@tec.mx
[3] Business School, Under Grant of FAIR, Tecnologico de Monterrey, San Pedro Garza Garca, Mexico
[4] National Central University, Taoyuan City, Taiwan
cwsshen@ncu.edu.tw

Abstract. Performance Evaluation provides details to all the stake holders and supports the life of the organization. There are various measures and magnitudes to understand and measure the current status ad predicts the future. Research on the applications of Financial ratios have been widely explored in terms of statistical applications. This study explores and applies Artificial intelligence (AI) in predicting by means of Artificial Neural Networks, Support Vector Regression, and linear regression. Investigates to support the technicalities with the decision support system architecture. The research methodology mainly consists of four steps broadly aligning with the knowledge discovery process in artificial intelligence. In addition, the paper directs future research towards artificial intelligence applications in predictive analytics, accounting and finance fields.

Keywords: Business intelligence · Artificial intelligence · Profitability ratios · Risk ratios · Predictive analytics.

1 Introduction

In an effort to improve the quality of businesses and investments, technological advancement has shifted decision-making from humans to artificially intelligent systems. Current environment is dynamically accelerated and systems are becoming more complicated, resulting in increasing competitiveness, necessitating a faster and more effective information decision system. Artificial Intelligence's Artificial Neural Network (ANN) and machine learning techniques are the finest options for creating a positive vision, delivering more value, and significantly improving the quality of business. Though the use of artificial intelligence

N. Nedjah et al. (Eds.): ICSPN 2021, LNNS 599, pp. 374–383, 2023.
https://doi.org/10.1007/978-3-031-22018-0_36

in business forecasting is not a new idea, its application in accounting and financial prediction is still in its infancy. There still exists a need to use machine learning techniques and its applications in financial applications.

Primary focus of this research study is the usage of quantitative data to suggest the qualitative outputs and emphasize the role of Artificial Intelligence in accounting and finance to predict the ratios from financial statements and evaluate the performance of the organization. This research will contribute in the literature with a new dimension of applications of predictions. Further it will assist to know the role of academic accounting and its future in accounting. The purpose is to establish a model for academic research, business, policymakers, and investors for application in real time.

2   Review of Literature

Business and academic research into the uses of intelligence systems has exploded in the past few years due to the progressive transformation in technology. With the COVID-19 AI growth was expedited and linked with industry restructuring. According to Financial Stability Board 2019 AI is explained as development and theory of computer related systems to complete tasks that have traditionally required the interference of human intelligence. ANNs are composed of a set of weights for each input connection and a bias adjustment, as well as a transfer function that translates the total of the weighted inputs and bias to identify the linear output value from the computing unit (node j).

In forecasting, predicting, clustering, classification and in pattern recognition problems, technology and techniques of Artificial Neural Network is used. ANN technology produces superior performance than other statistical techniques [1,2]. According to [3,4] usability of AI in accounting domains holds a vibrant research and a leadership role as compared to the earlier ideas of slow down growth in 1990s. According to study conducted by [5,8] accounting firms make substantial use of AI tools in their integrated audit support systems. [6] worked on targeting management accounting practices by business intelligence vendors. [7] reports emphasized on immediate and bigger understanding of concepts of machine- learning principles by the accounting major graduates. Importance of the recent advances of Artificial intelligence in the accounting profession are discussed connected to information technology with an emphasis on Data Analytics. Reports that are related to data of business and accounting recommends the students' needs for obtaining certification or training in the field of data analytics, statistics, use of R programming language and knowledge of the basics of machine learning that are related to both unstructured data and structured type of data. It is recommended to focus in the various skills of Artificial Intelligence. The recent trend in Machine learning is a key to success in the artificial intelligence sector, while it appears to be a target for wide usage in consultation, taxation and audit process, most likely in the management accounting process in most corporate and non-profit organizations.

AIS use in accounting is directed by the continuous monitoring and audit of data stream [9]. Accounting using AI techniques involves a complex process of data analytics since it deals with uncertainity. Artificial Neural Network (ANN) are used in this process. There is a level of uncertainty in predicting future events regarding financial data as it is involved with the recognition of patterns that will be used to forecast the future events. Some events like the changes in interest rare and movements of currency are considered as difficult exercises to predict in financial analysis although these are considered as critical exercises for the survival of the financial dealings in an organization. Accurate prediction of economic events, such as interest rate changes and currency movements currently rank as one of the most difficult exercises in finance considered to be critical for financial survival. ANNs provide a better solution to solve these problems as AI techniques deal well with large noisy data sets.

Artificial Neural Networks convert the complex form of data into mathematical functions that offer qualitative approaches for economic and business systems that cannot be offered by conventional quantitative techniques in econometrics and statistics. [10] enumerated the financial analysis tasks of prototype neural network-based decision aids for credit authorization screening, mortgage risk evaluation, project management and bidding strategy, and financials. According to [11] Artificial Neural Networks applications can be divided into three categories such as prediction, pattern classification, and the financial analysis and optimization and control. To determine the connection of pattern of one data to the next data is known as a temporal effect. A forecasting model from the previous data of time series can be built and forecasted using ANN approach. In order to deal with specific issues, recurrent ANN and Finite impulse response (FIR) are also being explored and developed.

ANNs use complex algorithms that are based on sending signals principles of neuron connecting to simulate the learning process of the human brain.A weightage is associated with each connection. The weightage can be associated with a positive or a negative value. A neuron is activated by positive weights and the activity of the neuron is inhibited by negative weights. Inputs (x1, x2,..., xi) are shown connected to neu-ron j with weights (w1j, w2j,..., wij) on each connection in Fig. 1. The neuron adds all the signals it receives, multiplying each signal by the weights associated with its connection.

According to [12] financial markets are affected by a number of economic, political and other factors in a very complex way therefore a precise prediction of these markets becomes difficult. According to the research findings, machine-learning processes give an excellent analytic tool with robustness and efficiency when an educator, researcher, financial service expert, lender, or policymaker must characterize and forecast a household's future financial condition. Support Vector Machines (SVM) and the Artificial Neural Networks (ANN) are more commonly used to predict the price movements of stocks. According to [13] investigation of the out-of-sample forecasting ability of recurrent and feed forward neural networks based on empirical foreign exchange rate data.

Artificial Neural Networks with the Backpropagation algorithm are a conventional classification and forecasting tool. Though numerous variants of Deep Neural Networks (DNN) are now popular, sophisticated analysis tools, the gradient descent approach employed in Feed forward and Back propagation algorithms remains the backbone of all neural network architectures. Both ANN and DNN have a variety of commercially significant uses. Methods and instruments for short time prediction of financial operations using neural network are bankruptcy forecasting [14].

A financial ratio may indicate distress when it is both higher than normal and lower than usual threshold [15]. In such cases, ANNs may provide a more accurate representation of the underlying relationships. This category assesses the relationship between input and output variables. The stronger the relationship, the higher the ability to predict the value of the output variable. Better the strength of link, the greater the capacity to anticipate the output variable's value. Integration of Machine Artificial intelligence is based on learning approaches which focus on extracting streams of patterns from older. The machine is put through this training data in order to predict the new results. Typically, there are two phases of research in machine learning to predict data.

There is interconnection between the basic layers of processing while there is a weightage assigned for each of the connection in the neural networks [16]. These weightages are changed and adjusted during the process of machine learning in the network. During the first phase of training, there is optimization of interconnections for the various layers of neurons. The process of optimization is further extended to the transfer function parameters between the neuron layers. This process minimizes the occurrence of errors. At the last layer of the network there is summation of the various signals into a single output signal. The output signal reflects the response of the network to the input data.

3   Research Methodology and Data Collection

Data comprises of annual financial statements for period from 2000–2021.The research methodology consists of four steps broadly aligning with the knowledge discovery process:

Step 1: This step indicates studying and analysing the Time Forecasting, and analysis.

Step 2: Identifying the best artificial intelligence techniques suited to the case, some of which are: neural networks, Support vector machine models, linear regression models

Step 3: This step further includes three sub-steps [17], namely data preprocessing, application of the algorithm, and pattern generation. Data preprocessing, normalizing data, and extracting features for it. Data pre-processing involves several techniques like data cleaning, data integration, data reduction [18].

Step 4: These steps align with the process and finally lead to the generation of patterns and results by applying various algorithms related to the main dimensions of performance.

Support Vector Regression and Linear Regression are utilized extensively for forecasting. In our research, we focus on developing an Artificial neural Network model. ANNs are built on the connection between neurons and signal passing principle in an way attempt to mimic the human brain's learning process through the use of complex algorithms. Each connection has either a positive or negative weight connected with it. Negative weights block the neuron whereas positive weights stimulate it. There are six steps in setting up an ANN. The input data and financial ratios are defined and delivered as a pattern to the ANN to signify the desired output. 80% of the sampled data is designated as the set for training or validation, whereas 20% is designated as the set for testing. The validation set employed contributes to the ANN's prediction power and suggests when training should be terminated. A neural network consists of an input layer, an output layer, and an additional hidden layer in between. These hidden layers illustrate the relationship between variables that is not linear. Each layer is composed of several neurons that are interconnected. With numerous layers of neurons, these networks may capture relatively complex phenomena. This training set procedure aids in the computation of output by minimizing forecasting mistakes caused by the iterations. The Algorithms used will modify the model parameters linking the nodes and weights. Each unit that is processing the output is relayed by each layer of the network.

Figure 1 depicts the whole methodology used to evaluate the three models. As previously indicated, the data is divided into 80% and 20% and then normalized. After the process of normalization, the data is input into LR, SVR, and ANN models. Meanwhile training the models with 80% of the data is completed and 20% of the remaining is utilized for evaluation of the models. The Mean Absolute error (MAE) and the root mean squared error (RMSE) errors are then made used to compare model performance. Estimations were performed on the data. R version 4.0.5 was used for this.

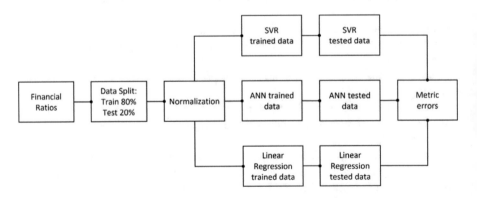

Fig. 1. The methodology applied in work

The section below provides a context for normalization, error measures, and the ANN model as an illustration.

The normalization stage ensures that features with smaller values do not predominate over those with larger values. In this way, the original matrix of financial ratios is transformed into a matrix of the range (0,1).Value after normalization is calculated as below:

$$Xnorms = \frac{X - Xmin}{Xmax - Xmin} \qquad (1)$$

After the normalization process, the 80% of training and 20% of testing data are separated .On the basis of the number of hidden layers and nodes, various ANN topologies were estimated. As transfer functions, the sigmoid and the hyperbolic tangent (Tan h) functions were applied. The activation function was used on the inputs that were weighted to calculate the outputs.

Figure 2 depicts a generic model of the ANN structure that has three inputs –X1, X2 and X3), one node, one hidden layer and one output (Y). Here f is the activation function.

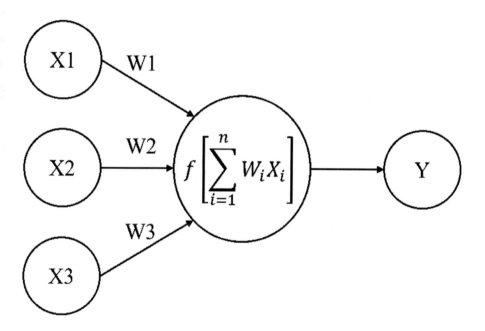

Fig. 2. ANN example

The logistic function is shown below

$$\varphi = \frac{x1}{1 + e^{-x}} \qquad (2)$$

The Tan h(x) function is shown in (3)

$$\tanh(x) = \frac{e^{2x} - 1}{e^{2x} + 1} \qquad (3)$$

Algorithms are compared for their performance analysis using RMSE and MAE as metrics . Equation (4) reflects the RMSE metric (Figs. 3 and 4).

$$RMSE = \sqrt{1/n \sum_{i=1}^{n} (x_i - \overline{x_i})^2} \qquad (4)$$

MAE matrix is shown below :

$$MAE = \sqrt{\frac{1}{n} \sum_{i=1}^{n} |x_i - \overline{x_i}|} \qquad (5)$$

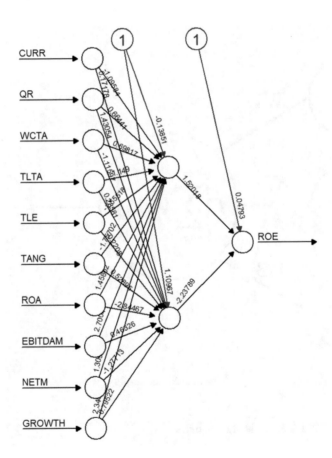

Fig. 3. Company 1

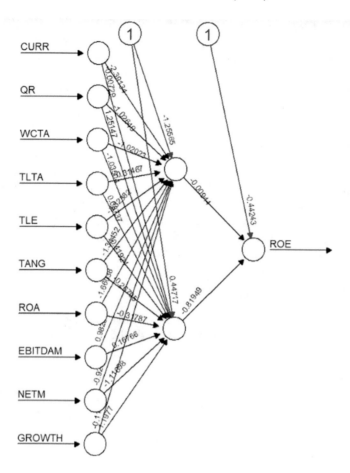

Fig. 4. .

4   Conclusion

The results demonstrate that the suggested ANN methodology outperforms previous statistical and financial techniques used to forecast performance evalua-
tions. Out of the three models used, SVM models have outperformed when there
is a non-linear performance. Using the statistical tools RMSE and MAE, empirical findings of ANN structures is validated in the tables in terms of algorithm
metric errors. The features of algorithm are considered to be ideal and outperform traditional regression. This study corresponds with the results that reveal
the prediction of performance of capital structure on the basis of ANN alone.
All ANN architectures are compatible with a single-layer, single-node structure with the same activation function which facilitates the passing of inputs to
the forecasting of capital structure output . ANN. models have outperformed
the econometric of forecasting for stock markets, financial risks, credit scores,

bankruptcy, and consumer choices [19]. The linear data outperformed LR and SVR. This result provides support for past studies that deemed ANN networks to be superior compared to other models. This work focused on predicting capital structure using input factors such as algorithm measures and financial error ratio measures in Artificial Neural Networks, Support Vector Machines and Linear Regression models.

4.1   Practical Implications

This research is one of the few to apply the ANN method for forecasting financial success that can be utilized by all parties involved. Financial statement and the capital structure is an area of significance for potential investors, auditing companies, financial situation to managers and other related personnel, investors and creditors, government regulators, financial institutions and analysts and others so that they can take timely measures to avoid losses.

4.2   Limitations and Future Scope

This study is limited with the application of financial ratios to single sector. This can be extended to other industries to identify the different behaviour of the industries.

Metrics Results

Company	Neural Network	SVR	Linear Regression
aljouf-rmse	0.1037961	0.0714185	0.0138223
aljouf-rse	0.2807046	0.1328952	0.0049779
aljouf-rae	0.4405598	0.2025537	0.0450133
aljouf-mae	0.0725317	0.0333475	0.0074108
jazan-rmse	0.0635795	0.0171325	0.0022059
jazan-rse	0.9349015	0.0678848	0.0011254
jazan-rae	0.9717091	0.2392354	0.0357036
jazan-mae	0.0407492	0.0100325	0.0014973

References

1. Altun, H., Bilgil, A., Fidan, B.C.: Treatment of multi-dimensional data to enhance neural network estimators in regression problems. Expert Syst. Appl. **32**(2), 599–605 (2007)
2. Al-Ayyoub, M., et al.: Accelerating 3D medical volume segmentation using GPUs. Multimed. Tools Appl. **77**(4), 4939–4958 (2018)
3. Brown-Liburd, H., Issa, H., Lombardi, D.: Behavioral implications of Big data's impact on audit judgment and decision making and future research directions. Acc. Horiz. **29**(2), 451–468 (2015)
4. Gupta, B.B., Badve, O.P.: GARCH and ANN-based DDoS detection and filtering in cloud computing environment. Int. J. Embed. Syst. **9**(5), 391–400 (2017)

5. Cheng, B., Titterington, D.M.: Neural networks: a review from a statistical perspective. Stat. Sci. 2–30 (1994)
6. Cortes, C., Vapnik, V.: Support-vector networks. Mach. Learn. **20**(3), 273–297 (1995)
7. Devadoss, A.V., Ligori, T.A.A.: Stock prediction using artificial neural networks. Int. J. Data Min. Techn. Appl. **2**(1), 283–291 (2013)
8. Agrawal, P.K., et al.: Estimating strength of a DDoS attack in real time using ANN based scheme. In: International Conference on Information Processing, pp. 301–310. Springer, Berlin, Heidelberg (2011, August)
9. Dowling, C., Leech, S.: Audit support systems and decision aids: current practice and opportunities for future research. Int. J. Acc. Inf. Syst. **8**(2), 92–116 (2007)
10. Elbashir, M.Z., Collier, P.A., Sutton, S.G.: The role of organizational absorptive capacity in strategic use of business intelligence to support integrated management control systems. Acc. Rev. **86**(1), 155–184 (2011)
11. Gan, C., Limsombunchao, V., Clemes, M.D., Weng, Y.Y.: Consumer choice prediction: artificial neural networks versus logistic models. J. Soc. Sci. **1**, 211–219 (2005)
12. Haykin, S.: Neural Netw., 2nd edn. Prentice hall, N.J. (1999)
13. Hill, T., Marquez, L., O'Connor, M., Remus, W.: Artificial neural networks for forecasting and decision making. Int. J. Forecast. **10**, 5–15 (1994)
14. Hill, T., O'Connor, M., Remus, W.: Neural network models for time series forecasts. Manage. Sci. **42**(7), 1082–1092 (1996)
15. Heo, W., Lee, J.M., Park, N., Grable, J.E.: Using artificial neural network techniques to improve the description and prediction of household financial ratios. J. Behav. Exp. Finance **25**, 100273 (2020)
16. Kuhn, J.R., Jr., Sutton, S.G.: Continuous auditing in ERP system environments: The current state and future directions. J. Inf. Syst. **24**(1), 91–112 (2010)
17. Krishnaswamy, C.R., Gilbert, E.W., Pashley, M.M.: Neural network applications in financial practice and education, 75–84 (2000)
18. Medsker, L.R.: Microcomputer applications of hybrid intelligent systems. J. Netw. Comput. Appl. **19**(2), 213–234 (1996)
19. Qi, M., Zhang, G.P.: Trend time series modelling and forecasting with neural networks. IEEE Trans. Neural Netw. **19**(5), 808–816 (2008)

GAN-Based Unsupervised Learning Approach to Generate and Detect Fake News

Pranjal Bhardwaj[1], Krishna Yadav[2(✉)], Hind Alsharif[3],
and Rania Anwar Aboalela[4]

[1] VIT Vellore, Vellore, India
pranjal.bhardwaj2019@vitstudent.ac.in
[2] National Institute of Technology Kurukshetra, Kurukshetra, India
krishna.nitkkr1@gmail.com
[3] Faculty of Computer and Information Technology, Umm AL-Qura University,
Makkah, Saudi Arabia
hhsharif@uqu.edu.sa
[4] Department of Information System, King Abdulaziz University, Rabigh, Saudi
Arabia
raboalela@kau.edu.sa

Abstract. Social media has grown into an increasingly popular means of disseminating information. Its massive growth has given evolution to fake news in misinformation and rumors, spreading very quickly. These days the generation of fake news is not only limited to the traditional method but is also extended to deep learning-based methods. The characteristics of fake news generated from these algorithms are very much identical to original news, which makes existing supervised machine learning algorithms difficult to detect these machine-generated fake news. Motivated by the problem, we have brought a fully unsupervised approach based on Autoencoder and GAN. With the help of an autoencoder, we have generated the high dimensional feature vector of news sentences which is later used by generators in GAN to create machine-generated fake news. The generated fake news is then identified with the real news with the help of a discriminator. We have tested our approach with the news dataset that contains about 30,000 news headlines. The obtained experimental results suggest that our approach is very reliable and can be very helpful in automating fake news detection.

Keywords: Generative adversarial networks · Autoencoders · Fake news detection · Fake news generation

© The Author(s), under exclusive license to Springer Nature Switzerland AG 2023
N. Nedjah et al. (Eds.): ICSPN 2021, LNNS 599, pp. 384–396, 2023.
https://doi.org/10.1007/978-3-031-22018-0_37

1   Introduction

Web-based media are these days one of the fundamental news hotspots for a vast number of individuals throughout the planet because of their minimal expense, simple access, and fast transmission. This case comes at the cost of the enormous danger of openness to 'fake news,' purposefully written to misdirect the users. Fake news is explicitly intended to sow a seed of uncertainty and fuel the current social elements by mishandling political inclinations. On Twitter, false news travels faster than true stories [1,2]. Not only does it spread faster, but fake news is also 70% more likely to be retweeted compared to real news. These days, fake news is not only limited to the social media networks but also have been extended to vehicular networks to misguide the travellers [3,5].

Fake news has been at the center of attention in recent years for having a broad adverse consequence on public events. A significant turning point of realization was the 2016 U.S. presidential election. It was believed that within the final three months leading up to the election, fake news favoring either of the two nominees was accepted and shared by more than 37 million times on Facebook [4,8]. Misinformation about COVID-19 also had additionally multiplied generally via web-based media in 2020, going from the hawking of so-called "fixes", for example, rinsing with lemon or salt-water and infusing yourself with bleach [6, 12]. The strategy film "Plandemic", appeared online on May fourth of 2020, gathering many perspectives and rapidly becoming quite possibly the most far and wide instances of Covid related deception [7,16]. For instance, the video advances risky well-being guidance, erroneously proposing that wearing a mask really "activates" the Covid. Fake news about the infection has likewise been effectively advanced by political elites, like US President Trump and Brazilian President Jair Bolsonaro. They erroneously guaranteed that hydroxychloroquine is "working in all spots", as a treatment against the infection [9]. In this manner, fake news and its viral dissemination have become a grave worry in the period of online media, where secrecy, client-created content, and geological distance may empower fake news sharing conduct. This makes the task of detecting fake news a crucial one.

These days with the growth of artificial intelligence, fake news generation is not only limited to human beings but is extended to intelligent machines. Authors at [10] discuss tools such as "Break your own news" to generate fake news. These tools require uploading photos, headlines, and article text to generate fake news and are very easy to use. The problem with this tool is that manually uploading all this information is time-consuming if we have to generate a large amount of fake news. With the success of deep learning models such as generative networks, different automated fake news generators have been developed. MaliGAN [11], LeakGAN [13], SeqGAN [14] are some of the available fake news generators based on generative models. The advantage of these approaches is that they are automated and can generate a large amount of fake news within a short amount of time.

Rumor and fake news identification strategies range from traditional supervised to unsupervised-based learning techniques. Starting approaches [15] attempted to distinguish fake news utilizing just linguistic features extracted from the content substance of reports. Existing work on fake news detection to a great extent depends on the utilization of broad recorded labeled datasets and supervised learning techniques that work over them. While supervised techniques are the normal initial move towards addressing any labeling task, they experience the ill effects of a few disadvantages. In the first place, supervised techniques require a significant amount of labeled data to learn meaningful models for the efficient detection of fake news. Getting manual explanations is labor-intensive as well as time-consuming, and crowdsourcing the information collection effort may involve quality weakening. Second, the usage of network and temporal features often lead to learning specific network and temporal behavioral patterns of individual users. Sometimes making the technique's effectiveness depends on the permanence of users' network positioning and their behavioral patterns. Third, methods that utilize content data would be influenced by topic drift other than being unable to sum up across natural languages.

Motivated by the automated fake news generation and its detection and observing the disadvantage of supervised approach in fake news detection, we have built an unsupervised-based approach for fake news generation as well as detection. We have used a combination of unsupervised algorithms such as autoencoder [30] and Generative Adversarial Network (GAN) to generate fake news and detect the machine-generated news with a very low amount of loss. Our developed approach first takes the unlabelled news dataset and generates high-dimensional feature vectors. These feature vectors are then fed into a generative model, i.e., LaTextGAN, and fake news is generated with the help of a generator. To detect the machine-generated news, we have simultaneously trained a discriminator in LaTextGAN. Given the use of unsupervised learning [31,32], our approach does not need data in its labeled form to either generate fake news or detect it which makes our approach very automated and reliable.

The rest of the paper is organized as follows: Sect. 2 discusses the most recent work available in the literature that discusses the use of various supervised and unsupervised learning methods for fake news detection. Section 3 gives a detailed overview of our proposed approach. Section 4 presents the results obtained from our approach. Finally, Sect. 5 concludes the paper.

2   Related Work

Existing work on fake news detection is mainly based on supervised methods. Authors at [17] evaluated the discriminative power of the previous features using several classic and state-of-the-art classifiers, including k-Nearest Neighbors (KNN), Naive Bayes (NB), Random Forests (RF), Support Vector Machine with RBF kernel (SVM), and XGBoost (XGB). The results correctly detected nearly all fake news in the data while misclassifying about 40% of true news. The data set used was also small. The limitation of this approach is that it is only using textual data to detect fabricated news from the tweets, and visual data is not being considered. The collective shortcoming of using supervised techniques is that these do not work efficiently on a new data set that was not used in the training process and might not yield true results. For such scenarios, unsupervised methods for fake news detection have been adopted recently. Authors at [18] model the distribution path of a report as a multivariate time series and identify fake news through propagation path classification with a combination of RNNs and CNNs. Authors at [19] inferred embeddings of social media user profiles and utilized an LSTM network over multiplication pathways to distinguish fake news. Some studies have shown that visual features like pictures or videos play a vital role in distinguishing fake news [20]. However, confirming the validity of multimedia content on online media has gotten less measure of examination. These features are still hand-crafted and can barely address complex appropriations of image content.

Deep neural networks have yielded immense success in learning image and textual representations. Authors at [21] have proposed a multimodal variational autoencoder that learns shared (visual + textual) representations to aid fake news detection. The model consists of three main components, an encoder, a decoder, and a fake news detector. However, the model is not extended to work on videos for fact-checking. Authors at [22] have tried to improve the accuracy of existing fake news detection using a Deep Convolutional Neural Network. In the proposed model, inputs are the word-embedding vectors. The model provides the same input word embedding vectors to all three parallel convolutional layers followed by a max-pooling layer. However, the approach is not extended to work with multi-labeled datasets and limits the analysis and broader generalization. At [23], authors have proposed an unsupervised fake news detection method based on autoencoder (UFNDA). They have extracted and fused Twitter's text content features, propagation features, image features, and user features in social networks, then used fake news as anomalous data and analyzed it with the proposed method UFNDA, and finally classified the test data by reconstructing errors. Although UFNDA performs well, there are still some improvements. In social networks, there are many features, such as the dissemination of news, and videos, that are important for fake news detection.

Recent developments in neural language models (LMs) have also raised concerns about their potential misuse for automatically spreading misinformation. In light of these concerns, several studies have proposed to detect machine-generated fake news by capturing their stylistic differences from human-written text. Authors at [24] have tried to show that stylometry is limited against machine-generated misinformation, whereas humans speak differently when trying to deceive, LMs generate stylistically consistent text, regardless of underlying motive. The model is effective in preventing impersonation, but limited in detecting LM-generated misinformation. Authors at [25] believe that developing robust verification techniques against generators like Grover (Generating aRticles by Only Viewing mEtadata Records) is critical. Grover can generate the rest of the article; humans find these generations to be more trustworthy than human-written disinformation. Counterintuitively, the best defense against Grover turns out to be Grover itself, demonstrating the importance of public release of strong generators. Another recent work [26] introduces the Neural-News dataset that contains both human and machine-generated articles with images and captions and proposes DIDAN (Detecting Cross-Modal Inconsistency to Defend Against Neural Fake News), an effective named entity-based model that serves as a good baseline for defending against neural fake news. Authors have empirically proved the importance of training with articles and nonmatching images and captions even when the adversarial generative models are known. However, the model does not evaluate generalization across different news sources, and a larger variety of neural generators.

3   Proposed Approach

Our proposed Autoencoder and GAN based architecture for fake news generation is shown in Fig. 1. The fake news generation is achieved in two parts. In the first part, an unlabelled news dataset is fed into the Long-Short Term Memory (LSTM) network based autoencoder. Autoencoders are artificial neural networks capable of learning dense representations of the input data, called latent representations or codings, without any supervision. Autoencoders also act as feature detectors, and they can be used for unsupervised pre-training of deep neural networks. In recent times, autoencoders have been very successful in generating the features from the unlabelled dataset. In our approach, encoder encodes the information from text and image into a latent vector and the decoder reconstructs back the original image and text from the latent vector. For faster and efficient training of autoencoders we used teacher forcing on our encoder and inference mode for our decoder.

In recent times, GAN has been very much successful in generating images; however, it's use in generation of text has been since text is in discrete form. Inspired by the work of [27], we have used LaTextGAN architecture. LaTextGAN is an architecture for the purpose of generating discrete text sequences. After generating a high-dimensional feature vector of news text from the autoencoder, the data is then fed into GANs and trained over it as shown in Fig. 1. The generator produces additional news sentences by generating more data points in the latent space which is later on used to decode the newly generated sentences. The generator proceeds to generate text which will try to fool the discriminator by reproducing similar news sentences. The discriminator then is used to distinguish between the machine-generated news and human-generated news fed as input. To tackle the vanishing gradient problem we used ResNet architecture for our Generative Adversarial Network. While using complex deep learning architecture like ResNets for generator and discriminator, traditional GAN training can prove to be less efficient. In such cases, WGAN can help the model to converge better [28]. Moreover, WGAN is more stable and less sensitive to choice of hyperparameter configurations.

In our approach, due to a very limited amount of training data, we decided to employ a skip-gram model to obtain the 200-dimensional word embeddings. The Skip-gram model is used to predict the context word from the given target word. The architecture of the skip-gram model is shown in Fig. 2. Let w(t) be the target word, and we are giving it as an input to the skip-gram model. The skip-gram model has one hidden layer which performs a matrix multiplication with the incoming target vector, i.e., w(t), and its weight matrix, i.e., w_h. The result of the dot product is then forwarded to the output layer, where the same phenomena happen by multiplying weight matrices with incoming vectors as in the hidden layer. The softmax activation is then applied to the output layer, which predicts different contextual output in the form of $w(t + 1)$, $w(t + 2)$, $w(t + 3)$.

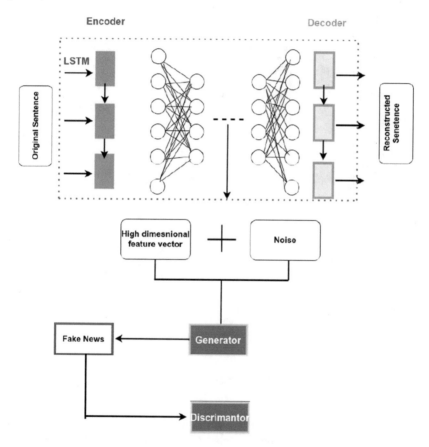

Fig. 1. Proposed autoencoder and GAN based architecture for fake news generation and detection

4   Experimentation and Result

Our experimentation was performed on a news summary dataset obtained from Kaggle [29]. The dataset contains 98,360 headlines of different news. The headlines were based on the content of the news article. Different news headlines in

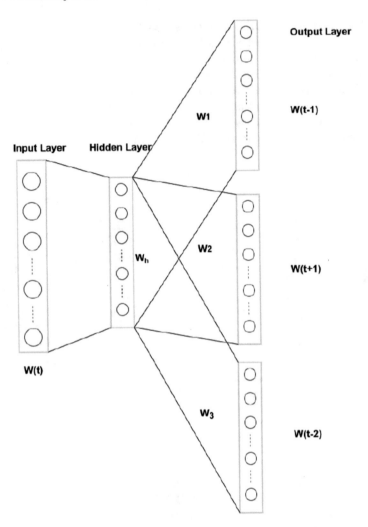

Fig. 2. Skip gram model

the dataset were extracted from the content available at The Hindu, The Indian Times, and The Guardian. The extracted news headlines were from February to August 2017.

Although the news dataset contains about 98,000 news headlines, due to the limitation of computational resources, we took only one-third of the article, i.e. 30,000. We used a randomize function to extract 30,000 from 98,000 news headlines randomly. Since the raw dataset contains several noises, to remove it, we performed several preprocessing steps. Since there were a lot of HTTP links, special characters, and emoticons present in the news headlines, we scanned through the headlines and removed such characters. Before feeding our data to NLTK for further analysis, tokenization of news headlines were necessary. For tokenization, we used Tweet Tokenizer. Tweet Tokenizer is different from the normal word tokenize as it keeps the hashtags intact. Additionally, Twitter Tokenizer also helps in extracting similar types of data in the news headlines that can be later on fed into unsupervised machine learning algorithms. Our news dataset contained a large number of numerical figures, which creates hindrances in training autoencoders. To solve this problem, inspired by the Natural Language Generation, we introduced unique tokens for numbers, i.e. "$< NUM >$" and for sentences, we generated further two tokens, i.e. "$< Start >$" and "$< End >$". Start represents the start of the sentence sequence whereas, the End represents the end of the sentence sequence. Furthermore, we removed all the words that did not appear as often compared to others by setting a specific frequency threshold. Removing these words helped us to remove the outliers in news headlines that can cause a higher reconstruction loss during the training of autoencoders. Since the news headlines were of variable sequence length, we applied padding. Padding helps to create an input sequence length of the same length. We further masked the input layer that will ignore the padded values. Masking helps to eliminate the impact of the padded word during training. A skip-gram model was then utilized to obtain 200-dimensional word embeddings. After implementing the skip-gram model, we obtained a vocab size of approximately 27,000, which helped in further training of autoencoders.

Once the dataset was preprocessed, it was necessary to generate the features from the unlabelled news dataset. To achieve that, we fed our preprocessed dataset into the autoencoder. We trained the autoencoder for 50 epochs, where we used Adam as an optimizer. In our autoencoder network, we adjusted the cell size of the LSTMs to 100 for the encoder and 600 for the decoders. The learning rate was set to $\alpha = 5^{10^{-4}}$ for autoencoders and $1^{10^{-4}}$ for WGAN. We plotted the loss obtained at each epoch during the training of the autoencoder, which is presented in Fig. 3. Figure 3 represents that we have achieved minimum reconstruction loss after 50 epochs. The minimum reconstruction loss makes our approach very much reliable in extracting features from news headlines.

Table 1 represents the generated output from the decoder. From the output sentences, we can conclude that although there is a dissimilarity between the input and output sentences; however, the overall sentiment of the news is being preserved. Authors at [13] suggest that it is very common in unsupervised-based text generation algorithms to have some inconsistency between the input and output texts.

Once the dataset with the features was obtained from the autoencoders, we fed the labeled dataset into GAN and trained GAN over 100 epochs. Figure 4 represents the loss graph obtained for the generator and discriminator. During the training of GAN, the generator and discriminator tries to beat each other, resulting in the increase of variance between their losses. In the training of GAN, the increase in variance suggests that the training of GAN is getting better as we move towards larger epochs. The same kind of scenario can be seen in Fig. 4. During the initial epochs, the loss of the generator and discriminator are gen-

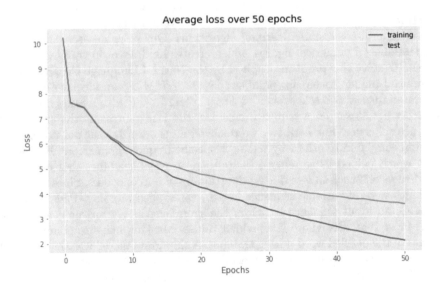

Fig. 3. Reconstruction loss graph of news headline

Table 1. Output sentence generated by autoencoder.

Input	Output
< *Start* > the supreme court has refused to stay the release of the punjabi film nanak shah fakir citing freedom of expression guaranteed by the constitution . this comes after shiromani gurdwara parbandhak committee banned the films release in punjab . the film , which is based on the life of guru nanak dev , was released all across india on april < *num* > , except in punjab . < *End* >	< *Start* > the supreme court has directed to entertain the national of the film cultural for dev zafar aur the struggle the . by the national . the comes after the akali parbandhak ((the national release on the and the film is which is scheduled on the life of the nanak dev , has released on the the in november < *num* > the < *End* >

erally very high, but as the training proceeds, the loss for both generator and discriminator is decreasing along with a significant amount of variance between them. This suggests that our model is converging as the epochs proceed. Additionally, the loss of the discriminator can be seen as very low, which suggests that the machine-generated fake news is being identified very correctly from the original news.

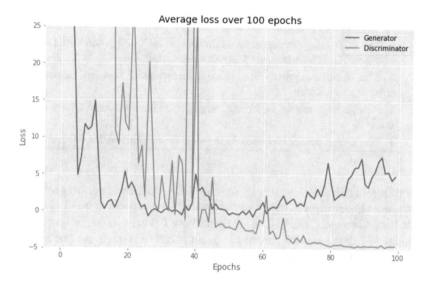

Fig. 4. Loss of generator and discriminator during training

Table 2 represents the fake news headlines generated by GAN. During the generation of fake news, it's very important to change the sentiment of the news that may deceive the false information to humans but at the same time, retaining the original information in the fake news is equally important. The words present in the generated fake news in Table 2 are the subset of the words of original news; however, the fake news is generated in such a way that it can deceive false information to the people. To compute the similarity between the machine-generated text and original text fed to the model, we computed the BLUE-4 score, which is described in Eq. 1. Computing similarity gives an idea about preserved originality between the original and machine-generated news.

Here, n corresponds to the n-gram length, pn is the modified n-gram precision scores and the entries in p correspond to the geometric averages of the modified n-gram precisions, and BP is the brevity penalty.

We calculated the scores with uniform values of weights, i.e., 0.25, for uni-, bi-, tri-, and four grams. Initially, we carried out our evaluation on 500 generated novel examples of fake news, which scored a BLUE-4 score of 0.4312. Further, we increased the number of input samples from 30,000 to 60,000 for autoencoders and calculated the BLUE-4 score on the output generated text. The increment

Table 2. Fake news generated from GAN.

Sample examples of generated text
1. The Supreme Court has criticised Sweden Manager of Britain
2. Union of Kerala Commission has announced a letter to the Tamil Nadu CM Jayalalitha
3. PM on Sunday asked government employees that developed a lie of Mahatma Gandhi Gandhi has been criticised by the university administration
4. The BJP led government has admitted that Tamil Nadu CM Lee was jailed in the memorial of Mysore memorial
5. The defence ministry has ordered an arrest of two Sri Lankan government officers
6. Punjab government on Sunday urged to remove external affairs of funds
7. The crown prince has sued Malaysia's prince for the Taj Mahal

resulted in a bigger training corpus for the GAN. At this time, we got a score of 0.5032 since the increase in input samples also increased the number of generated output text. Generally, a BLUE-4 score of above 0.3 is regarded as good. Further increment of input samples can result in an increase in output text and higher score; however, due to the limitation of computational infrastructures, we limited our input samples to only 60,000. Although a higher BLUE-4 score indicates a reasonable amount of similarity between the original text and machine-generated text, the developed discriminator model is able to distinguish between the two with very low loss that makes our developed approach for fake news detection very reliable.

5   Conclusion and Future Work

In this paper, we have developed an unsupervised-based approach to generate fake news using autoencoder and GAN. We have further described the use of discriminators in detecting machine-generated fake news. Additionally, we have also evaluated the similarity of generated fake news using the BLUE-4 score that shows the extent to which fake news can be generated, which is similar to the real news but can disseminate false information to the people. Our approach on generating fake news is not as syntactically correct as it should be. Syntactically incorrect fake news becomes very hard to understand for an audience with lower linguistic knowledge. These days with the development of several deep learning algorithms, sophisticated fake news can be generated to the large extent that may escape current fake news detection systems and is very much syntactically correct. In the future, we are planning to build a fake news generator which is very much syntactically sophisticated as well as an architecture to defend such syntactically correct fake news.

Acknowledegment. This research was partially funded by the Spanish Government Ministry of Science and Innovation through the AVisSA project grant number (PID2020-118345RB-I00).

References

1. Sahoo, S.R., et al.: Multiple features based approach for automatic fake news detection on social networks using deep learning. Appl. Soft Comput. **100**, 106983 (2021)
2. Tembhurne, J.V., Almin, M.M., Diwan, T.: Mc-DNN: fake news detection using multi-channel deep neural networks. Int. J. Semant. Web Inf. Syst. (IJSWIS) **18**(1), 1–20 (2022)
3. Gaurav, A., et al.: A novel approach for fake news detection in vehicular ad-hoc network (VANET). In: International Conference on Computational Data and Social Networks. Springer, Cham (2020)
4. Allcott, H., Gentzkow, M.: Social media and fake news in the 2016 election. J. Econ. Perspect. **31**, 211–236 (2017)
5. Srivastava, A.M., Rotte, P.A., Jain, A., Prakash, S.: Handling data scarcity through data augmentation in training of deep neural networks for 3D data processing. Int. J. Semant. Web Inf. Syst. (IJSWIS) **18**(1), 1–16 (2022)
6. van Der Linden, S., Roozenbeek, J., Compton, J.: Inoculating against fake news about COVID-19. Front. Psychol. **11**, 2928 (2020)
7. Cook, J., van der Linden, S., Lewandowsky, S., Ecker, U.K.H.: Coronavirus, 'Plandemic' and the seven traits of conspiratorial thinking. The Conversation (2020). Available online at: https://theconversation.com/coronavirus-plandemic-and-the-seven-traits-of-conspiratorial-thinking-138483. Accessed 15 May 2021
8. Sahoo, S.H., et al.: Hybrid approach for detection of malicious profiles in twitter. Comput. Electr. Eng. **76**, 65–81 (2019). ISSN 0045-7906. https://doi.org/10.1016/j.compeleceng.2019.03.003
9. Constine, J.: Facebook deletes Brazil president's coronavirus misinfo post. TechCrunch (2020). Available online at: https://techcrunch.com/2020/03/30/facebook-removes-bolsonaro-video/. Accessed 31 Mar 2021
10. Allcott, H., Gentzkow, M.: Social media and fake news in the 2016 election. Technical report, National Bureau of Economic Research (2017)
11. Che, T., Li, Y., Zhang, R., Hjelm, R.D., Li, W., Song, Y., Bengio, Y.: Maximum-likelihood augmented discrete generative adversarial networks. arXiv preprint arXiv:1702.07983 (2017)
12. Gupta, S., et al.: Detection, avoidance, and attack pattern mechanisms in modern web application vulnerabilities: present and future challenges. Int. J. Cloud Appl. Comput. (IJCAC) **7**, 1–43 (2017). https://doi.org/10.4018/IJCAC.2017070101
13. Guo, J., Lu, S., Cai, H., Zhang, W., Yu, Y., Wang, J.: Long text generation via adversarial training with leaked information. In: Thirty-Second AAAI Conference on Artificial Intelligence (2018)
14. Yu, L., Zhang, W., Wang, J., Yu, Y.: SeqGAN: sequence generative adversarial nets with policy gradient. In: Thirty-First AAAI Conference on Artificial Intelligence (2017)
15. Castillo, C., Mendoza, M., Poblete, B.: Information credibility on twitter. In: Proceedings of the 20th International Conference on World Wide Web, pp. 675–684. ACM (2011)

16. Agrawal, P.K., et al.: Estimating strength of a DDoS attack in real time using ANN based scheme. In: International Conference on Information Processing, Aug 2011, pp. 301–310. Springer, Berlin, Heidelberg (2011)
17. Reis, J.C.S., et al.: Supervised learning for fake news detection. IEEE Intell. Syst. **34**(2), 76–81 (2019)
18. Liu, Y., Brook Wu, Y.F.: Early detection of fake news on social media through propagation path classification with recurrent and convolutional networks. In: AAAI (2018)
19. Wu, L., Liu, H.: Tracing fake-news footprints: characterizing social media messages by how they propagate. In: Proceedings of the Eleventh ACM International Conference on Web Search and Data Mining, pp. 637–645. ACM (2018)
20. Wu, K., Yang, S., Zhu, K.Q.: False rumors detection on Sina Weibo by propagation structures. In: 2015 IEEE 31st International Conference on Data Engineering (ICDE), pp. 651–662. IEEE (2015)
21. Khattar, D., et al.: MVAE: multimodal variational autoencoder for fake news detection. In: The World Wide Web Conference (2019)
22. Kaliyar, R.K., et al.: FNDNet—a deep convolutional neural network for fake news detection. Cogn. Syst. Res. **61**, 32–44 (2020)
23. Li, D., et al.: Unsupervised fake news detection based on autoencoder. IEEE Access **9**, 29356–29365 (2021)
24. Schuster, T., et al.: The limitations of stylometry for detecting machine-generated fake news. Comput. Linguist. **46**(2), 499–510 (2020)
25. Zellers, R., et al.: Defending against neural fake news. arXiv preprint arXiv:1905.12616 (2019)
26. Tan, R., Plummer, B.A., Saenko, K.: Detecting cross-modal inconsistency to defend against neural fake news. arXiv preprint arXiv:2009.07698 (2020)
27. Donahue, D., Rumshisky, A.: Adversarial text generation without reinforcement learning. arXiv preprint arXiv:1810.06640 (2018)
28. Gulrajani, I., et al.: Improved training of Wasserstein GANs. arXiv preprint arXiv:1704.00028 (2017)
29. https://www.kaggle.com/sunysai12345/news-summary. Accessed 10 July 2021
30. D'Angelo, G., Palmieri, F.: Network traffic classification using deep convolutional recurrent autoencoder neural networks for spatial-temporal features extraction. J. Netw. Comput. Appl. **173**, 102890 (2021)
31. Garcia-Penalvo, F.J., et al.: Application of artificial intelligence algorithms within the medical context for non-specialized users: the CARTIER-IA platform. Int. J. Interact. Multimed. Artif. Intell. **6**(6), 46–53 (2021). https://doi.org/10.9781/ijimai.2021.05.005
32. Garcia-Penalvo, F.J., et al.: KoopaML: a graphical platform for building machine learning pipelines adapted to health professionals. Int. J. Interact. Multimed. Artif. Intell. (in press)

Metaverse: A New Tool for Real-Time Monitoring of Dynamic Circumstances in the Physical Production System

Pu Li[1], Ali Ali Azadi[2], Akshat Gaurav[3,4(✉)], Alicia García-Holgado[2], Yin-Chun Fung[5], and Zijian Bai[1]

[1] Software Engineering College, Zhengzhou University of Light Industry, Zhengzhou, China
lipu@zzuli.edu.cn, 1987002@email.zzuli.edu.cn
[2] University of Salamanca, Salamanca, Spain
ali.azadi@usal.es, aliciagh@usal.es
[3] Ronin Institute, Montclair, USA
akshat.gaurav@ronininstitute.org
[4] Lebanese American University, Beirut 1102, Lebanon
[5] School of Science and Technology, Hong Kong Metropolitan University, Hong Kong, China
ycfung@study.hkmu.edu.hk

Abstract. The notion of metaverses facilitates online communities where individuals may connect and socialise, highlighting the growing importance of online interactions in people's daily lives. The digital assets and values that the metaverse offers are useful in many fields since it is a simulation of the actual world with added enhancements.In this paper, we analyze the concept of metaverse in detail. Along with this, we analyze the evaluation of metaverse technology over the years. The results show that metaverse have a significant impact on the physical production system. Metaverse represent a major tool for real-time monitoring of dynamic circumstances in the physical environment. In addition to this, it is also evident that the in metaverse environment real-time monitoring of dynamic circumstances in the physical asset is possible.

Keywords: Metaverse · Digital avatar · Digital twin

1 Introduction

Blending real and virtual elements, the metaverse is a visual space [27]. The Metaverse is a collective online environment where users may navigate using an avatar they've created and engage in activities like socialising, doing business, and exploring new worlds with individuals who have similar interests [12,29]. Author Neal Stephenson originally used the term "metaverse" in his book Snow Crash, published in 1992. The story describes a future in which individuals live in a virtual reality environment and compete for social status by manipulating

© The Author(s), under exclusive license to Springer Nature Switzerland AG 2023
N. Nedjah et al. (Eds.): ICSPN 2021, LNNS 599, pp. 397–405, 2023.
https://doi.org/10.1007/978-3-031-22018-0_38

digital avatars. The term "metaverse" first emerged in print in the 1980s, and it has since appeared in a number of media formats. The term "metaverse" still hasn't been fully defined, although it's generally accepted as meaning "a parallel virtual universe to the actual world."

Metaverse is a virtual world that combines the real and the virtual, and it has a wide range of settings, non-player characters (NPCs), and avatars. The term "scene" is often used to describe several types of online locations. NPCs are non-interactable game elements that contribute to the game's atmosphere by acting independently of the player. The term "avatar" is often used to describe a player's digital persona while interacting with other players or computer agents in a virtual world [11,27]. Graphical approaches, such as the three-dimensional (3D) modelling of environments, non-player characters (NPCs), and player characters, provide the basis of the visual architecture of the metaverse, which is an integrated universe merging the real and virtual worlds. Virtual reality, augmented reality, and extended reality are the building blocks of the metaverse. One of the main benefits of virtual and augmented reality is the ability to create a convincing illusion of realism in computer-generated settings of metaverse [17,24]. The phrase "mixed reality," which is sometimes used interchangeably with "extended reality," describes one of the most cutting-edge trends in the field of information systems, processing, and management, and encompasses all VR/AR technologies. There has been a rise in the amount of literature on XR, namely on Augmented Reality (AR) and Virtual Reality (VR), demonstrating the far-reaching effects of this information systems breakthrough in many other domains [10,26].

There has been a lot of interest in the metaverse lately, both in academia and in business, but there are still many basic issues that need to be fixed before the metaverse can become a reality [3]. Even though various phrases and acronyms have evolved that seem to indicate a virtual world, there is still no universally agreed-upon definition for what exactly a virtual world is [18].

2   Research Methodology

A thorough literature review was conducted to chart the evolution of metaverse technology. The following procedures were followed while writing this paper: Choosing appropriate databases, selecting appropriate Queries, reading appropriate Research Articles, and assessing appropriate Studies were all part of the process. To reduce the risk of missing important articles during the search process, this research uses a subset of Scopus rather than the whole database since the Web-of-Science database indexes fewer journals. A limited number of publications were rejected from consideration since they did not fit the research focus. As an example, we do research only in computer science and not in any other discipline.

2.1   Data Source

Using the Scopus bibliographic database, the data was gathered in August 2022. The following two keywords were included in the search strategy to answer the research question.

– metaverse.

2.2   Search Query Selection

In order to get the information from Scopus database, we used the following query at different screening levels:

– **Stage 1**: included only papers written in English
  "(TITLE-ABS-KEY("metaverse")) AND (LIMIT-TO (LANGUAGE, "English"))"
– **Stage 2**: Include only papers from "computer science" domain,
  "(TITLE-ABS-KEY("metaverse")) AND (LIMIT-TO (LANGUAGE , "English")) AND (LIMIT-TO (SUBJAREA , "COMP"))) "
– **Stage 3** : Include only "Articles" and "conference papers",
  (TITLE-ABS-KEY("metaverse")) AND (LIMIT-TO (LANGUAGE , "English")) AND (LIMIT-TO (SUBJAREA , "COMP")) AND (LIMIT-TO (DOCTYPE , "cp") OR LIMIT-TO (DOCTYPE , "ar"))

3   Evaluation of Metaverse

In order to get information about the development of digital twin sector, we categorized this section into the following subsections.

3.1   Analysis Document Distribution

In this subsection, we give details about the scientific distribution of the research papers. To obtain the information of the article, we find highly cited articles related to the development of the digital twin sector. Figure 1a presents the number of papers related to digital twin over the years. From Fig. 1a it is clear that after 2017 there is an exponential growth in papers in the field of digital twin. Hence, we can say that this topic is still in its development phase.

3.2   Analysis of Country Distribution

The distribution of researchers by nations is also a significant and beneficial component in analyzing the development in the field of digital twin. This metric indicates the effectiveness of researchers in a country. Figure 1d represents the distribution of countries according to the total number of publications. From Fig. 1d it is clear that *China, Germany, United States, Russia, and UK* researchers are actively working in the field of digital twin.

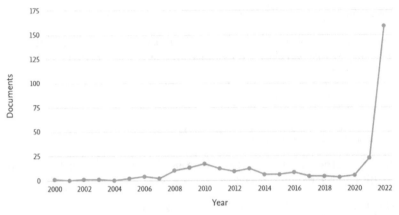

(a) Document Distribution

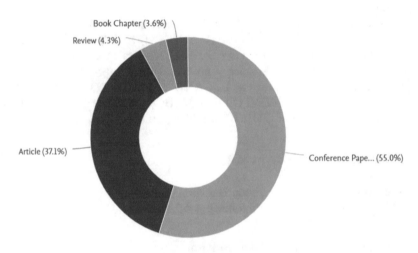

(b) Different Document Types

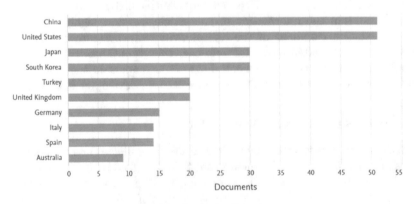

(c) Most Productive Countries

Fig. 1. Analysis of literature

(d) Important Keywords

Fig. 1. (*continued*)

3.3   Trending Research Topics

We analyze the current research directions and trends in this subsection. To begin the study, we examine the distribution of the keywords of the paper. Keywords are a representation of the paper's principal subject. Figure 1d represents the frequency of keywords in our research database; as the frequency of the keyword increases, its size in Fig. 1d increases.

4   Theoretical and Practical Implications

4.1   Limitation of Digital Avatars

Specifically, people didn't think their virtual avatars looked like them when their actual face appearance was used, but they did when their facial behaviours were used. Given this, it's possible that designs won't be limited to resembling humans in any way. In the future, researchers may test whether or not people can recognise resemblance between animal-inspired avatars and real people when the avatars mimic real people's facial expressions.

Similarity and attraction were shown to have a linear, positive connection, according to the study authors. However, there was no discernible difference in attractiveness between the different virtual avatars. Possible explanation: Just one exposure is needed for interaction.

However, when the data is broken down into categories, it's clear that routine expressions had a significant impact on the aggregate of likeability and similarity ratings. The findings imply that extended engagement with an individual avatar might create attraction due to the interplay between perceived resemblance and like.

Moreover, the personalised virtual avatars did not significantly affect the level of immersion [22].

Table 1. Highely cited papers

Paper	DOI	Total Citations	TC per Year	Normalized TC
DAVIS A, 2009, J ASSOC INF SYST [5]	10.17705/1jais.00183	199	14.2143	9.5815
DIONISIO JDN, 2013, ACM COMPUT SURV [6]	10.1145/2480741.2480751	92	9.2	8.4923
KUMAR S, 2008, COMPUTER [16]	10.1109/MC.2008.398	82	5.4667	6.7769
CHOI H-S, 2017, INT J INF MANAGE [4]	10.1016/j.ijinfomgt.2016.04.017	49	8.1667	3.6981
BOURLAKIS M, 2009, ELECTRON COMMER RES [2]	10.1007/s10660-009-9030-8	40	2.8571	1.9259
PARK S-M, 2022, IEEE ACCESS [21]	10.1109/ACCESS.2021.3140175	34	34	31.2486
DUAN H, 2021, MM - PROC ACM INT CONF MULTIMED [7]	10.1145/3474085.3479238	30	15	8.4146
JAYNES C, 2003, PROC WORKSHOP VIRTUAL ENVIRON, EGVE [14]	10.1145/769953-769967	24	1.2	1
GARRIDO-IÑIGO P, 2015, INTERACT LEARN ENVIRON [8]	10.1080/10494820.2013.788034	21	2.625	2.2105
OLIVER IA, 2010, MMSYS - PROC ACM SIGMM CONF MULTIMEDIA SYST [20]	10.1145/1730836.1730873	21	1.6154	5.1
NEVELSTEEN KJL, 2018, COMPUT ANIM VIRTUAL WORLDS [19]	10.1002/cav.1752	19	3.8	2.7143
HASSOUNEH D, 2015, J ELECTR COMMER RES [13]		19	2.375	2
GONZÁLEZ CRESPO R, 2013, EXPERT SYS APPL [9]	10.1016/j.eswa.2013.06.054	16	1.6	1.4769
SIYAEV A, 2021, SENSORS [23]	10.3390/s21062066	15	7.5	4.2073
ZIMMERMANN R, 2008, [28]	10.1145/1459359.1459400	15	1	1.2397
XI N, 2022, INF SYST FRONT [26]	10.1007/s10796-022-10244-x	14	14	12.8671
BARRY DM, 2015, PROCEDIA COMPUT SCI [1]	10.1016/j.procs.2015.08.181	14	1.75	1.4737
OWENS D, 2009, IT PROF [5]	10.1109/MITP.2009.35	14	1	0.6741
ULLRICH S, 2008, PROC INT CONF ADV COMPUT ENTERTAIN TECHNOL, ACE [25]	10.1145/1501750.1501781	14	0.9333	1.157
KANEMATSU H, 2014, PROCEDIA COMPUT SCI [15]	10.1016/j.procs.2014.08.224	13	1.4444	4.875

The Metaverse is an amalgamation of media that necessitates a number of methods for efficient and high-quality content production, generation, transmission, and visualisation. In addition, several fields, including networking, AI, VR hardware, etc., will need to collaborate to deliver a complete answer [3].

4.2 Documents Distribution

In this part, we discuss the academic dissemination of the publications. Not all studies that make it into print are equally significant or informative. We thus locate highly referenced publications about the growth of mataverse technologies to glean the article's content. Table 1 provides a breakdown of these types of articles. The citation counts, average citations, and normalised citations for each work are shown in Table 1.

5   Conclusion

It is possible to think of the Metaverse as a hypothetical future version of the Internet that provides users with a persistent online virtual environment in which they may work, play, and engage socially in an environment that seems very real to them. At present, metaverse is still in its infancy, and as such, there is no established structure for its visual creation and exploration. Furthermore, there has been no effort to carefully synthesise the technological foundation for its entire visual development and investigation, nor have graphics, interaction, and visualisation been investigated outside of the metaverse's setting. The investigation into the present condition of their respective disciplines is the closest to our own. In this context, this paper presents a general overview of the metaverse sector. We analyze the development of metavers sector along with its limitations and challenges.

Acknowledgment. The works described in this paper are supported by The Program for Young Key Teachers of Henan Province under Grant No. 2021GGJS095; The Project of collaborative innovation in Zhengzhou under Grant No. 2021ZDPY0208; The National Natural Science Foundation of China under Grant No. 61802352.

References

1. Barry, D.M., Ogawa, N., Dharmawansa, A., Kanematsu, H., Fukumura, Y., Shirai, T., Yajima, K., Kobayashi, T.: Evaluation for students' learning manner using eye blinking system in metaverse. Procedia Comput. Sci. **60**, 1195–1204 (2015)
2. Bourlakis, M., Papagiannidis, S., Li, F.: Retail spatial evolution: paving the way from traditional to metaverse retailing. Electronic Commer. Res. **9**(1), 135–148 (2009)
3. Chen, S.C.: Multimedia research toward the metaverse. IEEE MultiMedia **29**(1), 125–127 (2022)

4. Choi, H.S., Kim, S.H.: A content service deployment plan for metaverse museum exhibitions- centering on the combination of beacons and HMDs. Int. J. Inf. Manage. **37**(1), 1519–1527 (2017). https://doi.org/10.1016/j.ijinfomgt.2016.04.017

5. Davis, A., Murphy, J., Owens, D., Khazanchi, D., Zigurs, I.: Avatars, people, and virtual worlds: foundations for research in metaverses. J. Assoc. Inf. Syst. **10**(2), 90–117 (2009). https://doi.org/10.17705/1jais.00183

6. Dionisio, J., Burns Iii, W., Gilbert, R.: 3d virtual worlds and the metaverse: current status and future possibilities. ACM Comput. Surv. **45**(3) (2013). https://doi.org/10.1145/2480741.2480751

7. Duan, H., Li, J., Fan, S., Lin, Z., Wu, X., Cai, W.: Metaverse for social good: a university campus prototype, pp. 153–161 (2021). https://doi.org/10.1145/3474085.3479238

8. Garrido-Inigo, P., Rodrguez-Moreno, F.: The reality of virtual worlds: pros and cons of their application to foreign language teaching. Interact. Learn. Environ. **23**(4), 453–470 (2015). https://doi.org/10.1080/10494820.2013.788034

9. Gonzalez Crespo, R., Escobar, R., Joyanes Aguilar, L., Velazco, S., Castillo Sanz, A.: Use of ARIMA mathematical analysis to model the implementation of expert system courses by means of free software Opensim and Sloodle platforms in virtual university campuses. Expert Syst. Appl. **40**(18), 7381–7390 (2013). https://doi.org/10.1016/j.eswa.2013.06.054

10. Gupta, B.B., Badve, O.P.: GARCH and ANN-based DDos detection and filtering in cloud computing environment. Int. J. Embed. Syst. **9**(5), 391–400 (2017)

11. Gupta, B.B., Misra, M., Joshi, R.C.: An ISP level solution to combat DDos attacks using combined statistical based approach. arXiv preprint arXiv:1203.2400 (2012)

12. Gupta, S., et al.: Detection, avoidance, and attack pattern mechanisms in modern web application vulnerabilities: present and future challenges. Int. J. Cloud Appl. Comput. (IJCAC) **7**(3), 1–43 (2017)

13. Hassouneh, D., Brengman, M.: Retailing in social virtual worlds: developing a typology of virtual store atmospherics. J. Electron. Commer. Res. **16**(3), 218–241 (2015)

14. Jaynes, C., Seales, W., Calvert, K., Fei, Z., Griffioen, J.: The metaverse—a networked collection of inexpensive, self-configuring, immersive environments, pp. 115–124 (2003). https://doi.org/10.1145/769953-769967

15. Kanematsu, H., Kobayashi, T., Barry, D.M., Fukumura, Y., Dharmawansa, A., Ogawa, N.: Virtual stem class for nuclear safety education in metaverse. Procedia Comput. Sci. **35**, 1255–1261 (2014)

16. Kumar, S., Chhugani, J., Kim, C., Kim, D., Nguyen, A., Dubey, P., Bienia, C., Kim, Y.: Second life and the new generation of virtual worlds. Computer **41**(9), 46–53 (2008). https://doi.org/10.1109/MC.2008.398

17. Lee, H., Woo, D., Yu, S.: Virtual reality metaverse system supplementing remote education methods: based on aircraft maintenance simulation. Appl. Sci. **12**(5), 2667 (2022)

18. Nevelsteen, K.J.: Virtual world, defined from a technological perspective and applied to video games, mixed reality, and the metaverse. Comput. Animation Virtual Worlds **29**(1), e1752 (2018)

19. Nevelsteen, K.: Virtual world, defined from a technological perspective and applied to video games, mixed reality, and the metaverse. Comput. Animation Virtual Worlds **29**(1) (2018). https://doi.org/10.1002/cav.1752

20. Oliver, I., Miller, A., Allison, C.: Virtual worlds, real traffic: interaction and adaptation, pp. 305–316 (2010). https://doi.org/10.1145/1730836.1730873

21. Park, S.M., Kim, Y.G.: A metaverse: taxonomy, components, applications, and open challenges. IEEE Access **10**, 4209–4251 (2022). https://doi.org/10.1109/ACCESS.2021.3140175

22. Park, S., Kim, S.P., Whang, M.: Individual's social perception of virtual avatars embodied with their habitual facial expressions and facial appearance. Sensors **21**(17), 5986 (2021)

23. Siyaev, A., Jo, G.S.: Towards aircraft maintenance metaverse using speech interactions with virtual objects in mixed reality. Sensors **21**(6), 1–21 (2021). https://doi.org/10.3390/s21062066

24. Tewari, A., et al.: Secure timestamp-based mutual authentication protocol for IoT devices using RFID tags. Int. J. Semant. Web Inf. Syst. (IJSWIS) **16**(3), 20–34 (2020)

25. Ullrich, S., Prendinger, H., Ishizuka, M.: Mpml3d: agent authoring language for virtual worlds. In: Proceedings of the 2008 International Conference on Advances in Computer Entertainment Technology, pp. 134–137 (2008)

26. Xi, N., Chen, J., Gama, F., Riar, M., Hamari, J.: The challenges of entering the metaverse: an experiment on the effect of extended reality on workload. Inf. Syst. Front. 1–22 (2022)

27. Zhao, Y., Jiang, J., Chen, Y., Liu, R., Yang, Y., Xue, X., Chen, S.: Metaverse: perspectives from graphics, interactions and visualization. Visual Inf. (2022)

28. Zimmermann, R., Liang, K.: Spatialized audio streaming for networked virtual environments, pp. 299–308 (2008). https://doi.org/10.1145/1459359.1459400

29. Zyda, M.: Let's rename everything "the metaverse." Computer **55**(3), 124–129 (2022)

Security of Android Banking Mobile Apps: Challenges and Opportunities

Akash Sharma[1]([✉]) [ID], Sunil K. Singh[2] [ID], Sudhakar Kumar[2] [ID],
Anureet Chhabra[2] [ID], and Saksham Gupta[2] [ID]

[1] Chandigarh College of Engineering and Technology, Panjab University,
Chandigarh, India
sharmaaakash080@gmail.com
[2] Department of CSE, Chandigarh College of Engineering and Technology,
Panjab University, Chandigarh, India
sksingh@ccet.ac.in, sudhakar@ccet.ac.in

Abstract. As of 2021, there are 14.91 billion mobile devices, which shows that cell phones continues to be a vital part of present day human's existence and the Android Operating system is the most used OS in these cell phones. Even though there are alot of Apps available to use but Application utilization and cell phone penetration are as yet developing at a consistent rate, with practically no indications of dialing back soon. Apps on our mobile devices make our lives easier by providing multiple features such quick payments and more. Nowadays, there are a lot of banks that offer mobile banking services using apps and increase in online payments using smartphones has been observed. Due to the massive and dynamic nature of mobile banking apps, they pose a risk of security breaches. Because vulnerabilities can lead to huge financial losses, we have presented a comprehensive empirical studies of the security risks of global banking apps in order to provide useful insights and improve security. These vulnerabilities may lead to serious financial losses due to data-related weaknesses in banking apps. In this paper, we looked at mobile banking app vulnerabilities as well as a security difficulties with mobile internet banking applications, and then studied a few security strategies to address the relevant security issues.

Keywords: Mobile banking application · Vulnerability · Authentication · Weakness · Threats

1 Introduction

Android gadgets have acquired notoriety after some time because of the accessibility of essentially a wide range of utilizations, including games, banking/monetary applications, music and video web based applications, and shopping applications like Amazon, to specify a couple of instances of generally utilized applications with a large number of downloads. Android additionally has

© The Author(s), under exclusive license to Springer Nature Switzerland AG 2023
N. Nedjah et al. (Eds.): ICSPN 2021, LNNS 599, pp. 406–416, 2023.
https://doi.org/10.1007/978-3-031-22018-0_39

various valuable highlights like calling, messaging, and area based administrations. As indicated by Google, there are now 2.8 billion dynamic Android clients, with 2.78 million applications accessible on the Google Play Store in 2021, up from 1 million out of 2013. Android has a portion of the overall industry of over 75 % [1,2]. Android devices, on the other hand, are plagued with malware and offer a number of privacy and security risks, some of which are particularly easy to attack, as we'll show below. These malicious programs can gather client login passwords, private data, digital payments, client areas and IP addresses, and other possibly horrendous ways of behaving [4] (Fig. 1).

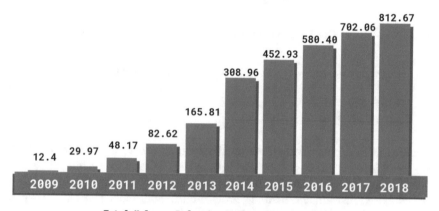

Total Malware Infection Growth Rate (In Millions)

Fig. 1. Graph on malware infection growth rate in each year

Mobile banking refers to financial services and transactions that are carried out via mobile devices. Data transfer will be required for mobile banking security, which is crucial for safeguarding users' data and preventing hackers from accessing and stealing it. Banking apps are among the most secure and data-sensitive apps on the market. Cashless mobile payment has had a significant impact on traditional financial services, beginning with the first ATM and concluding with e-banking. Authentication is also important since it guarantees that only authorized people have access to the data. In order to obtain data quickly, it is also necessary to avoid cumbersome authorization [5]. Users frequently believe that banking applications provide secure transactions and an easy-to-use interface because they believe that all interactions between local banking apps and distant bank servers are encrypted (e.g., over HTTPS). Regrettably, this assumption isn't always correct. After studying a number of real-world banking apps, we discovered new sorts of defects that are difficult to spot using standard industrial and open-source technologies [7] (Fig. 2).

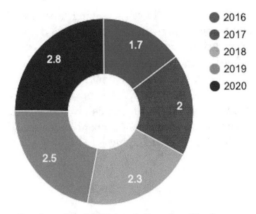

Number of Android apps (values in million)

Fig. 2. Pie chart representing android apps growth in each year

2   Vulnerability Analysis

Malware authors focus their efforts on the Android ecosystem, introducing malicious code into consumers' apps and attempting to steal everything they want. They may swiftly get clients' login passwords and information, for example, via internet-based installment/banking programmes, and, more tragically, they can use these clients to disseminate malware to other clients, completing an incredible chain. Piggybacking is the process of repackaging an Android application after altering its code and content to include a malicious payload that may compromise the overall security of the Android device [6,8]. These payloads might be anything from simple adware to complex malware. According to some research, the majority of malware on Android is merely repackaged variations of legitimate applications, although other applications contain more embedded code as compared to the initial application. The functionality, components, and user interface of these piggybacked applications are indistinguishable from the original programming, yet they include dangerous code [3,10].

According to Ariel Sanchez, programmers may exploit security flaws in mobile financial applications to steal money from customers. According to Sanchez, "70% of the applications did not use alternative validation procedures, for example, multifaceted confirmation, which might further expand the security perspective and kill specific risk that could trigger data assaults." Mobile banking apps, which are broadly utilised by clients on their cell phones and tablets, have sparked concerns about exchange security all over the world. Thus, in previous years, the European Central Bank urged the introduction of a slew of new portable instalments security requirements that installment specialist organisations (PSPs, for example, banks and versatile instalment arrangement suppliers' (MPSPs) should adhere to [9,11,13].

2.1   The Risk Involved in Information Transferring Over the Network

There may be various different security offers. One of the most genuine security blemishes in versatile banking is moving a client's information starting with one portable organization then onto the next using any operating system like linux [12,14]. All encoded information ought to be unscrambled in this example for the wellbeing of effortlessness. Whenever cell phones demand pages from an organization server in versatile banking, a fundamental development is made to exhibit that the solicitations are coming from the Wireless Transport Security Layer (WTSL) standard. At a WAP (Wireless Application Protocol) section, the requests are decoded. Whenever the solicitations have been decoded, they are shipped off the norm [15,17].

Encryption technology is utilized to secure data in today's mobile environment. According to Ghosh, the structure has lately become vulnerable to assaults. A component of the existing encryption computations has been cracked by software developers. As a result, development is not without flaws. However, there is no all-encompassing managerial design in place to ensure and assess security risks. As innovation improves and more people get confused, no one can guarantee client security, thus clients should be cautious while utilizing small financial innovation[16,18,20].

2.2   Two Factor Authentication Problem

Two-factor approval isn't frequently utilized while doing sensitive exchanges on cell phones. As per reviews, clients ordinarily use static passwords rather than two-factor confirmation while doing essential web exchanges on their cell phones [19]. Passwords can be speculated, neglected, got on paper and taken, or tuned in on while involving static passwords for check [21]. Customary passwords and PINs give a lower level of safety than two-factor affirmation, which might be vital for delicate transactions. Two-factor approval expects clients to approve their character utilizing somewhere around two isolated "factors": something they know, something they have, or something they are. Phones can be used as an ensuing variable in some two-factor approval methods. Pass codes can be produced on a PDA or shipped off a telephone through instant message. On the off chance that two-factor confirmation isn't utilized, unapproved clients could gain admittance to touchy data and misuse mobile phones [22,23].

3   Tools for Detecting Vulnerabilities and Their Difficulties

There are different procedures accessible to recognize fake android applications, however these state of the art work [DroidMoss] [DNADroid] are basically restricted to perceiving repackaged applications as opposed to recognizing piggybacked applications, in spite of the way that these issues are practically the same, this collaboration isn't great for piggybacked applications. Moreover, these work expect the applications on Google Play Store to be special, and checking repackaged applications isn't great for piggybacked applications [24]. These state of the art works basically do pairwise code assessment between the first and test application and endeavor to lay out differentiation and closeness to recognize repackaged applications (essentially a language structure based separating is done in the motor), however we are at present talking in the large numbers, and it is overall testing to work on this scale on the grounds that the above strategy is neither speedy nor scalable [25].

The following are the most prominent App Security Testing tools that are used throughout the world; many of them are also open-source.

1. Zed Attack Proxy
2. ImmuniWeb® MobileSuite
3. Android Debug Bridge
4. WhiteHat Security
5. Micro Focus
6. AndroBugs
7. Mobile Security Framework (MobSF)
8. QARK.

Indeed, even these out-of-the-box administrations and contraptions, in any case, face many difficulties. These are a couple of the monetary unequivocal weakness assortment restrictions [26, 27].

1. They generally search for blemishes and security gambles in nonexclusive projects instead of specific financial applications. Subsequently, the models they use to separate issues in banking applications are challenging to find data related blemishes.
2. They regularly go to express design based separation in the wake of separating endless weak applications, making it difficult to check the genuine data stream.
3. Banking or money related applications are challenging to test on numerous clients simultaneously while guaranteeing consistent live data and continuous live trades since they disregard a critical number of bank and client security and shows.

4   Mobile Banking Apps Security and Threats

4.1   Issues Involved with Wireless Application Protocol(WAP)

Access to the internet has become a lot easier thanks to new technology. Clients connect their mobile phones to WAP and GPRS to access a wide range of financial services, such as transferring money from one account to another, and making installment payments for purchases [28].

Web banking becomes more useful with WAP. Cell phones are used in financial administrations for electronic communication, but when encryption is used for it, it is hampered for the protection of sensitive information between banks and clients because accessing sensitive data requires very powerful gadgets and a lot of storage space [29]. Secure web banking requires complex encryption cycles and robust PC frameworks. Nonetheless, mobile devices have a limited processing capability, making them unsuitable for using complicated cryptographic frameworks. WAP security is critical from start to finish, but it's challenging to provide since the information isn't jumbled at the door during the convention process [30]. While using cell phones to make decisions, online data will be available on the door, making it possible for the assailant to access the information [31].

4.2 Virus/Malware Attacks in Mobile Banking or Financial Applications

PC malware, web toxic ventures, and TrojanZeus Trojan designated versatile bank clients all exist. Attackers often utilize the Zitmo disease to forestall SMS banking. Notwithstanding disease, developers normally use Zeus to get sufficiently close to portable trade affirmation numbers or passwords [32].

As indicated by Kaspersky Lab, Svpeng is a sort of malware that targets cell phones. The spyware, which targets Android gadgets, searches for explicit versatile financial applications on the telephone prior to locking it and mentioning cash to open it. Speng penetrates a mobile phone with a social designing plan including texting. The spyware chases after programs from a specific arrangement of monetary establishments after it has wormed its direction into a gadget. The Trojan likewise contains code that might be utilized to encode reports, permitting it to muddle information put away on the telephone and afterward request cash to unravel it [33].

The security of versatile banking applications is threatened by Trojans, rootkits, and diseases. An assortment of bank applications have been affected by Zitmo, Perkel/Hesperbot, Wrob, Bankum, ZertSecurity, DroidDream, and Keylogger infections. These infections have been modified by digital thieves to target mobile phones so that they can access monetary records and become resistant to security efforts. Below are some common malware that affect portable financial applications. [34].

1. Through instant messaging discussions, Zitmo fraudsters attack and steal TAN codes from clients.
2. JavaScript Infusion (JS) is used in this case on a PC in order to demand a cell phone number and afterward send a Trojan through SMS. Trojan camouflages itself as a security program [35].
3. Bankum is a financial application that pretends to be a real financial application but in fact is a Trojan clone that poses as a Google Play application and replaces browser-installed financial applications.
4. ZertSecurity - emulates bank login, gathers certifications, and introduces rootkits.

5. Using the rage against the rageagainstthecage exploit, DroidDream allows you to root your gadget, gather information, install other applications, and order other applications remotely [36].
6. Keyloggers transmit keystrokes and other relevant data in addition to acting as an outer control center.

5   Mobile Banking Applications

5.1   Starbuck App

The Starbuck app lets you pay for items, enjoy your Starbuck rewards, and more. Both Android and iPhone versions of the app are available. It allows users to pay with their phones at the register. Using a MasterCard or a bank account, a customer can reload a Starbucks gift card immediately.

How secure are Starbuck versatile application exchanges?

Notwithstanding its accommodation, the degree of assurance given by Starbuck's security engineering is very moderate. At the point when an interloper accesses a client's Starbucks account on the web, a couple of things could occur. A client had a beverage at Starbuck in Sugar Land, Texas, in one of the cases. He kept on getting Paypal alerts about reloading his Starbuck card. He likewise got a few texts from Starbuck. As indicated by current realities of the case, Starbuck didn't look for affirmation from the client prior to making the transaction [37].

5.2   Bank of China Application (Hong Kong)

BOCHK, an organization that is comparable to Maybank in its mobile banking and speculation administrations, has developed a mobile financial app. It is possible for customers to manage their finances using their mobile devices from anywhere in the world, so the app is accessible from anywhere.

What is the gamble factor for Bank of China versatile financial applications?

Since the BOCHK adaptable monetary application plays out the whole cycle on HTTP, there might be a gamble of man-in-the-center attack. This program ought to empower complete monetary organizations, yet additionally a part that makes demands for electronic trade records, as well as opening and shutting different electronic portion trade capacities and setting quite far. Therefore, social planning might be considered an immediate result of the application's control of the sureness [38].

6   Research Model Analysis

With the end goal of this paper, we will think about a basic model of portable banking or money related applications for additional conversation. Compact banking or financial applications are basic for android application security since they handle a great deal of exchanges and are straightforwardly taking care of assets, making them more defenseless against digital attack. Existing exploration has neglected to sufficiently analyze the security dangers of globalized structures, given the extraordinary potential for monetary calamity banking or money related applications, and to guarantee the wide-going security of these significant and generally utilized sorts of Android applications. Credit-just convenient portion has considerably modified customary monetary administrations. Clients recognize that all associations between nearby banking applications and far off bank data bases are protected (e.g., through HTTPS) and that all banking or money related applications give a solid installment choice and a simple to-utilize UI. Deplorably, ceaseless examinations and exploration have uncovered that specific ensured banking applications incorporate imperfections or holes that are challenging to distinguish utilizing industry and open-source techniques [39].

Prateek Panda distributed a noteworthy investigation named "Security Report of Top 100 Mobile Banking Applications" in 2015, and as per his discoveries, 85% of the applications bombed essential security tests, and most of them have 4–6 imperfections that developers may effortlessly take advantage of. In spite of the way that it has been six years and applications are as yet advancing as far as security, the reality stays that numerous monetary or banking applications are not quite as secure as we accept, and clients might confront huge monetary misfortunes assuming nothing is done to work on the security of these often utilized applications[40].

As per many examinations that inspected security weaknesses in banking applications top to bottom, by far most applications in Europe and North America had recently 0.27 information spillage concerns per application. Asia has the most security hardships, with a normal of 6.4 shortcomings per application, because of the popularity for credit only installment frameworks. The flaws of various financial apps vary by market and country. Even versions of similar apps have problems. Security levels are lower for subsidiaries than for parent banks. As demonstrated by the Citibank application for South Korea, this remark is right.

7   Security Measures

The accompanying five safety efforts ought to be carried out as indicated by [41,42]:

1. Customary Access Control: Attempts to access devices using standard methods, for example, passwords and screens that lock after idle time.
2. Application Provenance: Before being made alter-safe, each application is branded with the creator's personality (utilizing a computerized signature). It allows a client to select whether or not to utilize the application based on the creator's personality.
3. In case of a setback or a theft, encryption guarantees that data on the gadget is protected.

4. The craving to restrict admittance to delicate information or frameworks on a gadget by an application using separation.
5. Assent Based Access Control furnishes every application with a bunch of approvals that limit their admittance to device data and a bunch of systems that match those approvals, hence, keeping them from performing exercises past those approvals.

8 Conclusion

The most popular smartphone operating system is Android. To protect the user's privacy and personal information, it's critical to improve the security of an Android OS. In this paper we had gone through security risks and vulnerabilities that are associated with mobile banking apps, and the security methods to address the difficulties. This research had presented the analysis of various banking apps as well as studied the different tools available to test the security of the payment done through these app. This study concludes that in order to improve app security and enable future technologies, mobile banking apps require a foundation. This implies that versatile apps and their security system are future-proof, requiring less assets to preserve within the long run.

References

1. Wu, L., Grace, M., Zhou, Y., Wu, C., Jiang, X.: The impact of vendor customizations on Android security. In: Proceedings of the ACM Conference on Computer and Communications Security, pp. 623–634,(2013). https://doi.org/10.1145/2508859.2516728
2. Pousttchi, K., Schuring, M.: Assessment of today's mobile banking applications from the view of customer requirements. In: Proceedings of the 37th Annual Hawaii international Conference (2004)
3. Shashank Gupta et al. 2015. PHP-sensor: a prototype method to discover workflow violation and XSS vulnerabilities in PHP web applications. In Proceedings of the 12th ACM International Conference on Computing Frontiers (CF '15). Association for Computing Machinery, New York, NY, USA, Article 59, pp. 1–8. https://doi.org/10.1145/2742854.2745719
4. Yuan, H., Tang, Y., Wenjuan, S., Liu, L.: A detection method for android application security based on TF-IDF and machine learning. PloS One **15**, e0238694 (2020). https://doi.org/10.1371/journal.pone.0238694
5. Sanchez, A.: Security Flaws in mobile banking apps identified by researcher. Out-Law.com, 13 Jan 2014. [Online]. Available: www.out-law.com/en/articles/2014/january/security-flaws-inmobile-banking-apps-identified-by-researcher/. Accessed 25 Oct 2015
6. Gupta, S., et al.: Detection, avoidance, and attack pattern mechanisms in modern web application vulnerabilities: present and future challenges. Int. J. Cloud Appl. Comput. (IJCAC) **7**, 1–43 (2017). https://doi.org/10.4018/IJCAC.2017070101
7. Zhang, Y., Lee, W.: Intrusion detection in wireless ad-hoc networks. In: Hulme, G. (ed)Services Seeks to Bring e-Business to Small Businesses, p. 21. In Informationweek.com (2000)

8. de la Puente, S.A.J.F.: Virus attack to the PC bank. In: Security Technology Proceedings

9. Sharma, R., Sharma, T.P., Sharma, A.K.: Detecting and preventing misbehaving intruders in the Internet of vehicles. Int. J. Cloud Appl. Comput. (IJCAC) **12**(1), 1–21 (2022)

10. Singh, S., Kaur, K., Aggarwal, A.: Emerging trends and limitations in technology and system of ubiquitous computing. Int. J. Adv. Res. Comput. Sci. **5**, 174–178 (2014)

11. Crosman, P.: First major mobile banking security threat hits the US. American Banker, 13 June 2014. [Online]. Available: www.americanbanker.com/issues/ 179_114/first-major-mobilebanking-security-threat-hits-the-us-1068100-1.html. Accessed 26 Oct 2015

12. Ling, Z., Hao, Z.J.: An intrusion detection system based on normalized mutual information antibodies feature selection and adaptive quantum artificial immune system. Int. J. Semant. Web Inf. Syst. (IJSWIS) **18**(1), 1–25 (2022)

13. Webroot: The risks and rewards of mobile banking apps. Webroot, United States (2014). [18] Zhang, Y., W. Lee, W.: Intrusion detection in wireless ad-hoc networks. In: ACM/IEEE MobiCom (2000)

14. Singh, S.K.: Linux Yourself: Concept and Programming (1st ed.). Chapman and Hall/CRC. https://doi.org/10.1201/9780429446047 (2021)

15. Filiol, E., Irolla, P.: Security of Mobile Banking ... and of Others, BlackHat Asia, pp. 1–22 (2015)

16. Kumar, S., Singh, S.K., Aggarwal, N., Aggarwal, K.: Evaluation of automatic parallelization algorithms to minimize speculative parallelism overheads: an experiment. J. Discrete Math. Sci. Crypt. **24**(5), 1517–1528 (2021)

17. Brdesee, H.S., Alsaggaf, W., Aljohani, N., Hassan, S.U.: Predictive model using a machine learning approach for enhancing the retention rate of students at-risk. Int. J. Semant. Web Inf. Syst. (IJSWIS) **18**(1), 1–21 (2022)

18. Cooney, M..: 10 common mobile security problems to attack. PC world. [Online]. Available www.pcworld.com/article/2010278/10-common-mobile-securityproblems-to-attack.html. Accessed 26 Oct 2015

19. Zhou, W., Zhou, Y., Grace, M., Jiang, X., Zou, S.: Fast, scalable detection of "Piggybacked" mobile applications. https://doi.org/10.1145/2435349.2435377 (2013)

20. Zhou, Z., et al.: A statistical approach to secure health care services from DDoS attacks during COVID-19 pandemic. Neural Comput. Applic. (2021). https://doi. org/10.1007/s00521-021-06389-6

21. Mallat, N., Rossi, M., Tuunainen, V.K.: Mobile banking services. Commun. ACM **47**(May), 42–46 (2004)

22. Nie, J.; Hu, X.: Mobile banking information security and protection methods. Comput. Sci. Softw. Eng. (2008)

23. Gupta, A., Singh, S.K., Chopra, M., Gill, S.S.: An inquisitive prospect on the shift toward online media, before, during, and after the COVID-19 pandemic: a technological analysis (2022)

24. Verma, P., Charan, C., Fernando, X., Ganesan, S. (eds.): Advances in data computing, communication and security. Lect. Notes Data Eng. Commun. Technol. **106**

25. Chopra, M., Singh, S.K., Sharma, A., Gill, S.S.: A comparative study of generative adversarial networks for text-to-image synthesis. Int. J. Softw. Sci. Comput. Intell. (IJSSCI) **14**(1), 1–12 (2022). https://doi.org/10.4018/IJSSCI.300364

26. Singh, A., Singh, S.K., Mittal, A.: A review on dataset acquisition techniques in gesture recognition from Indian sign language. Adv. Data Comput. Commun. Secur. 305–313 (2022). https://doi.org/10.1007/978-981-16-8403-6_20
27. Narendiran, C., Rabara, S.A., Rajendran, N.: Public key infrastructure for mobile banking security. In: Global Mobile Congress, pp. 1–6 (2009)
28. Dai, W., Tang, Y.: Research on security payment technology based on mobile e-Commerce. In: e-Business and Information System Security, pp. 1–4 (2010)
29. Singh, S.K., Kumar, A., Gupta, S., Madan, R.: Architectural Performance of WiMAX over WiFi with Reliable QoS over wireless communication. Int. J. Adv. Networking Appl. (IJANA) 03(01), 1016–1023 (2011). [EISSN: 0975–0282]
30. Nachenberg, C.: A window into mobile device security. Technical report, Symantec (2011)
31. Singh, I., Singh, S.K., Singh, R., Kumar, S.: Efficient loop unrolling factor prediction algorithm using machine learning models. In: 2022 3rd International Conference for Emerging Technology (INCET), pp. 1–8 (2022). https://doi.org/10.1109/INCET54531.2022.9825092
32. Aggarwal, K., Singh, S.K., Chopra, M., Kumar, S., Colace, F.: Deep learning in robotics for strengthening industry 4.0.: opportunities, challenges and future directions. In: Nedjah, N., Abd El-Latif, A.A., Gupta, B.B., Mourelle, L.M. (eds.) Robotics and AI for Cybersecurity and Critical Infrastructure in Smart Cities. Studies (2022)
33. Singh, S.K., Singh, R.K., Bhatia, M.: Design flow of reconfigurable embedded system architecture using LUTs/PLAs. In: 2012 2nd IEEE International Conference on Parallel, Distributed and Grid Computing, pp. 385–390 (2012). https://doi.org/10.1109/PDGC.2012.6449851
34. Gupta, S., Singh, S.K., Jain, R: Analysis and optimisation of various transmission issues in video streaming over Bluetooth . Int. J. Comput. Appl. 11(7), 44–48 9 (2010)
35. Chopra, M., et al.: Analysis and prognosis of sustainable development goals using big data-based approach during COVID-19 pandemic. Sustain. Technol. Entrepreneurship. www.sciencedirect.com/science/article/pii/S2773032822000128 (2022)
36. Sharma, S.K., Singh, S.K., Panja, S.C.: Human factors of vehicle automation. Auton. Driving Adv. Driver-Assistance Syst. (ADAS), 335–358 6 (2021)
37. Singh, R., Rana, R., Singh, S.K.: Performance evaluation of VGG models in detection of wheat rust. Asian J. Comput. Sci. Technol. 7(3), 76–81 5 (2018)
38. Ghosh, Security and Privacy for E-Business
39. Singh, S.K., Singh, R.K., Bhatia, M.P.S., Singh, S.P.: CAD for delay optimization of symmetrical FPGA architecture through hybrid LUTs/PLAs (2013)
40. Meghanathan, N., Nagamalai, D., Chaki, N. (eds.): Advances in Computing and Information Technology. In: Advances in Intelligent Systems and Computing, vol. 178. Springer, Berlin, Heidelberg (2012). https://doi.org/10.1007/978-3-642-31600-5_57
41. Kaur, P., Singh, S.K., Singh, I., Kumar, S.: Exploring convolutional neural network in computer vision based image classification. https://ceur-ws.org/Vol-3080/21.pdf (2022)
42. Tsai, C., Chen, C., Zhuang, D.: Secure OTP and biometric verification scheme for mobile banking. In: 2012 Third FTRA International Conference on Mobile, Ubiquitous, and Intelligent Computing, pp. 138–141 (2012). https://doi.org/10.1109/MUSIC.2012.31

Author Index

© The Editor(s) (if applicable) and The Author(s), under exclusive license
to Springer Nature Switzerland AG 2023
N. Nedjah et al. (Eds.): ICSPN 2021, LNNS 599, pp. 417–418, 2023.
https://doi.org/10.1007/978-3-031-22018-0

Printed in the United States
by Baker & Taylor Publisher Services